Intelligent Methods in Signal Processing and Communications

D. Docampo
A. R. Figueiras-Vidal
F. Pérez-González
Editors

1997
Birkhäuser
Boston • Basel • Berlin

D. Docampo
F. Pérez-González
Departamento Tecnologías
de las Comunicaciones
E.T.S.I. Telecomunicación
Universidad de Vigo
36200 Vigo, Spain

A.R. Figueiras-Vidal
Departamento de Senales, Sistemas
y Radiocomunicaciones
E.T.S.I. Telecomunicación
Universidad Politécnica de Madrid
Ciudad Universitaria
28040 Madrid, Spain

Library of Congress Cataloging-in-Publication Data

Intelligent methods in signal processing and communications / Domingo
Docampo, Fernando Pérez-González, Anibal Figueiras-Vidal, editors.
p. cm.
Includes bibliographical references.
ISBN 0-8176-3960-8 (hc : alk. paper). -- ISBN 3-7643-3960-8 (alk. paper)
1. Signal processing. 2. Neural networks (Computer science)
3. Multisensor data fusion. I. Docampo, Domingo, 1954- .
II. Pérez-González, Fernando, 1967- III. Figueiras-Vidal, A. R.
(Anibal R.), 1950-
TK5102.9.I545 1997
621.382' 2' 028563--dc21 97-183
CIP

Printed on acid-free paper

ISBN 0-8176-3960-8
ISBN 3-7643-3960-8
Typeset by the Editors in LaTeX.
Printed and bound by Hamilton Printing, Rensselaer, NY.
Printed in the U.S.A.

9 8 7 6 5 4 3 2 1

Contents

Contributors

C.T. Abdallah, EECE Department, University of New Mexico. Albuquerque, New Mexico 87131-1356. U.S.A.

Y.S. Abu-Mostafa, California Institute of Technology. Mail Stop 136-93. Pasadena, CA 91125, USA

S.C. Ahalt, Department of Electrical Engineering, The Ohio State University. Columbus, Ohio 43210, U.S.A.

M. Ansorge, Institute of Microtechnology, University of Neuchatel. Rue de Tivoli 28. 2003 Neuchatel, Switzerland.

F. Argenti, Dipartamento di Ingegneria Elettronica, Universitá di Firenze. Via di Santa Marta, 3. 50319 Firenza, Italy.

H. Aydınoğlu, School of Electrical and Computer Engineering, Georgia Institute of Technology. Atlanta, Georgia 30332-0250, U.S.A.

J. Bracamonte, Institute of Microtechnology, University of Neuchatel. Rue de Tivoli 28. 2003 Neuchatel, Switzerland.

B. Brogelli, Dipartamento di Ingegneria Elettronica, Universitá di Firenze. Via di Santa Marta, 3. 50319 Firenza, Italy.

J. Cid-Sueiro, Depto. de Teoria de la Señal y Comunicaciones e I.T. E.T.S.I. Telecomunicación, Universidad de Valladolid. c/ Real de Burgos s/n. 47011 Valladolid, Spain.

E. Del Re, Dipartamento di Ingegneria Elettronica, Universitá di Firenze. Via di Santa Marta, 3. 50319 Firenza, Italy.

D. Docampo, Departamento Tecnologías de las Comunicaciones, E.T.S.I. Telecomunicación, Universidad de Vigo. 36200 Vigo, Spain.

A.R. Figueiras-Vidal, DSSR, ETSI Telecomunicación, Universidad Politécnica de Madrid. Ciudad Universitaria. 28040 Madrid, Spain.

J. Ghattas, Depto. de Teoria de la Señal y Comunicaciones e I.T. E.T.S.I. Telecomunicación, Universidad de Valladolid. c/ Real de Burgos s/n. 47011 Valladolid, Spain.

L.J. Griffiths, Department of Electrical and Computer Engineering, University of Colorado. Boulder, CO 80309-0425. U.S.A.

M. Hayes, School of Electrical and Computer Engineering, Georgia Institute of Technology. Atlanta, Georgia 30332-0250, U.S.A.

S. Haykin, Communications Research Laboratory, McMaster University. 1280 Main Street West, Hamilton, Ontario, Canada, L8s 4K1.

J.R. Hernández, Departamento Tecnologías de las Comunicaciones, E.T.S.I. Telecomunicación, Universidad de Vigo. 36200 Vigo, Spain.

D. Hush, EECE Dept., University of New Mexico. Albuquerque, New Mexico 87131-1356. U.S.A.

M. P. Kennedy, Department of Electronic and Electrical Engineering, University College Dublin. Dublin 4, Ireland.

S.Y. Kung, Department of Electrical Engineering, Princeton University. Princeton, NJ08544. U.S.A.

S.-H. Lin, Epson Palo Alto Lab. 3145 Porter Drive, Suite 104, Palo Alto, CA94304. U.S.A.

C. Mosquera, Departamento Tecnologías de las Comunicaciones, E.T.S.I. Telecomunicación, Universidad de Vigo. 36200 Vigo, Spain.

F. Pellandini, Institute of Microtechnology, University of Neuchatel. Rue de Tivoli 28. 2003 Neuchatel, Switzerland.

F. Pérez-González, Departamento Tecnologías de las Comunicaciones, E.T.S.I. Telecomunicación, Universidad de Vigo. 36200 Vigo, Spain.

W. Pierson, Department of Electrical Engineering, The Ohio State University. Columbus, Ohio 43210, U.S.A.

J.L. Sancho-Gómez, DSSR, ETSI Telecomunicación, UPM. Ciudad Universitaria. 28040 Madrid, Spain.

I.W. Sandberg, Department of Electrical and Computer Engineering,

The University of Texas at Austin. Austin, Texas 78712. U.S.A.

J. Sill, Computation and Neural Systems Program, MC 136-93
California Institute of Technology. Pasadena, CA 91125, U.S.A.

B. Ulug, Department of Electrical Engineering,
The Ohio State University. Columbus, Ohio 43210, U.S.A.

S. Verdú, Department of Electrical Engineering, Princeton University.
Princeton, NJ 08544, U.S.A.

Preface

From June 24 to 26, 1996 a Workshop —the fourth in a series— was celebrated at the Parador Nacional de Baiona, Spain. Its topic, Intelligent Methods for Signal Processing and Communications, led the Workshop to an objective additional to the scientific and technical discussion: to prepare the work to be done around signal processing and intelligence facilities for communication terminals, which is the subject of an European cooperative research project, COST#254, recently started.

This opportunity served to invite six distinguished speakers to offer plenary talks; besides, 51 refereed contributions were also presented in a poster format. The papers tried either to summarize the state-of-the-art and trends in some relevant topics in the field of the Workshop, or to open new basic research avenues to be further developed from their initial conceptions to results which can be applied in future communication terminals or (other) signal processing applications; without disregarding the consideration of techniques and tools needed for the corresponding implementations.

This book contains extended versions of the six invited lectures plus eight selected regular contributions: most of them making emphasis on fundamental concepts and formulations, but having a clear potential for useful applications.

Among the invited talks, there are two which overview two important aspects of present day communications: "Adaptive Antenna Arrays in Communications", by L.J. Griffiths, and "Multiuser Demodulation for CDMA Channels", by S. Verdu. The paper "Intelligent Signal Detection", by S. Haykin, gives a powerful and practical approach for designing radar and sonar detectors. S.-H. Lin and S. Y. Kung suggest an interesting option for security purposes in "Biometric Identification for Access Control". The contributions by I.W. Sandberg, "Multidimensional Nonlinear Myopic Maps, Volterra Series, and Uniform Neural-Network Approximations", and by J. Sill and Y. Abu-Mostafa, "Monotonicity Theory", are advances in approximation and classification theories which can help to extend applications and to improve performances.

The same basic character corresponds to four of the contributed regular papers: "Analysis and Synthesis Tools for Robust SPR Discrete Systems", by C. Mosquera et al.; "Boundary Methods for Distribution Analysis", authored by J. L. Sancho et al.; "Constructive Function Approximation: Theory and Practice", written by D. Docampo et al.; and the work by J. Cid-Sueiro et al. entitled "Decision Trees Based on Neural Networks". While M. P. Kennedy explores spread spectrum transmission and security uses of the pervasive theory of chaos in his "Applications of Chaos in Communications". A new point of view on the largely used filter banks is given in "Design of Near Perfect Reconstruction Non-Uniform Filter Banks" by F. Argenti and coauthors. M. H. Hayes explains a method to include a very interesting capability in terminals and other machines: "A Genetic Algorithm for Synthesizing Lip Movements from Speech". And, finally, J. Bracamonte et al. address the implementation aspects of communication coders from a general perspective in "Design Methodology for VLSI Implementation of Image and Video Coding Algorithms".

We believe that these articles constitute an updated and balanced version between fundamentals and applicability in the field of intelligent processing and communications. Needless to say, this is not editors' merit, but a result of the efforts ot the authors and of many colleagues who have helped along years to reach a high scientific and technical level at the Baiona Workshops; in particular, V. Cappellini, K. Fazekas, M. Kunt, and M. Najim. To all them, our sincere gratitude.

The support of the following institutions is gratefully acknowledged: CI-CYT, Xunta de Galicia, Universidad de Vigo, Spanish Branch of the IEEE. We also wish to thank the United States Air Force European Office of Aerospace Research and Development and the Office of Naval Research, Europe, for their contribution to the success of the conference.

D. Docampo
A. R. Figueiras-Vidal
F. Pérez-González

Vigo, Spain
October, 1996

1

Adaptive Antenna Arrays in Mobile Communications

L. J. Griffiths

ABSTRACT This paper presents a discussion of the applications of adaptive array processing methods in mobile communication systems. A brief history of the development of adaptive arrays is presented together with a summary of current methodologies used for implementation. A "best estimate" of what adaptive applications are likely in communication systems in the future is provided together with a description of the significant technical challenges. The paper is intended as an overview and does not contain detailed mathematical development.

1.1 Introduction

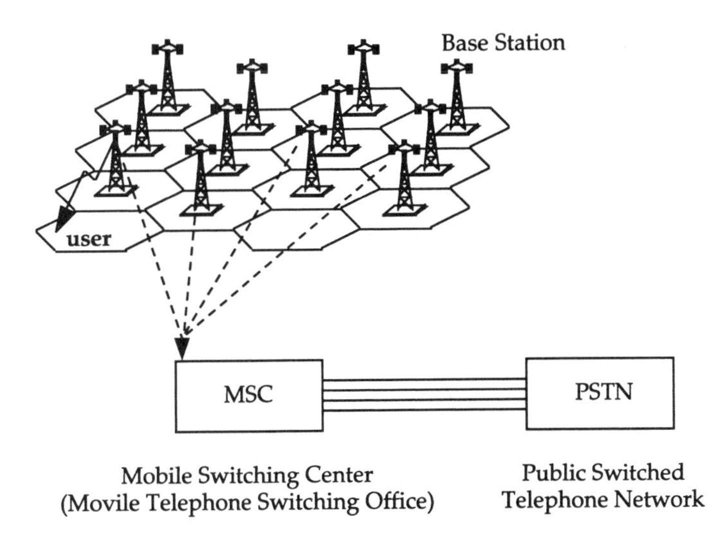

FIGURE 1.1. Cellular telephone system.

Figure 1.1 illustrates a basic cellular telephone system. Many excellent references can be found which describe these systems. T.S. Rappaport's excellent book, [1], for example, contains descriptions of mobile cellular systems concepts, design, and analysis. An earlier book by the same author

[2] contains selected readings on the topic drawn from a variety of sources. Another good source for background material on cellular systems appeared in the June 1991 *IEEE Communications Magazine* with excellent review articles by D.J. Goodman [3], A.D. Kucar [4], and others. [1].

In most cellular systems, towers are used as base stations that transmit to (and receive from) mobile users located within the cellular boundaries. In the simple illustration in Fig. 1.1, individual cells are represented by hexagonal shapes. The base stations, in turn, are connected to the public telephone system through transmissions to the Mobile Switching Center (MSC). These transmissions are illustrated using dashed lines.

Figure 1.2 illustrates two methods for the placement of base stations. In the center-excited system, each base station radiates with an omnidirectional pattern so as to reach all users within a defined radius. In the edge-excited geometry, the stations are placed at the vertices of the hexagonal cells and three different antennas with patterns of approximate width 120° are employed to reach the cells defined by the intersection point. The frequency reuse pattern for cellular systems is designed to minimize

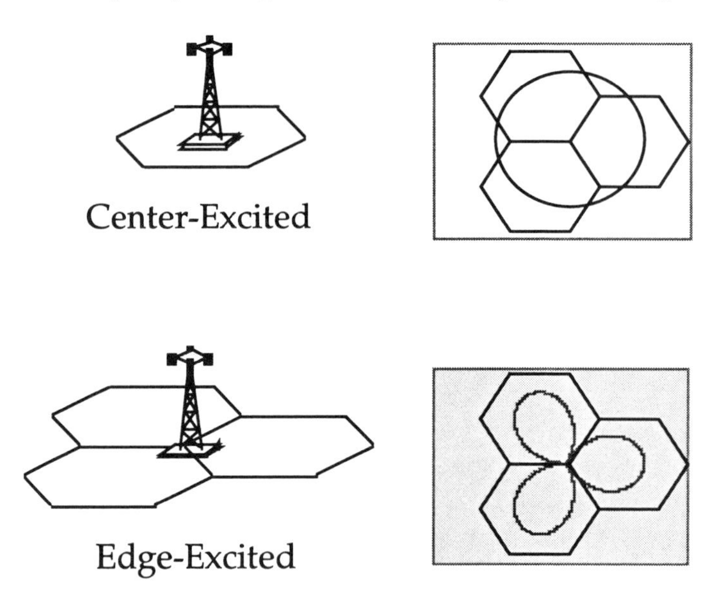

FIGURE 1.2. Center and edge-excited cellular geometries.

interference and is shown in figure 1.3. The letters in each cell represent transmission on identical frequencies. (Figures 1.1, 1.2, and Fig. 1.3 are taken from reference [1].)

In mobile systems that operate within crowded urban environments, interference is the major limiting factor in the performance of cellular radio systems. This interference arises from several sources. Primary among these is the radiation produced by multiple callers in the same cell or by calls

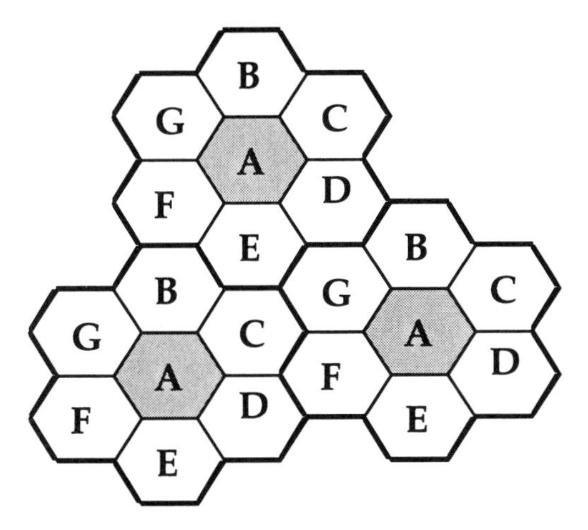

FIGURE 1.3. Frequency reuse patterns in cellular systems.

made to adjacent cells. Other sources include base stations operating at the same frequency and non-cellular transmissions or interference. The capacity of the overall system is limited by this interference. Allocation of different users to non-overlapping frequency bands minimizes interference but simultaneously limits the number of users within a given cell.

Perhaps the most troublesome source of interference in cellular systems is co-channel interference caused by frequency reuse. Since all base stations operate at the same power, the interference is independent of transmitter power. Because the source is "in-band", the only potential method for reducing its effect is through spatial processing. In fact, the level of co-channel interference, is critically dependent upon the spatial distribution of the towers with respect to the users and to the type of antenna pattern used by the towers. Since overall system capacity is ultimately limited by the amount of co-channel interference, there is significant need to employ "smart" antennas which enable reductions in the amount of this interference. The purpose of this paper is to discuss potential applications of adaptive array processing in this environment. In particular, the degree to which adaptive processing at the base station can provide improvements is the subject of this paper.

1.2 Adaptive Arrays in Base Station Antennas

One obvious approach to reducing in-band interference in cellular systems is to employ beam shape processing in the base station antennas. There are, however, some fundamental constraints that must be addressed. First, existing base stations employ relatively small numbers of array elements. Four element arrays are typical and rarely are more than six elements used.

A second limitation is that any spatial processing technology must fit into the existing very large installed system base. In effect, the process must be viewed as "add-on" to the existing structure both in terms of hardware and software. The implementation of spatial processing techniques cannot impact the mobile radio design due to the large number of receivers currently in the field. The overall system will have added complexity as a result of the adaptive technology. This complexity must provide an increased effectiveness which justifies the overall cost. Tradeoff studies which predict effectiveness as a function of cost for these systems are complicated to implement at a level which produces meaningful results. Factors which must be taken into account include multipath effects with small differential relative delays and non-planar wavefront propagation. In most adaptive array studies, these factors are assumed to have negligible effect on the overall system. The same, however, cannot be said for the cellular mobile environment.

A significant amount of work has been carried out in the area of base station beam processing systems. Such systems are referred to as "SMART" array systems. There have been two general approaches. In the area of multi-beam and/or sectored transmitting arrays, the work contained in [5], [6], and [7] is representative. These systems are not adaptive in the sense normally employed in adaptive array literature. Instead, they take advantage of fixed, multi-beam narrowband arrays which concentrate the transmitted power into smaller spatial regions than that employed by traditional base station systems.

The second approach employs adaptive array elements in which the gain/phase weighting at each element is controlled by a feedback algorithm. The work contained in references [7], [8], [9], and [10] (and in the contained references) describes this approach. Figure 1.4 illustrates the basic approach used in a typical adaptive array system. The array element

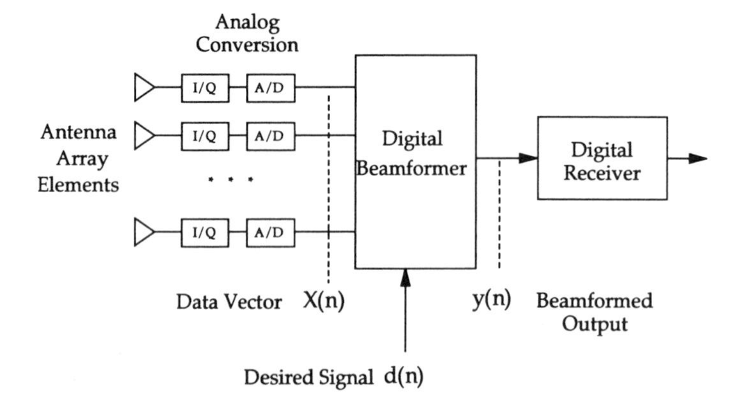

FIGURE 1.4. Basic digital beamforming adaptive array structure.

outputs are sampled and passed to the digital beamformer to produce the digital beamformed output. Final reception is then carried out digitally. One beamformer is required for each beam. Thus, the digital beamformer shown in Fig. 1.4 typically has multiple outputs representing the multiple beams pointed at the different users. The overall idea is to separate transmissions to users spatially and thus employ the same transmission frequency to multiple users. In this fashion, the overall capacity of the system is increased. A desired signal is used to identify the user for each beam. Typically, this signal would be the known handshake waveform employed in the existing cellular system. However, the details of how this waveform is derived and employed represent significant complications to the overall system which are beyond the scope of this paper.

1.3 Adaptive Array Details

The purpose of this section is to describe the basic methodology used in adaptive arrays and to illustrate the performance of small (four element) adaptive systems in a simulated communication environment. Figure 1.5 shows the details interior to the digital beamformer in Fig. 1.4. The weights w_n are complex, representing both gain and phase terms, and $\mathbf{W}_a$ is the vector of array weights,

$$\mathbf{W}_a = \begin{bmatrix} w_1 \\ w_2 \\ \vdots \\ w_N \end{bmatrix} . \tag{1.1}$$

The weights determine the shape of the transmitted beampattern in response to the adaptive algorithm contained in the "Weight Vector Computation" system. The objective of the algorithm is to minimize the average value ξ^2 of the complex error term $\epsilon(n)$. Thus $\mathbf{W}_a$ minimizes the term,

$$\xi^2 = \sum_n [|\epsilon(n)|^2] = \sum_n [|d(n) - y(n)|^2] . \tag{1.2}$$

The weights which minimize ξ^2 are termed the least-squares solution $\mathbf{W}_{LS}$ and given by,

$$\mathbf{W}_{LS} = \mathbf{R}_{XX}^{-1}\mathbf{P}_{Xd} , \tag{1.3}$$

where the correlation matrix $\mathbf{R}_{XX}$ and cross-correlation vector $\mathbf{P}_{Xd}$ are given by,

$$\mathbf{R}_{XX} = \sum_n [\gamma(n)\mathbf{X}(n)\mathbf{X}^\dagger(n)] , \tag{1.4}$$

$$\mathbf{P}_{Xd} = \sum_n [\gamma(n)\mathbf{X}(n)d(n)] . \tag{1.5}$$

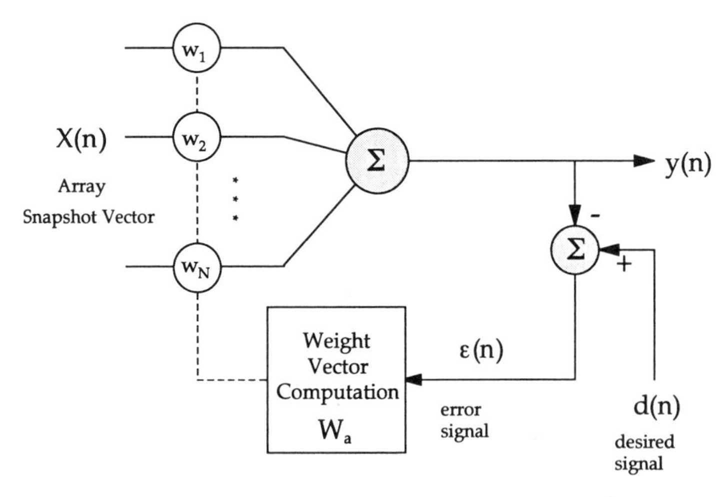

FIGURE 1.5. Details for digital adaptive beamforming system.

Weight vectors determined by this sequence of operations are said to be determined by "sample matrix inversion" or SMI techniques. Different methods are used to compute the summations in (1.4) and (1.5) depending upon how the summation limits are generated and the form of the weighting term $\gamma(n)$ which is referred to as the "forget factor". For example, in block-adaptive processing, the summation is taken over the past K_0 samples and $\gamma(n) = 1$. In recursive least-squares approaches, the summation is over the infinite past and $\gamma(n)$ is an exponential term with increasing powers for older data samples. A vast amount of literature on this subject is available. The interested reader is referred to [11] and related references for further information and details.

A wide variety of algorithms are available for updating the weights in the adaptive array shown in Fig. 1.5. One of the most widely used, and certainly the simplest, it the LMS algorithm originally proposed by Widrow [12]. Additional references on the algorithm can be found in reference [13]. The LMS recursion for the weights is,

$$\mathbf{W}_a(n+1) \;=\; \mathbf{W}_a(n) \;+\; \mu\, \mathbf{X}(n)\, \epsilon^{\dagger}(n)\;, \tag{1.6}$$

where the error term, $\epsilon(n)$ is,

$$\epsilon(n) \;=\; d(n) \;-\; y(n)\;. \tag{1.7}$$

The adaptive step size parameter μ is normalized by the average power level in the adaptive processor. One approach is,

$$\mu \;=\; \frac{\alpha}{p\{\mathbf{X}(n)\}}\;, \quad 0 < \alpha < 1\;, \tag{1.8}$$

with a power estimate $p\{\mathbf{X}(n)\}$ given by,

$$p\{\mathbf{X}(n)\} = \sum_{n} \gamma(n)\mathbf{X}^{\dagger}(n)\mathbf{X}(n) . \tag{1.9}$$

The advantage of the normalization in (1.8)-(1.9) is that convergence is assured for positive values of α which are less than unity. In addition, the fractional missadjustment noise [12] caused by adaptation can be shown to be equal to α. In typical applications, $\alpha = 1$ which results in a 10% increase in error due to adaptation.

1.4 LMS Adaptive Array Examples

In order to demonstrate the adaptive procedures described above, a simple example involving synthetic communication signals was constructed. Figure 1.6 shows the ideal beampatterns which result for this case when perfect covariance information is available. In this example, the signals were mod-

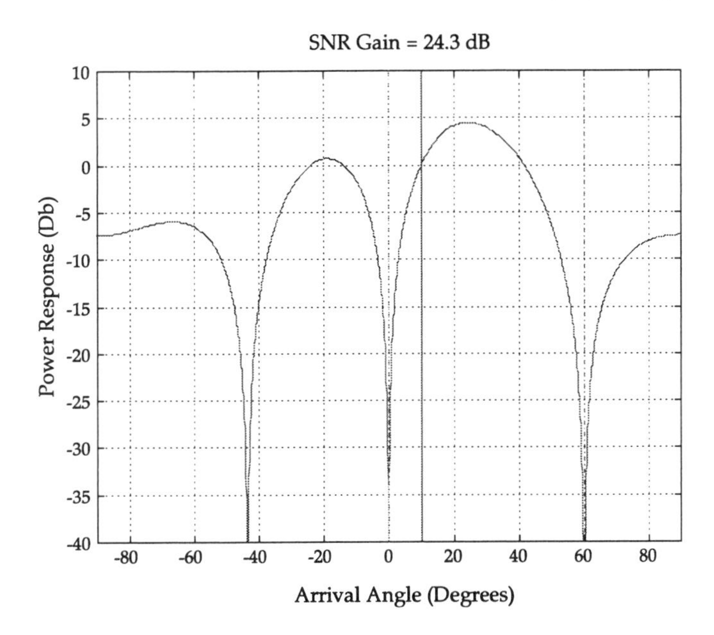

FIGURE 1.6. Adaptive beampattern achieved with exact covariance matrix.

eled as AMPS (Advanced Mobile Phone System) transmissions that were assumed to be incident on a four element base station array with inter-element separation of $\lambda/2$. Three independent signals were assumed to be incident on the array from angles (relative to boresight) of 0°, 10°, and 60° with the 10° signal representing the desired signal for the simulation.

The LMS algorithm was implemented on the data using a training signal consisting of the handshake signal for the 10° arrival. Two different values of α were employed and the results obtained are displayed in Fig. 1.7. Note

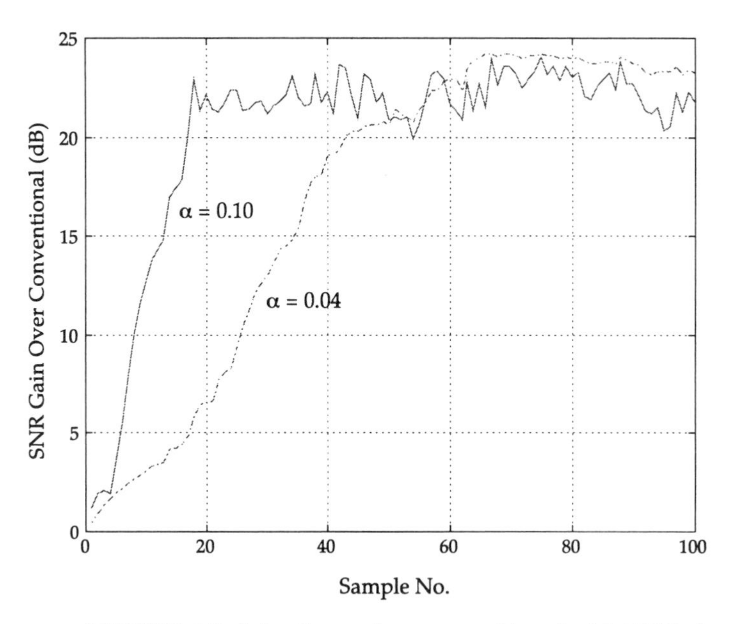

FIGURE 1.7. Adaptive performance achieved with LMS algorithm.

that the smaller step size takes longer to converge and results in an overall performance level that is better in the steady-state condition. Even at the slower convergence rate, approximately 70 samples are required to achieve convergence. Given the 30KHz AMPS one-way bandwidth and the corresponding sampling frequency of 60KHz, this represents a time increment of approximately 1.2 msec. Comparison of the time-adaptive results in Fig 1.7 with the exact solution in Fig. 1.6 shows that the performance penalty incurred with use of LMS is less than 1 dB. Clearly, for this simple example, LMS offers a practical approach to achieving spatial filtering at the base station.

1.5 Desired Signal Availability

The results achieved with the LMS array in the previous section were conditioned on the assumption that a desired signal could be generated at the base station using *a prior* knowledge of the handshake signal. However, this represents a complication in the implementation of the system. Alternative methods, which do not require a specific desired waveform to achieve convergence are also available. The purpose of this section is to summarize three alternative adaptive approaches that do not require an externally-generated desired signal waveform.

P-Vector Adaptation

The first approach to adaptation without a replica of the desired signal is applicable to cases in which the arrival angle of the desired signal is approximately known *a priori*. In this case, the LMS algorithm can be modified [14] to reflect the given information. Assuming that the known arrival angle is represented by $\boldsymbol{\theta_s}$, a simple model for the array snapshot vector is,

$$\mathbf{X}(n) = s(n)\mathbf{V}(\boldsymbol{\theta_s}) + \mathbf{Z}(n) , \qquad (1.10)$$

where $s(n)$ is the desired signal time waveform and $\mathbf{Z}(n)$ represents all interference and noise contributions. The latter are assumed to be uncorrelated with the desired signal. Under these conditions, the cross-correlation vector $\mathbf{P}_{Xd}$ in (1.5) is,

$$\begin{aligned} \mathbf{P}_{Xd} &= \mathbf{E}\left[\mathbf{X}(n)s^{\dagger}(n)\right] \\ &= \sigma_s \mathbf{V}(\boldsymbol{\theta_s}) , \end{aligned} \qquad (1.11)$$

where σ_s is the desired signal power level and $\mathbf{V}(\boldsymbol{\theta_s})$ is the array steering vector for direction $\boldsymbol{\theta_s}$. Thus, the cross-correlation vector is equal to the steering vector, within a scale factor. Assuming that the element response for the array is known or, equivalently, that the array has been calibrated, then $\mathbf{V}(\boldsymbol{\theta_s})$ is also known for each steering angle $\boldsymbol{\theta_s}$.

A modified LMS algorithm, termed the "P-Vector" algorithm results from replacing the term that depends upon $s(n)$ in the LMS algorithm, (1.6). Specifically the update term in LMS is,

$$\mathbf{X}(n)\, \epsilon^{\dagger}(n) = \mathbf{X}(n)d^{\dagger}(n) + \mathbf{X}(n)y^{\dagger}(n) . \qquad (1.12)$$

Since $d(n) = s(n)$ by assumption, the average value of the first term in this equation is proportional to $\mathbf{V}(\boldsymbol{\theta_s})$. The following approximation is then valid,

$$\mathbf{X}(n)\, \epsilon^{\dagger}(n) \approx \mathbf{V}(\boldsymbol{\theta_s}) + \mathbf{X}(n)y^{\dagger}(n) . \qquad (1.13)$$

The P-Vector algorithm results from the use of this approximation in the LMS algorithm, i.e.,

$$\mathbf{W}_a(n+1) = \mathbf{W}_a(n) + \mu \left[\mathbf{V}(\boldsymbol{\theta_s}) - \mathbf{X}(n)y^{\dagger}(n)\right] . \qquad (1.14)$$

Experiments with the P-Vector approach illustrate that it has a performance that is comparable with the LMS algorithm under ideal conditions. However, under strong desired signal conditions, the method is sensitive to errors between the assumed steering vector $\mathbf{V}(\boldsymbol{\theta_s})$ and the actual phase characteristics of the arriving signal. It is then possible to have a situation in which the algorithm discriminates against the desired signal arrival. A common occurrence is when the estimated arrival angle is in error. The degree to which this is an issue with the P-Vector approach is directly proportional to the relative strength of the desired signal.

Decision Feedback Adaptation

An alternative approach to adaptation without a desired signal replica is to generate one from the beamformed output. It is applicable to digital modulation systems such as PCS. With reference to Fig. 1.4, assuming that the array has been initially pointed in the direction of the desired signal, then the digital receiver output should contain the desired waveform. If this receiver includes a detector which generates the binary sequence corresponding to the desired signal, then an approximation to the desired waveform for the adaptive array can be constructed by encoding this binary sequence back into the appropriate modulated signal. This approach is called decision feedback or blind equalization.

Decision Feedback Adaptation

The third method is useful in modulation schemes in which the desired signal waveform has angle modulation, such as occurs in phase or frequency modulation methods. Under these conditions, the modulus of the desired signal is constant,

$$|s(n)| = k , \tag{1.15}$$

where k is independent of the sample interval. If the modulus of the beamformed output $|y(n)|$ is not constant, then residual interference and/or noise is present. The difference between the output modulus and a constant can then be used as an adaptive feedback signal. The resulting constant modulus algorithm (CMA) [15] is given by,

$$\mathbf{W}_{cma}(n+1) = \mathbf{W}_{cma}(n) + \mu \, \mathbf{X}(n) \, \tilde{\epsilon}^{\dagger}(n) , \tag{1.16}$$

where the error term, $\tilde{\epsilon}(n)$ is,

$$\tilde{\epsilon}(n) = \frac{y(n)}{|y(n)|} - y(n) . \tag{1.17}$$

As in the LMS case, the step size is normalized by the power of the input data samples. Details of this process can be found in reference [15].

The CMA algorithm was implemented on the data described previously for use with the LMS algorithm. Two different values of α were employed and the results obtained are displayed in Fig. 1.8. In this simulation, the power normalization approach used with the LMS algorithm was also employed. Comparison with the LMS case shows comparable performance.

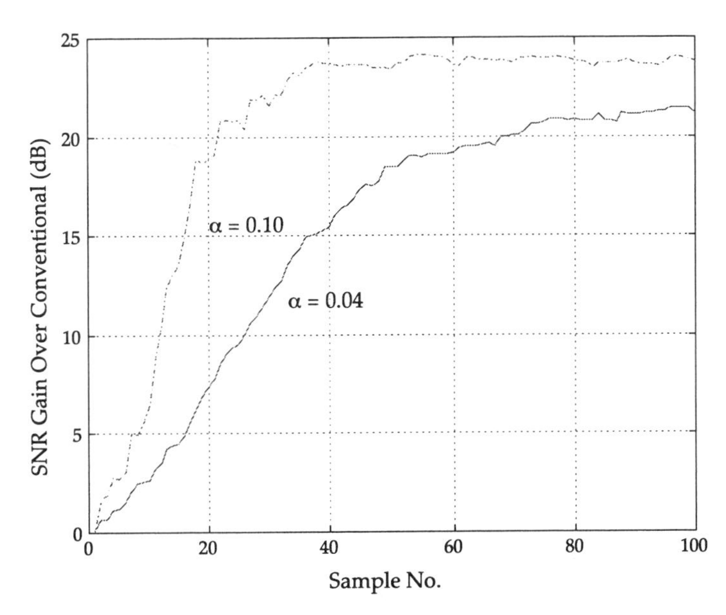

FIGURE 1.8. Adaptive performance achieved with CMA algorithm.

Note, however, since the only requirement in the CMA feedback term is that the output modulus should be a constant, the CMA algorithm may "lock" onto any constant modulus signal in the field of view of the array. In particular it is possible that the method will form a beam on a weak multipath of the desired signal.

In general, if the initialization weights have a beam maximum that is "close" to the arrival angle for the desired signal, it is likely that CMA will lock onto this signal. This is particularly true when the signal amplitude is relatively high. In effect, CMA operates as a "blind" method which can be used to successively extract constant modulus signals from an arbitrary input using serial CMA processors [15].

1.6 Discussion and Observations

This paper has presented a brief discussion of the advantages of using adaptive array processing in mobile communication systems and has identified candidate algorithms for this application. Adaptation offers the potential to increase system capacity by allowing base stations to operate with reduced interference. It may be possible, for example to have transmissions from two or more different users operating on the same frequency to a single base station. In order to accomplish this spatial separation, the base station must be able to simultaneously form beams in the directions of specific users while simultaneously placing nulls in the directions of others.

In order to accomplish this improvement, the algorithm must be relatively simple to implement. Given the large installed base, any highly complex procedure for achieving spatial processing that requires significant hardware and/or software changes is not likely to be implemented. Adaptive processing must be provided as an "add-on" feature to existing systems.

The digital LMS, P-Vector, and CMA algorithms discussed here are all extremely simple to implement and the simulations given demonstrate that significant advantage may accrue through the use of these methods. However, in order to achieve the predicted results, each element in the base station must have an analog to digital converter at each element. Existing implementations use analog beamforming. Effective conversion to a digital system at the element level will require specialized hardware. Advantages provided by this conversion, however, will accrue even if adaptive array processing is not employed. For example, fixed sector beam steering is much easier to implement with a digitally-based system.

Finally, the relatively small number of elements used in the current base station configuration limits the ability of the adaptive system to provide spatial nulling. A four element adaptive array can null up to three spatial sources while maintaining a beam on the desired signal. However, the presence of uncorrelated noise in the system may well limit the effective number of sources to two. One method that may increase the flexibility of the processor is to consider the use of space-time adaptive processing (STAP) [16], [17]. In this approach, adaptive weights are applied to successive time samples at each element. This increases the spatio-temporal degrees of freedom significantly. For example, using five weights per element increases the degrees of freedom from four to twenty. For the case in which spatial sources occupy the full system bandwidth, the added degrees of freedom will not enable nulling of more spatial sources. However, if there are spectral differences between the sources, additional nulling is possible.

1.7 References

[1] T. S. Rappaport, *Wireless Communications: Principles and Practice*, Prentice Hall PTR, Upper Saddle River, New Jersey 07458, 1996.

[2] T. S. Rappaport (Ed.), *Cellular Radio & Personal Communications: Selected Readings*, IEEE Press, New York, ISBN: 0-7803-2283-5, 1995.

[3] D.J. Goodman, "Trends in cellular and cordless communications," *IEEE Communications Magazine*, pp. 31-39, June 1991.

[4] A.D. Kucar, "Mobile radio – An overview," *IEEE Communications Magazine*, pp. 72-85, June 1991.

[5] S.C. Swales, M.A. Beach, D.J. Edwards, and J.P. McGeehn, "The performance enhancement of multibeam adaptive base station antennas for cellular land mobile radio stations," *IEEE Trans. Veh. Tech.*, Vol. VT-39, pp56-67, February 1990.

[6] G.K. Chan, "Effect of sectorization on the specturm efficiency of cellular radio systems," *IEEE Trans. Veh. Tech.*, Vol. VT-41, August 1992.

[7] S.P. Stapelton and G.S. Quon, "A cellular base station phased array antenna system," *Proc. VTC-93*, Vol. I, (Secaucus, NJ), pp 93-96, May 1993.

[8] J.H. Winters, "Signal acquisition and tracking with adaptive arrays in wireless systems," *Proc. 43rd Veh. Tech. Conf.*, Vol I, pp85-88, November 1993.

[9] A.F. Naguib and A. Paulraj, "Performance of cellular CDMA with M-ary orthogonal modulation and cell site antenna arrays," *Proc. ICC-95*, (Seattle, WA), June 1995.

[10] J.H. Winters, J. Saltz, and R.D. Gitlin, "The impact of antenna diverityon the capacity of wireless communication systems," *IEEE Trans. Commun.*, Vol COM-41, pp 1840-1751, April 1994.

[11] D.H. Johnson and D.E. Dudgeon, *Array Signal Processing: Concepts and Techniques*, Prentice Hall PTR, englewood Cliffs, New Jersey 07632, 1993.

[12] B. Widrow and S.D. Stearns, *Adaptive Signal Processing*, Prentice Hall PTR, englewood Cliffs, New Jersey 07632, 1985.

[13] S. Haykin, *Adaptive Filter Theory*, Prentice Hall PTR, englewood Cliffs, New Jersey 07632, 1986.

[14] L.J. Griffiths, "A simple adaptive algorithm for real-time processing in antenna arrays," *Proc. IEEE*, Vol. 57. No. 7, pp. 1696-1713, October 1969.

[15] J.J. Shynk, A.V. Keerthi, and A. Mathur, "Steady-state analysis of the multistage constant modulus array," *IEEE Trans. Signal Proc.*, Vol. 44. No. 4, pp. 948-962, July 1995.

[16] R. DiPietro, "Extended factored space-time processing for airborne radar systems," *Proc. 26th Asilomar Conf. on Signals, Systems, and Computing*, Pacific Grove, CA, pp. 425-430, October, 1992.

[17] J. Ward, *Space-Time Adaptive Processing for Airborne Radar,* Lincoln Laboratory Technical Report 1015, Lexington, MA, December 1994.

[18] L. J. Griffiths and C.W. Jim "An alternative aproach to linearly constrained adaptive beamforming," *IEEE Trans. Ant. Prop.*, Vol. AP-30, pp. 27-34, July 1982.

[19] L. J. Griffiths, K. M. Buckley. "Quiescent Pattern Control in Linearly Constrainted Adaptive Arrays," *IEEE Trans. ASSP*, vol. ASSP-35. no. 7, pp. 917-926, July 1987.

2

Demodulation in the Presence of Multiuser Interference: Progress and Misconceptions

Sergio Verdú

ABSTRACT The last ten years have witnessed the appearance of a large number of works in signal processing applied to multiuser communications, and in particular to the demultiplexing of overlapping digital streams, such as Code-Division Multiple-Access (CDMA) channels. This chapter reviews a sampling of the vast literature on *multiuser detection* and critically examines some of the misconceptions that may have arisen during its development.

2.1 Introduction

Sharing of channel resources is dictated by wireless multiaccess of bursty, uncoordinated, mobile users. The major approaches to channel sharing followed by first-generation digital mobile telephony are Time-Division Multiple-Access (TDMA) and Code-Division Multiple-Access (CDMA). In TDMA, time is partitioned into slots assigned to each incoming digital stream in round-robin fashion. This requires that the geographically-separated transmitters and the receiver maintain time-synchronism. Provision for dynamic reassignment of idle time-slots is crucial to avoid inefficiency of channel utilization. Demultiplexing is carried out by simply switching on to the received signal at the appropriate epochs. Ideally, no multiuser (also known as *co-channel*) interference arises and adjacent time-slots are completely transparent to each demodulated stream; in practice, imperfect time-synchronism and non-ideal received waveforms translate into multiuser interference. In first-generation wireless digital telephony TDMA systems such as GSM [17] and IS-54 [113], this non-ideal effect is combatted inserting guard-times, which decreases the efficiency of channel utilization. In contrast to TDMA, CDMA lets transmitters occupy simultaneous time-frequency slots. If time-synchronism can be maintained and the overall data rate is not too high relative to channel bandwidth, CDMA signals (called spreading codes or signature waveforms) can be designed that overlap in both time and frequency, while preserving the orthogonality enjoyed by TDMA signals. Again, orthogonality is desirable because it avoids multiuser interference. Allowing overlap in both time and frequency has the advantage that the each individual transmitter can use spread-spectrum signaling (bandwidth much larger than symbol rate), which is not only more robust against channel impairments (such as multipath and fading) but makes them easier to combat. This reason, as well as information theoretic arguments, are adduced by the proponents of the CDMA-based IS-95 North-American standard for first-generation digital-telephony [114]. In such a system, orthogonal CDMA is used in the downlink channel (from base station to mobile users), whereas non-orthogonal asynchronous CDMA is used in the uplink channel. Non-orthogonal CDMA has two major advantages: (1) the mobile transmitters are not required to be synchronized, (2) a higher number of users can be accommodated in the same bandwidth. A major challenge arising from non-orthogonal CDMA is multiuser interference. In the Qualcomm [138] embodiment of IS-95, multiuser interference is combatted through the use of fast and accurate power control and the use of highly redundant error control codes.

Multiuser detection studies the demodulation of one or more digital signals in the presence of multiuser interference. Those strategies apply when multiuser interference is present by design (non-orthogonal CDMA) as well as to TDMA, orthogonal CDMA, or orthogonal frequency division multiplexing (OFDM), where co-channel interference arises from channel non-ideal effects and out-of-cell transmissions.

Misconception 2.1 *Direct-Sequence CDMA is the only multiplexing strategy for which multiuser detection can be applied.*

The following sections are organized according to the main strategies that have been proposed for digital demodulation in the presence of multiaccess interference.

2.2 Single-user Matched Filter

When CDMA was first proposed five decades ago [105], [104], the presence of multiaccess interference at the receiver was simply neglected, on the grounds that its statistical properties would be similar to additive white Gaussian noise, and therefore a single-user matched filter should be near-optimal to combat such interference. Figure 2.1 shows a bank of single-user matched filters. Each branch operates independently and neglects the presence of multiaccess interference in the channel. The designers of the IS-95 system in the late eighties took the same approach. The only signal processing technique used in the Qualcomm demodulator of IS-95 signals is the *rake*: an adaptive matched filter developed in the late fifties [96] for demodulating a single-user spread-spectrum HF teletype signal subject to multipath. Very accurate and fast *power control* is required in the IS-95

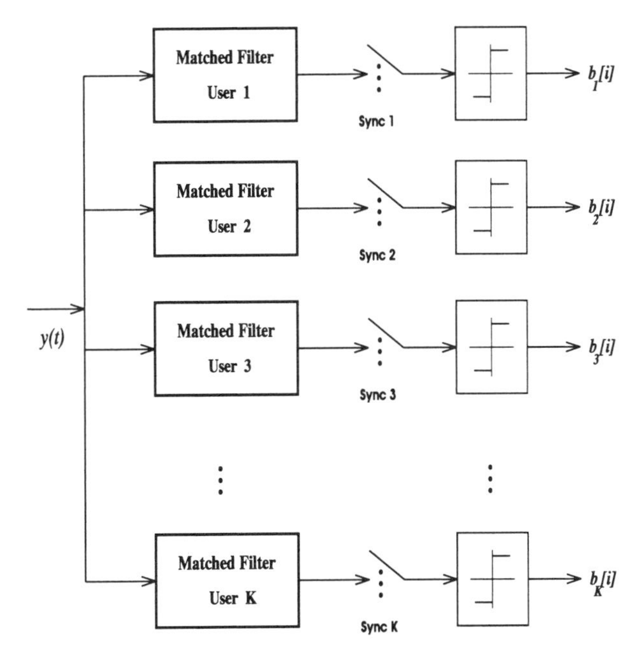

FIGURE 2.1. Bank of Single-user Matched Filters

system to keep the levels of the received signals to within a fraction of one dB. Among the benefits brought about by the use of power control we can cite:

- In mobile urban environments with typical cell sizes, received power imbalances can be as high as 90 dB (the so-called *near-far problem*). It is very challenging to implement robust signal processing, time-acquisition, and demodulation for such enormous dynamic ranges.

- Power control contributes to save energy (very important for portable transmitters) by expending no more energy than that required by the receiver to achieve a given performance level.

- Power control alleviates the effects of fading.

- Power control reduces out-of-cell interference.

Those benefits accrue regardless of the multiplexing strategy used; however, they are most crucial in non-orthogonal CDMA demodulated with single-user matched filters. Unfortunately, not all wireless applications lend themselves to fast and accurate power control: satellite-mobile systems (due to long round-trip delays); peer-to-peer communication and systems with mobile base stations; multimedia CDMA where discordant rate and quality requirements of various information sources make equal received powers ill-advised [7]. Furthermore, reliance on very accurate and fast power control greatly increases the susceptibility of the system to jamming and to the deleterious effects of malfunctions of power control subsystems. Thus, even though power control is beneficial, it is important to have available demodulation strategies offering some degree of resistance to near-far effects.

Often, analyses of power-controlled CDMA systems using single-user matched filters incur in the following over-simplification:

Misconception 2.2 *The bit error rate of a single-user matched filter in the presence of many equal-power interferers is accurately approximated by assuming that the multiaccess interference at the output of the matched filter is a Gaussian random variable (e.g. [97] and [149]).*

Unless the signal-to-noise ratio is very low, the Gaussian approximation is unreliable even in perfect power-controlled situations with many users and mild individual interference. This is illustrated in Figure 2.2 (also, e.g. [117]). Uncritical use of the Gaussian approximation leads to misconceptions on the performance of the single-user matched filter such as the following (quoted verbatim from [125]):

Misconception 2.3 *The capacity [of CDMA with single-user matched filtering] is limited by the total power of the interference. The number of interferers is irrelevant, as is the actual received power of each interferer.*

This statement is refuted by a very simple analysis of the error probabilities of the single-user matched filter with one interferer and with an infinite number of interferers with the same total power as the single interferer. The reason for the inaccuracy of the Gaussian approximation is that the error is accumulated in the tails of the distribution which are responsible for the low bit-error-rate behavior. Several works have studied better approximations to the error proability of the single-user receiver in the presence of multiaccess interference [25], [98], [70], [85], [102], [35], [36], [117].

Misconception 2.4 *The single-user matched filter achieves quasi-optimal performance whenever its output can be approximated as a Gaussian random variable.*

Even if the scenario is such that a *bona-fide* application of the central limit theorem is feasible (e.g. fixed overall interference power and a number of equal-power interferers growing to infinity), the single-user matched filter is far from optimal. The flaw in the argument of Misconception 2.4 is to restrict attention to demodulation strategies based only on the output of the matched filter of the desired user, which is not a sufficient statistic (unless the multiplex is orthogonal). The outputs of the matched filters of the interfering users give valuable information for the demodulation of the bit stream of interest. So much so, that in the hypothetical case when no background noise is present, error-free demodulation can be achieved (under mild conditions on the signature waveforms) using all matched filter outputs. In contrast, in a model with an infinite number of equal-power

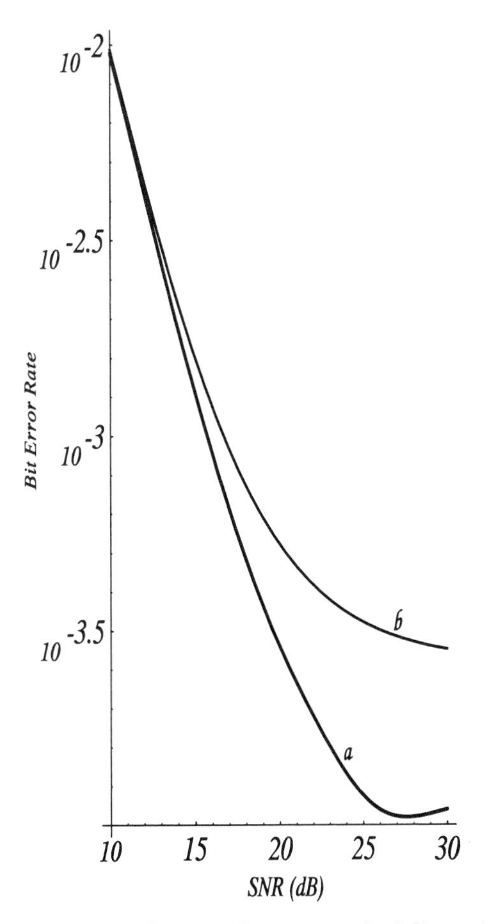

FIGURE 2.2. Bit-error-rate of the single-user matched filter with 10 equal-energy users and identical crosscorrelations $\rho_{kl} = 0.08$; (a) exact, (b) Gaussian approximation.

interferers the bit-error-rate of the single-user matched filter achieves a nonzero limit as the background noise goes to zero.

2.3 Optimum Multiuser Detection

The optimum multiuser detector employs the same front-end as in Figure 2.1: a bank of matched filters, but rather than demodulating each stream independently a combinatorial optimization algorithm demodulates all streams simultaneously [127], [132]. Such a combinatorial optimization problem was shown [127] and [128] to belong to the class of NP-complete problems, for which polynomial algorithms (in the number of users) exist if and only if they can be found for longstanding combinatorial problems such as the traveling salesman problem and integer linear programming. It is generally believed that such polynomial algorithms do not exist. Interestingly, whether the users are synchronous or asynchronous has very little impact on the asymptotic complexity of optimum multiuser demodulation. This is thanks to the possibility of using the Viterbi dynamic programming algorithm [132] for maximum likelihood detection and backward-forward dynamic programming [127], [137] for minimum probability of error detection. Prior to [133], [127], [132], the applicability of the Viterbi algorithm for the demodulation of asynchronous multiuser channels had been hinted in [103] and [64]. The vector versions of the optimum multiuser detector in [16] and [133] do not exploit fully the structure of the asynchronous channel and (for the same complexity) can accommodate only a fraction of the number of users of the algorithm in [132].

Misconception 2.5 *The optimum multiuser detector is useless in practice since its complexity is exponential in the number of interferers.*

With the computational power of a personal computer it is possible to demodulate optimally streams of nine 8 kbps transmitters. It is impossible to demodulate optimally, say, 20 or more such users. Such impossibility of *optimal* performance in large user populations does not imply, however, that the structure of the maximum likelihood detector is useless, as it naturally lends itself to various approximations. In particular, neglecting all but a fixed number of interferers may be a sensible strategy (e.g. [119]); for example, non-ideal effects in TDMA often limit the number of effective interferers to two in-cell and two out-of-cell interferers. Moreover, extensive progress has been made in reduced-state implementations of the Viterbi algorithm that achieve near-optimum performance.

In contrast to the bank of single-user matched filters, the optimum multiuser detector requires knowledge of received signal powers. The application of that algorithm to mobile users with imperfect power control, requires the estimation of the received amplitudes. This problem was treated first in [127] in the case when training sequences are available. Other works on this topic include [93], [77], [111], [84], [77], [146], [51], [49] [108], and [109].

The analysis of minimum probability of error of multiuser signals in additive noise is due to [132]. A main part of that analysis was based on a novel technique of *indecomposable vectors* which exploited the Euclidean structure of the problem to get tight bounds on bit error rate. The results of this analysis (cf. Figure 2.3) refuted the following (cf. Misconception 2.4):

Misconception 2.6 *Significant performance gains over the single-user matched filter are only possible for small numbers of interferers or in near-far situations.*

Moreover, the results in [132] and [135] showed that the near-far problem is not an inherent shortcoming of CDMA, but of the single-user matched

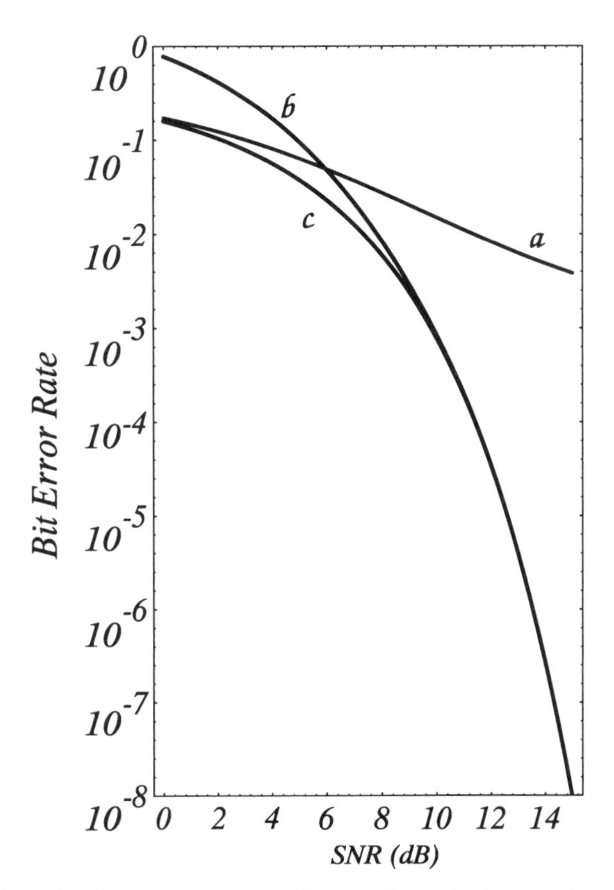

FIGURE 2.3. Bit-error-rate in a 15-user channel with equal energy users and $\rho_{kl} = 0.09$; (a) single-user matched filter, (b) upper bound to minimum bit-error-rate, (c) lower bound to minimum bit-error-rate.

filter. This gave rise to:

Misconception 2.7 *Multiuser detection eliminates the need for power control.*

The aforesaid desirable features of power control are not negated at all by the use of multiuser detection. In fact, a modicum of power control is necessary if acquisition and synchronization are carried out in the traditional approach which neglects the presence of multiaccess interference (e.g. [124] and references therein). As in demodulation, the structure of the multiaccess interference can be exploited to increase robustness, near-far resistance and acquisition/synchronization speed [50] [73], [6], [15]. Having said that, multiuser detection does indeed lower considerably the requirements for power control in terms of both accuracy and tracking speed. A related misconception is:

Misconception 2.8 *Unequal received powers are best when using multiuser detection.*

As with several other misconceptions that we will review, the origin of Misconception 2.8 can be attributed to the fact that multiuser detection is often equated with the particular technique of successive cancellation (Misconception 2.20). Even in the simplest setting of two users, it is shown in [131] that for most multiuser detectors and signature waveforms of interest, perfect power control is optimal if the sum of the powers is fixed.

2.4 Linear Multiuser Detection

The gaps in performance and complexity between the optimum multiuser detector and the single-user matched filter have spurred considerable interest in suboptimal solutions. A substantial body of literature has been devoted to the study of linear multiuser detectors. A simple way to view those detectors is to consider the bank in Figure 2.1 with matched filters modified to take into account the multiaccess interference.

Misconception 2.9 *Multiuser detection applies to multipoint-to-point channels where the receiver is interested in demodulating all users in the channel.*

As we mentioned, the phrase *multiuser detection* is generally used as a synonym of demodulation in the presence of multiaccess interference, whether the receiver is interested in demodulating one, a group, or the totality of active transmitters. Linear multiuser detectors can naturally focus exclusively on the user or users of interest by simply including as many branches in the bank of modified matched filters as users are to be demodulated.

A natural linear method [83] to mitigate the effects of multiuser interference in Direct-Sequence Spread Spectrum is to approximate the additive interference as a stationary Gaussian process with a power spectral density which depends exclusively on the chip waveforms (and the number of users and their strength). This results on a simple modification of the single-user matched filter where chip matched filtering is done with respect to a pseudo-chip waveform. A related solution was proposed in [127] and [95] as the solution to a locally-optimum detection problem. The benefits of this approach are limited since no attempt is made to exploit (or learn) the spreading codes of the interferers.

The first linear multiuser detector (other than the single-user matched filter in Section 2) analyzed was the *decorrelating detector* [72]. The essence of this receiver is to invert the linear transformation consisting of the linear modulators and matched filters. In the asynchronous case, there is not a unique decorrelating detector. The length of the impulse response can be chosen by the designer (depending on the desired performance/complexity tradeoff). The solutions range from the infinite-impulse response decorrelator in [71] to the one-shot decorrelator of [134] and [56], including intermediate solutions such as [142] and [54].

It was shown in [72] and [71] that the decorrelating detector has the same *near-far resistance* as the optimum detector. Near-far resistance is a worst-case measure of performance in the low background noise region. Interestingly, this property is achieved by the decorrelating detector (and by the adaptive multiuser detectors discussed below) without requiring any knowledge of the received power of the individual users.

Misconception 2.10 *Multiuser detection requires knowledge of the received interfering powers. It is not robust against imperfect knowledge of those parameters.*

Often the individual crosscorrelations between the signature waveforms are relatively small. In that case, a first-order approximation of the decorrelating detector achieves excellent performance [76] and is very easy to implement. This approximation was proposed in [63] and [129]. (Other approximations can be found in [26] and [148].) Like the optimum multiuser detector [132], the first-order approximation of the decorrelating detector avoids any algebraic operations with the signal crosscorrelations. The crosscorrelations can even be arbitrarily time-varying, making this multiuser detector suitable for CDMA channels where the duration of the spreading codes is (much) longer than the duration of each symbol. The first-order approximate decorrelator, as well as others proposed in the literature [55], refute the following (e.g.[125]):

Misconception 2.11 *Multiuser detection requires time-invariant channel conditions; it can only be used when the spreading code spans only one symbol interval.*

The IS-95 CDMA format [114] uses spreading codes whose length is much larger than the spreading gain (number of chips per symbol). Extremely long spreading codes are easily generated by shift registers. However, the overwhelming majority of the benefits of spread spectrum depend directly on the spreading gain rather than periodicity of the spreading code. Reliable privacy is obtained through cryptography rather than long spreading codes. The advantage of using spreading codes whose length is much larger than the spreading gain is easily outweighed by the burden that it imposes on adaptive multiuser detection strategies. A purported advantage of extremely long pseudo-random spreading sequences (so called R-CDMA in [125]) is the following, quoted verbatim from [125]:

Misconception 2.12 *R-CDMA is fundamentally a minimax approach to system design. It is well-known from information theory that when the variance of an additive interferer is constrained, the worst-case additive interference to any user is white Gaussian noise [22]. Hence a robust design* *methodology is to make multi-user interference look like AWGN through careful system design. This is accomplished by assigning users extremely long pseudo-random spreading sequences.*

The recourse to a minimax result (one actually due to Shannon [107]) is uncalled-for in this situation. Minimax results are applicable to *non-cooperative* games, such as those that model jamming or channels with statistical uncertainty. In this case, the system designer has influence over not only the modulation/coding strategy but also the multiaccess "noise". It does not make much sense to try to make that noise be the most harmful to the communication system under design. In fact, information theory teaches us exactly the opposite lesson: the less uncertainty the receiver has about the noise, the higher the capacity of the channel. Indeed, the important performance gains achievable by multiuser detection stem from exploiting the considerable structure of the multiaccess noise.

Another purported advantage of extremely long pseudo-random spreading sequences [125] is the ability of the conventional single-user matched filter to take advantage of the burstiness of active transmitters (due to non-continuous voice activity in telephony, for example).

Misconception 2.13 *Voice activity cannot be easily exploited in CDMA systems where the spreading sequence spans a single bit interval [125].*

The benefits of reduced interference accrue to the conventional single-user detector regardless of the length of the spreading sequence. Moreover, a base station receiver knows the activity status of each user in the cell and therefore can easily reconfigure the multiuser detector to take into account only those users active in the channel. On the other hand, in a decentralized situation a decorrelating detector which does not monitor the activity of interferers would not be able to take advantage of voice activity factors. Adaptive algorithms that can be applied to reconfigure the decorrelating detector to take into account only those users that are truly active have been studied in [80] and [81]. Moreover, the linear multiuser detector discussed in the next paragraph is naturally suited to take advantage of activity factors [37].

Another linear multiuser detection technique which has proven very successful is Minimum Mean-Square (MMSE) multiuser detection. This technique can be seen as a compromise between the multiaccess interference rejection capabilities of the decorrelating detector and the white-noise rejection capabilities of the single-user matched filter. It was proposed in [147], [75], [100], [99] and [101]. The bit error rate of the MMSE multiuser detector has been analyzed in [94]. The average multiuser capabilities of optimally near-far resistance receivers with Direct-Sequence Spread Spectrum with random signature waveforms have been succinctly demonstrated in [74] and [3]. The variability in performance due to random signature sequences has been studied in [40] finding conclusions which disagree with the claims in [125]. Simulations in realistic environments are reported in [89], [40] and [24]. Unlike the decorrelating detector, the MMSE linear multiuser detector does not require that the signature waveforms be linearly independent. This well-known fact contrasts with the following claim (verbatim from [125]):

Misconception 2.14 *Linear multiuser receivers require that the waveforms assigned to the users be linearly independent.*

The fact that the optimal multiuser detector, the decorrelator and the MMSE multiuser detector have direct forerunners in single-user communications subject to intersymbol interference may give the erroneous impression that the body of results in multiuser detection is a direct conceptual extension of single-user digital communication theory. However, a number of major results obtained for multiuser channels have no counterparts in the classical theory; for example, the optimal near-far resistance property of the decorrelator and the NP-completeness of optimum multiuser detection. Furthermore, many of the multiuser detectors reviewed in the sequel do not have natural counterparts for single-user channels. Undoubtedly, research in multiuser digital communication benefits enormously from a thorough understanding of the single-user theory. Conversely, synchronous CDMA channels (often with just two users) are often better pedagogical and conceptual tools than classical single-user channels subject to intersymbol interference.

An important advantage of MMSE multiuser detectors is that they lend themselves to *adaptive* implementations. A survey of the state-of-the-art in adaptive multiuser detection in 1994 can be found in [130]. Adaptive methods for demodulation in the presence of multiuser interference are either training-sequence-based ([75], [100], [79]), or blind ([38], [21], [39]). The dimensionality of the adaptive vector in those algorithms ranges from the spreading gain to a small multiple thereof. Thus the complexity of those adaptive detectors is independent of the number of interferers.

Misconception 2.15 *The complexity of a multiuser detector invariably grows with the number of interferers.*

The blind adaptive multiuser detector in [39] converges to the optimally near-far resistant MMSE linear multiuser detector requiring no more prior knowledge than the conventional detector. Thus, it is possible to combat interference caused by users whose signature waveforms, amplitudes, and timing are a-priori unknown. Thus, those adaptive detectors prove the following misconception wrong.

Misconception 2.16 *To combat an interfering user it is necessary to know its signature waveform, acquire its timing and despread it.*

Moreover, the detectors in [39] and [41] contradict:

Misconception 2.17 *Adaptive multiuser detection requires transmission of training sequences every time there is a significant change in channel conditions, such as powering on of an interferer.*

The adaptive detectors referenced so far are fully decentralized in that they do not incorporate knowledge of these signature waveforms of the interferers. In a base station demodulating all the active users in its cell, a cooperative approach is possible where the signature waveforms of all users in the cell can be used for initialization purposes and to improve the speed of convergence. Except for that aspect, which does not affect the steady-state solution of the algorithm, out-of-cell interference receives no different treatment than intra-cell interference: the effect of an interferer on the MMSE solution depends on its received power. Thus, the MMSE detector does not neglect the presence of out-of-cell interference, which is usually a substantial percentage of the overall multiaccess interference [140], [139]:

Misconception 2.18 *Multiuser detection cannot cope with out-of-cell interference.*

Another criticism that has been leveled against adaptive algorithms for demodulation in the presence of multiuser interference is that [125] "they need a high SNR to converge, and yet in a fading channel the SNR may exhibit momentary fluctuations of 10 dB or more." Analytical convergence results have been shown for the detectors in [75] and [39]. More important, there is experimental evidence of convergence of adaptive algorithms subject to the fading conditions of a typical mobile urban environment. A prime example is the rake adaptive algorithm whose speed is sufficient to track the channel variations encountered by IS-95 CDMA signals. Efforts in testing and optimizing adaptive multiuser detectors in realistic time-varying channel conditions have begun only recently. Encouraging simulation results can be found in [40], [37] and [23] and the references therein.

Considerable attention has been paid to combine multiuser detection with rake receivers [152], [153], [155], [19], [156], [60], [8], [90], [44], [47], [46], [48]. For example, a bank of rake receivers adapt themselves to the received single-user spreading waveforms and internal crosscorrelators (of the rake impulse responses) supply the (time-varying) crosscorrelations to the multiuser demodulators. In this way, the ability of rake adaptation to cope with multipath channels, well-established in connection with single-user demodulation [59], carries over to multiuser demodulation. In particular, in contrast to the following erroneous assertion, those combined rake-multiuser detectors do not treat each time-delayed version of each interfering waveform as a separate interferer.

Misconception 2.19 *Unlike R-CDMA, where multipath energies if harnessed effectively result in negligible capacity degradation, in D-CDMA interference cancellation is affected because any particular interferer will ap-*

pear in multiple dimensions, thereby increasing the perceived number of users [125].

This misconception confuses the dimensionality of the space spanned at any given time by the received waveforms (upper bounded by the number of users) with the dimensionality of the space spanned by all signature waveforms and all their time-delayed replicas.

2.5 Decision-based Multiuser Detection

A natural strategy in multiuser detection is *successive cancellation*: demodulate one user first (treating everyone else as noise); remodulate that signal with the recovered data; subtract it from the received process and repeat the procedure with all other users. Although this strategy is quite different from those we have discussed so far, some think it is the only known way to combat multiaccess interference:

Misconception 2.20 *Multiuser detection is successive cancellation.*

Successive cancellation is a proof technique in information theory [9], where it is used to show achievability of the capacity region of the multiaccess capacity region of the additive Gaussian channel. That scenario applies to error-control coded communication with arbitrarily low error probability. Otherwise, successive cancellation is suboptimal. In the context of multiuser detection of CDMA channels, successive cancellation has been examined in [62], [151], [10], [150], [91], [151],[52], [90], [34], [11]. References [138] and [140] question the implementability and robustness of this technique. Essentially, successive cancellation shows performance gains over linear multiuser detectors only when there are large imbalances in received powers. This fact together with Misconception 2.20 has led some authors to believe:

Misconception 2.21 *The power control required by multiuser detection is more complex than that required by the single-user matched filter.*

Reference [125] goes as far as claiming that even MMSE multiuser detection requires a more complex power control algorithm than single-user matched filtering. This claim contrasts with the fact that the sensitivity of the bit-error-rate of MMSE multiuser detection to variations in received powers is much lower than that of the single-user matched filter (e.g. the sensitivity of the bit error rate of the MMSE detector to received powers vanishes as the Gaussian background noise vanishes). Power control in conjunction with MMSE multiuser detection has been considered in [68].

Decision-directed multiuser detectors which overcome some of the shortcomings of successive cancellation are *multistage detectors* and *decision-feedback multiuser detectors.* Multistage detectors were proposed in [122] and [120] for synchronous and asynchronous channels, respectively; various embodiments and applications have been analyzed (among others) in [151], [43], [86]. Decision-feedback multiuser detectors have been proposed and analyzed in [147], [1], [129], [18], [2], [12], [13], [92], [145].

2.6 Noncoherent Multiuser Detection

In the development of multiuser detection, there has been much emphasis on linear modulation models. This emphasis is epitomised in the work on linear multiuser detection which exploits the rich structure of those models. Such emphasis may have led to:

Misconception 2.22 *Multiuser detection requires linear modulation and coherent demodulation.*

But multiaccess interference can be combatted even if it is not linearly modulated: consider, for example, the successive cancellation scheme, which requires no structure on the modulation format. The optimal detector has also been derived in non-coherent scenarios [136].

More importantly, there is now a substantial body of literature on multiuser detection for differentially coherent modulation and for fading channels. A tutorial discussion of this and other topics on multiuser detection can be found in [14]. A few of the more salient contributions are: [121], [118], [123], [45], [90], [52], [154], [155], [153], [156]. Contributions to noncoherent multiuser detection for Frequency-Hopping CDMA include [116], [115], [31] and [58].

2.7 Multiuser Detection combined with Array Processing

It is natural to exploit not only the time-domain structure of the multiaccess interference but the fact that users are not co-located. Array processing can be used to improve the signal-to-interference ratio. However, in nonorthogonal multiplexing, array processing (even with unlimited complexity, precision and speed) cannot eliminate the need to cope with multiaccess interference, because the because transmitters can be arbitrarily close.

Although [125] gives no reasons for the following puzzling assertion, it is wrong to treat multiuser detection and array processing as alternative mutually-exclusive options to deal with the same problem.

Misconception 2.23 *The benefits of sectorization are very hard, if not impossible, to achieve with CDMA if the spreading sequence spans one symbol interval.*

The integration of time-domain and space-domain signal processing techniques is currently the object of many research efforts. Existing results already offer much promise: [65], [4], [77], [78], [47], [45] [44], [112], [32], [66], [82], [42], [69], [57], [61], [110], [157], [87], [144].

The counterpart of knowing the signature waveform of the desired user at the receiver is the knowledge of its direction of arrival, which in a multipath environment generalizes to the knowledge of a directivity pattern. Obviously, in the spatial domain there is even more need than in the time domain for a robust blind adaptive approach to obtain the desired direction of arrival, as in that case no nominal information is available. In the same way that multiuser detection effectively exploits the fine structure of the digitally-modulated multiuser interference, it is important to exploit this structure in adaptive array processing. For example, the fact that the several paths due to one user contain the same data and are uncorrelated with paths of other users, leads to novel results [143] on the capabilities of beamforming.

2.8 Multiuser Detection with Error Control Coded Data

Much of the analysis of multiuser detectors has focused on performance measures such as bit-error-rate, asymptotic efficiency and near-far resistance under the assumption that the transmitted symbols are independent. In wireless systems, the data is error control coded, and thus, the assumption of independent data symbols no longer applies. Although the design of multiuser detectors has focused on hard-decision demodulators, in practice, it is likely that decision variables will not be thresholded and soft-decisions will be fed to the error-control decoder. This philosophy is called "user-separating demodulation" in [101]. Naturally, the bit-error-rate for independent data still provides a good indication of performance for the system designer, and in particular asymptotic multiuser efficiency [135] is a good measure of the effectiveness of the signal separation process.

Some of the major advances in multiuser detection (obtained under the assumption of unencoded data) have already been considered in conjunction with error correcting decoders. Succesive cancellation with orthogonal convolutional decoding was considered in [141] in a model where all the code-division multiplexing is on the side of high redundancy error control coding and no signature waveforms are used. The Viterbi algorithm of the optimum multiuser detector can be naturally adapted to the decoding of convolutionally encoded data at the expense of an increase in the dimensionality of the trellis [28], [27], [29]. In addition to demodulated data, the synchronous multiuser detector of [88] produces reliability values that are fed to a soft-decision Viterbi decoder (see also [33]). The capacity of channels achievable with the decorrelating detector in a frequency-selective Rayleigh fading environment has been considered in [30]. A constructive scheme that uses error control coding and decorrelator outputs is proposed in [5].

Simple linear codes are used in [106] in a fashion akin to letting time diversity play the role of space diversity. The combination of Turbo codes and multiuser detection for a CDMA mobile radio system has been explored in [53]. The superior bandwidth efficiency of trellis-coded modulation for additive noise channels has led to the investigation of multiuser detection with trellis decoders in [20] and [126]. The latter work shows that the error floor of conventional detection decreases exponentially with the product of space diversity and the minimum Hamming distance of the coded-modulation scheme. Excellent numerical results have been reported for trellis codes used in conjunction with adaptive MMSE multiuser detectors in [89] and [67]. This is particularly encouraging because at first glance one might expect that when error control coding is used, the benefits of using multiuser detection would decrease because of the tolerable lower signal-to-noise ratios. However, to be fair one must fix the spectral efficiency of the system (information rate/bandwidth) in which case, adding redundancy translates into smaller spreading gains for the signature sequences and consequently to higher levels of multiuser interference, enhancing the gains achievable by multiuser detection.

Misconception 2.24 *Multiuser interference is best combatted with error control coding rather than multiuser detection.*

But why should error control coding and multiuser detection be mutually exclusive approaches to deal with interference? There is no question that in most channels reliable digital communication requires error control coding and that disregarding the structure of the interference prior to decoding is distinctly suboptimal. Consider the case of single-user intersymbol interference where, after decades of research and development, the need for equalization can hardly be questioned. Admittedly, asynchronism, varying-user populations and fading make adaptive multiuser detectors more challenging to implement robustly in wireless channels than adaptive equalizers for the telephone channel. Moreover, since multiuser detection was at its infancy when first-generation digital telephony systems were designed, it is understandable that those systems relied exclusively on error control coding to combat multiaccess interference. However, as [125] states: "there is a widespread perception that wireless CDMA systems should be designed in order to exploit interference cancellation techniques." Indeed, increasing needs in spectral and power efficiency, rapid technological advances in the speed and accuracy of signal processors and the impressive growth in worldwide research and development activities in multiuser detection make it unlikely that future wireless systems (TDMA or CDMA) will be designed disregarding the availability of effective signal processing techniques to combat co-channel interference.

Acknowledgments: This work was partially supported by the United States Army Research Office under ARO DAAH04-96-1-0379

2.9 References

[1] M. Abdulrahman, *Cyclostationary crosstalk suppression by decision feedback equalization on digital subscriber loops*, IEEE Journal on Selected Areas in Communications, 10, No. 3 (April 1992), pp. 640–649.

[2] M. Abdulrahman, A. Sheikh, and D. Falconer, *Decision feedback equalization for CDMA in indoor wireless communications*, IEEE Journal on Selected Areas in Communications, (May 1994), pp. 698–706.

[3] I. Acar and S. Tantaratana, *Performance analysis of the decorrelating detector for DS spread-spectrum multiple-access communication systems*, Proc. 1995 Allerton Conf. Communications, Control, Computing, (Oct. 1995), pp. 1073–1082.

[4] B. G. Agee, *Solving the near-far problem: exploitation of spatial and spectral diversity in wireless personal communications*, preprint.

[5] P. Alexander, L. Rasmussen, and C. Schlegel, *A linear receiver for the coded asynchronous multiuser CDMA channel*, Proc. of the 33rd Annual Allerton Conference on Communication, Control, and Computing,, 3 (October 1995), pp. 29–38.

[6] S. E. Bensley and B. Aazhang, *Subspace-based delay estimation for CDMA communication systems*, Proc. 1994 International Symposium on Information Theory, (June 1994), p. 138.

[7] V. Bhargava, *High rate data transmission in mobile and personal communications*, 1995 PIMRC, (Sept. 1995).

[8] J. Blanz, A. Klein, M. NaBhan, and A. Steil, *Performance of a cellular hybrid c/tdma mobile radio system applying joint detection and coherent receiver antenna diversity*, IEEE Journal on Selected Areas in Communications, 12 (May 1994), pp. 568–579.

[9] T. Cover, *Some advances in broadcast channels*, in Advances in Communication Systems, Academic, New York, 1975, pp. 229–260.

[10] P. Dent, B. Gudmundson, and M. Ewerbring, *CDMA-ic: A novel code division multiple access scheme based on interference cancellation*, IEEE Globecom, (1992), pp. 98–102.

[11] D. Divsalar, M. Simon, and D. Raphaeli, *A new approach to parallel interference cancellation for CDMA*, preprint, (1996).

[12] A. Duel-Hallen, *Decorrelating decision-feedback multiuser detector for synchronous CDMA*, IEEE Trans. Communications, 41 (Feb. 1993), pp. 285–290.

[13] ——, *A family of multiuser decision-feedback detectors for asynchronous code-division multiple-access channels*, IEEE Trans. Communications, COM-43 (Feb./Mar./Apr. 1995), pp. 421–434.

[14] A. DUEL-HALLEN, J. HOLTZMAN, AND Z. ZVONAR, *Multiuser detection for CDMA systems*, IEEE Personal Communications, 2, No. 2 (April 1995), pp. 46–58.

[15] S. M. E.G. STROM, S. PARKVALL AND B. OTTERSTEN, *Propagation delay estimation in asynchronous direct-sequence code-division multiple access systems*, IEEE Trans. Communications, (January 1996), pp. 84–93.

[16] W. V. ETTEN, *Maximum likelihood receiver for multiple channeltransmission systems*, IEEE Trans. Commun., COM-24 (February 1976), pp. 276–283.

[17] EUROPEAN TELECOMMUNICATIONS STANDARDIZATION INSTITUTE, *Group Speciale Mobile or Global System for Mobile Communication (GSM) Recommendation*, Sophia Antipolis, France, 1988.

[18] D. FALCONER, M. ABDULRAHMAN, N. LO, AND B. PETERSEN, *Advances in equalization and diversity for portable wireless systems*, Digital Signal Processing, 3 (1993), pp. 148–162.

[19] U. FAWER AND B. AAZHANG, *A multiuser receiver for code division multiple access communications over multipath channels*, IEEE Trans. Communications, (Feb. 1995), pp. 1556–1565.

[20] ——, *Multiuser reception for trellis-based code division multiple access communications*, Proc. of MILCOM'94, (Oct. 1994), pp. 977–981.

[21] K. FUKAWA AND H. SUZUKI, *Orthogonalizing matched filter (omf) detection for DS-CDMA mobile radio systems*, Proc. 1994 Globecom, 1 (Nov. 28-Dec. 1, 1995), pp. 385–389.

[22] R. G. GALLAGER, *Information Theory and Reliable Communication*, Wiley, New York, 1968.

[23] R. D. GAUDENZI, F. GIANNETTI, AND M. LUISE, *Advances in satellite CDMA transmission for mobile and personal communications*, Preprint, (1996).

[24] ——, *An advanced linear adaptive interference-rejection receiver fo CDMA satellite communications*, Proc. 5th ESA Int. Workshop on Digital Signal Processing Techniques Applied to Space Communications, 1 (September 1996), p. 11.19.

[25] E. GERANIOTIS AND M. PURSLEY, *Error probability for direct-sequence spread-spectrum multiple-access communications - part II: Approximations*, IEEE Trans. Commun., COM-30 (May 1982), pp. 985–995.

[26] V. GHAZI-MOGHADAM, L. NELSON, AND M. KAVEH, *Parallel interference cancellation for CDMA systems*, Proc. 1995 Allerton Conf., (Oct. 1995).

[27] T. GIALLORENZI AND S. WILSON, *Multiuser ML sequence estimator for convolutionally coded asynchronous DS-CDMA systems*, IEEE Trans. Communications, (Aug. 1996), pp. 997–1008.

[28] ——, *Trellis-based multiuser receivers for convolutionally coded CDMA systems*, Proc. of the 31st Allerton Conference on Comm., Control and Computing, (Oct. 1993).

[29] ——, *Suboptimum multiuser receivers for convolutionally coded asynchronous DS-CDMA systems*, IEEE Trans. Communications, (September 1996), pp. 1183–1196.

[30] D. GOECKE AND W. STARK, *Throughput optimization in multiple-access communication systems with decorrelator reception*, Proc. 1996 IEEE Int. Symposium on Information Theory and Its Applications, 2 (September 1996), pp. 653–656.

[31] D. GOODMAN, P. HENRY, AND V. PRABHU, *Frequency-hopped multilevel FSK for mobile radio*, The Bell System Technical Journal, 59 (September 1980), pp. 1257–1275.

[32] S. GRAY AND J. C. PREISIG, *Multiuser detection in mismatched multiple-access channels*, Proc 1994 Conf. Information Sciences and Systems, (Mar. 1994).

[33] P. HOEHER, *On channel coding and multiuser detection for DS/CDMA*, Proc. of the 2nd International Conference on Universal Personal Communications, (1993), pp. 641–646.

[34] J. HOLTZMAN, *DS/CDMA successive interference cancellation*, in Code Division Multiple Access Communications, Kluwer Academic, 1995, pp. 161–182.

[35] J. HOLTZMAN, *A simple accurate method to calculate spread-spectrum multiple access error probabilities*, IEEE Trans. on Communications, 40 (Mar. 1992), pp. 461–464.

[36] J. HOLTZMAN, *On calculating DS/SSMA error probabilities*, Proc. 1992 Int. Symp on Spread Spectrum Techniques and Applications, (Nov-Dec 1992), pp. 23–26.

[37] M. HONIG, *Performance of adaptive interference suppression for DS-CDMA with a time-varying user population*, Proc. 1996 Int. Symposium on Spread Spectrum Techniques and Applications, (Sep. 1995).

[38] M. HONIG, U. MADHOW, AND S. VERDÚ, *Blind adaptive interference suppression for near-far resistant CDMA*, Proc. 1994 Globecom, 1 (Dec. 1994), pp. 379–384.

[39] ——, *Blind adaptive multiuser detection*, IEEE Trans. on Information Theory, 41 (July 1995), pp. 944–960.

[40] M. HONIG AND W. VEERAKACHEN, *Performance variability of linear multiuser detection for DS-CDMA*, Proc. Vehicular Technology Conference, (May 1996).

[41] M. L. HONIG, *A rescue operation to enhance the near-far adaptability of linear MMSE detectors for DS-CDMA*, IEEE Information Theory Workshop, (1995), pp. 3.4–3.5.

[42] S. HOSUR, A. TEWFIK, AND V. MOGHADAM, *Adaptive multiuser receiver schemes for antenna arrays*, Proc. of the 6th IEEE Int. Symp. on Personal, Indoor and Mobile Radio Communications (PIMRC), 3 (Sept. 1995), pp. 940–944.

[43] A. HOTTINEN, H. HOLMA, AND A. TOSKALA, *Performance of multistage multiuser detection in a fading multipath channel*, Proc. of the 6th IEEE Int. Symp. on Personal, Indoor and Mobile Radio Communications (PIMRC), 3 (Sept. 1995), pp. 960–964.

[44] H. HUANG, *Combined Multipath Processing, Array Processing, and Multiuser Detection for DS-CDMA Channels*, PhD thesis, Princeton University, January 1996.

[45] H. HUANG AND S. SCHWARTZ, *Noncoherent multiuser detection using array sensors*, Proc. Asilomar Conference on Signals, Systems and Computing, (1994), pp. 863–867.

[46] H. HUANG AND S. SCHWARTZ, *A comparative analysis of linear multiuser detectors for fading multipath channels*, Proc 1994 IEEE Globecom, (Dec. 1994).

[47] H. HUANG, S. SCHWARTZ, AND S. VERDÚ, *Combined multipath and spatial resolution for multiuser detection: Potentials and problems*, Proc. 1995 IEEE Int. Symp. Information Theory, (Sept. 17-22, 1995), p. 380.

[48] H. HUANG AND S. VERDÚ, *Linear differentially coherent multiuser detection for multipath channels*, Wireless Personal Communications, (Jan. 1997).

[49] R. ILTIS AND L. MAILAENDER, *An adaptive multiuser detector with joint amplitude and delay estimation*, IEEE J. on Selected Areas in Communications, 12, No. 5 (June 1994), pp. 774–785.

[50] R. ILTIS AND L. MAILAENDER, *Multiuser code acquisition using parallel decorrelators*, Proc. Conference on Information Sciences and Systems, (Mar. 1994), pp. 109–114.

[51] R. ILTIS AND L. MAILAENDER, *A symbol-by-symbol multiuser detector with joint amplitude and delay estimation*, Proceedings of the Asilomar Conference on Signals, Systems and Computers, (Oct. 1992), pp. 108–112.

[52] L. JALLOUL AND J. HOLTZMAN, *Performance analysis of DS/CDMA with noncoherent m-ary orthogonal modulation in multipath fading channels*, IEEE J. on Selected Areas in Communications, 12, No. 5 (June 1994), pp. 862–870.

[53] P. JUNG, M. NASSHAN, AND J. BLANZ, *Application of turbo-codes to CDMA mobile radio system using joint detection and antenna diversity*, Proc. of VTC'94, (June 1994), pp. 770–774.

[54] M. JUNTTI AND B. AAZHANG, *Linear finite memory-length multiuser detectors*, IEEE Global Telecommunications Conference, (Nov. 13-17, 1995).

[55] M. JUNTTI, B. AAZHANG, AND J. LILLEBERG, *Linear multiuser detection for R-CDMA*, Proc. 1996 IEEE Global Telecommunications Conference, (Nov. 18-22, 1996).

[56] A. KAHIWARA AND M. NAKAGAWA, *Crosscorrelation cancellation in ss/ds block demodulator*, IEICE Transactions, (Sep. 1991), pp. 2596–2602.

[57] S. KANDALA, E. SOUSA, AND S. PASUPATHY, *Multi-user multi-sensor detectors for CDMA networks*, IEEE Trans. Communications, (Feb. 1995), pp. 946–957.

[58] S. KANG, Y. HASUKA, AND R. KOHNO, *Multiuser detection for MFSK/FH-CDMA in a multi-cellular environment*, Proc. 1996 IEEE Int. Symposium on Information Theory and Its Applications, 2 (September 1996), pp. 538–541.

[59] C. KCHAO AND G. STUBER, *Performance analysis of a single cell direct sequence mobile radio system*, IEEE Trans. on Communications, COM-41 (Oct. 1993), pp. 1507–1516.

[60] A. KLEIN AND P. BAIER, *Linear unbiased data estimation in mobile radio systems applyinf CDMA*, IEEE Journal on Selected Areas on Communications, SAC-11 (Sep. 1993), pp. 1058–1066.

[61] R. KOHNO, *Spatial and temporal filtering for co-channel interference in CDMA*, in Code Division Multiple Access Communications, Kluwer Academic, 1995, pp. 117–146.

[62] ——, *Pseudo-noise sequences and interference cancellation techniques for spread spectrum systems–spread spectrum theory and techniques in Japan*, IEICE Trans., E.74 (May 1991), pp. 1083–1092.

[63] R. KOHNO, M. HATORI, AND H. IMAI, *Cancellation techniques of co-channel interference in asynchronous spread spectrum multiple access systems*, Electronics and Communications, 66-A (1983), pp. 20–29.

[64] R. KOHNO, H. IMAI, AND M. HATORI, *Cancellation techniques of co-channel interference and application of Viterbi algorithm in asynchronous spread spectrum multiple access systems*, 1982 Symposium on Information Theory and its Applications, (Oct. 1982), pp. 659–666. in Japanese.

[65] R. KOHNO, H. IMAI, M. HATORI, AND S. PASUPATHY, *Combination of an adaptive array antenna and a canceller of interference for Direct Sequence Spread Spectrum multiple access system*, IEEE Journal on Selected Areas in Communications, (1990), pp. 675–681.

[66] R. KOHNO, N. ISHII, AND M. NAGATSUKA, *A spatially and temporally optimal multi-user receiver using an array antenna for DS/CDMA*, Proc. of the 6th IEEE Int. Symp. on Personal, Indoor and Mobile Radio Communications (PIMRC), 3 (Sept. 1995), pp. 950–954.

[67] R. KOHNO, P. RAPAJIC, AND B. VUCETIC, *An overview of adaptive techniques for interference minimization in CDMA systems*, Wireless Personal Communications, 1, No. 1 (1994), pp. 3–21.

[68] P. KUMAR AND J. HOLTZMAN, *Power control for a spread spectrum system with multiuser receivers*, Proc. of the 6th IEEE Int. Symp. on Personal, Indoor and Mobile Radio Communications (PIMRC), 3 (Sept. 1995), pp. 955–959.

[69] W. LEE AND R. PICKHOLTZ, *Maximum likelihood multiuser detection with use of linear antenna arrays*, preprint, (1996).

[70] J. LEHNERT AND M. PURSLEY, *Error probability for binary direct-sequence spread-spectrum communications with random signature sequences*, IEEE Trans. Communciactions, COM-35 (1987), pp. 87–98.

[71] R. LUPAS AND S. VERDÚ, *Near-far resistance of multiuser detectors in asynchronous channels*, IEEE Trans. Communications, COM-38 (Apr. 1990), pp. 496–508.

[72] ——, *Linear multiuser detectors for synchronous code-divsion multiple-access channels*, IEEE Trans. Information Theory, IT-35 (Jan. 1989), pp. 123–136.

[73] U. MADHOW, *Blind adaptive interference suppression for acquisition and demodulation of direct-sequence CDMA signals*, Proc. 1995 Conf. Information Sciences and Systems (CISS '95), (Mar. 1995).

[74] U. MADHOW AND M. HONIG, *Performance analysis of MMSE detectors for direct sequence CDMA assuming random signature sequences*, IEEE Transactions on Information Theory, (1995). submitted.

[75] ——, *MMSE interference suppression for direct-sequence spread spectrum CDMA*, IEEE Trans. Communications, 42 (Dec. 1994), pp. 3178–3188.

[76] N. MANDAYAM AND S. VERDÚ, *Analysis of an approximate decorrelating detector*, Wireless Personal Communications, (1997). to appear.

[77] S. MILLER, *Detection and Estimation in Multiple-Access Channels*, PhD thesis, Princeton University, Princeton, NJ, Oct. 1989.

[78] S. MILLER AND S. C. SCHWARTZ, *Integrated spatial-temporal detectors for asynchronous gaussian multiple-access channels*, IEEE Trans. Communications, (Feb. 1995), pp. 396–411.

[79] S. L. MILLER, *An adaptive direct-sequence code-division multiple-access receiver for multi-user interference rejection*, IEEE Trans. Communications, 43 (Apr. 1995), pp. 1746–1755.

[80] U. MITRA AND H. POOR, *Adaptive receiver algorithms for near-far resistant CDMA*, IEEE Trans. Communications, (Feb. 1995), pp. 1713–1724.

[81] ——, *Analysis of an adaptive decorrelating detector for synchronous CDMA channels*, IEEE Trans. Communications, (February 1996), pp. 257–268.

[82] V. MOGHADAM AND M. KAVESH, *Interference cancellation using antenna arrays*, Proc. of the 6th IEEE Int. Symp. on Personal, Indoor and Mobile Radio Communications (PIMRC), 3 (Sept. 1995), pp. 936–940.

[83] A. MONK, M. DAVIS, L. MILSTEIN, AND C. HELSTROM, *A noise-whitening approach to multiple access noise rejection-part I: Theory and background*, IEEE J. on Selected Areas in Communications, 12, No. 5 (June 1994), pp. 817–827.

[84] T. MOON, Z. XIE, C. RUSHFORTH, AND R. SHORT, *Parameter estimation in a multi-user communication system*, IEEE Trans. Communications, (Aug. 1994), pp. 2553–2560.

[85] R. K. MORROW AND J. S. LEHNERT, *Bit-to-bit error dependence in slotted DS/SSMA packet systems with random signature sequences*, IEEE Trans. on Communications, COM-37 (Oct. 1989), pp. 1052–1061.

[86] S. MOSHAVI, E. KANTERAKIS, AND D. SCHILLING, *Multistage linear receivers for DS-CDMA systems*, Intl. Journal of Wireless Information Networks, 3 (Jan. 1996), pp. 1–18.

[87] O. MUÑOZ-MEDINA AND J. FERNANDEZ-RUBIO, *Optimum spatial-temporal detector for rician multiple-access channels*, Proc. 5th ESA Int. Workshop on Digital Signal Processing Techniques Applied to Space Communications, 1 (September 1996), p. 11.32.

[88] M. NASIRI-KENARI AND C. RUSHFORTH, *An efficient soft-decision decoding algorithm for synchronous CDMA communications with error-control coding*, Proc. IEEE International Symp. on Info. Theory, (June/July 1994), p. 227.

[89] I. OPPERMANN, B. VUCETIC, AND P. RAPAJIC, *Capacity of digital cellular CDMA system with adaptive receiver*, 1995 IEEE Int. Symp. Information Theory, (1995), p. 110.

[90] P. PATEL AND J. HOLTZMAN, *Analysis of a simple successive interference cancellation scheme in DS/CDMA system*, IEEE Journal Selected Areas on Communications, (June 1994), pp. 796–807.

[91] ——, *Analysis of a DS/CDMA successive interference cancellation scheme using correlations*, Proceedings of Globecom '93, (Nov. 29 - Dec. 2 1993), pp. 76–80.

[92] B. PETERSEN AND D. FALCONER, *Suppression of adjacent-channel, cochannel, and intersymbol interference by equalizers and linear combiners*, IEEE Trans. Communications, (Dec. 1994), pp. 3109–3118.

[93] H. POOR, *On parameter estimation in DS/SSMA formats*, Proc. 1988 Int. Conf. Advances in Communications and Control Systems, 1 (Oct. 1988), pp. 98–109.

[94] H. Poor and S. Verdú, *Probability of error in MMSE multiuser detection*, IEEE Trans. Information Theory, (1997).

[95] ——, *Single-user detectors for multiuser channels*, IEEE Trans. Communications, 36 (Jan. 1988), pp. 50–60.

[96] R. Price and P. Green, *A communication technique for multipath channels*, Proc. IRE, 46 (Mar. 1958), pp. 555–570.

[97] M. Pursley, *Performance evaluation for phase-coded spread-spectrum multiple-access communication – part I: System analysis*, IEEE Trans. on Communications, COM-25 (Aug 1977), pp. 795–799.

[98] M. Pursley, D. Sarwate, and W. Stark, *Error probability for direct-sequence spread-spectrum multiple-access communications - part I: Upper and lower bounds*, IEEE Trans. Commun., COM-30 (May 1982), pp. 975–984.

[99] P. Rapajic and B. Vucetic, *Linear adaptive transmitter-receiver structures for asynchronous CDMA systems*, European Trans. Telecommunications, (Jan/Feb 1995), pp. 21–27.

[100] ——, *Adaptive receiver structures for asynchronous CDMA systems*, IEEE Journal on Selected Areas in Communications, (May 1994), pp. 685–697.

[101] M. Rupf, F. Tarkoy, and J. Massey, *User-separating demodulation for code-division multiple-access systems*, IEEE J. on Selected Areas in Communications, 12, No. 5 (June 1994), pp. 786–795.

[102] J. Sadowsky and R. Bahr, *Direct sequence spread spectrum multiple acces communication with random signature sequences*, IEEE Trans. Information Theory, IT-37 (May 1991), pp. 514–527.

[103] K. Schneider, *Optimum detection of code division multiplexed signals*, IEEE Trans. Aerosp. Electron. Syst., AES-15 (January 1979), pp. 181–185.

[104] R. Scholtz, *The evolution of spread-spectrum multiple-access communications*, in Code Division Multiple Access communications, S. G. Glisic and P. Leppänen, eds., Kluwer, 1995, pp. 3–28.

[105] ——, *The origins of spread-spectrum communications*, IEEE Trans. on Communications, COM-30 (May 1982), pp. 822–854.

[106] N. Seshadri, A. Calderbank, and G. Pottie, *Channel coding for cochannel interference suppression in wireless communication systems*, Proc. 1995 IEEE Vehicular Technology Conference, (1995).

[107] C. E. SHANNON, *A mathematical theory of communication*, Bell Sys. Tech. J., 27 (Jul.-Oct. 1948), pp. 379–423, 623–656.

[108] Y. STEINBERG AND H. POOR, *Multiuser delay estimation*, Proc. 1993 Conf. Inform. Sci. Syst., (1993).

[109] ——, *Sequential amplitude estimation in multiuser communications*, IEEE Trans. Information Theory, 38 (Jan. 1994), pp. 11–20.

[110] M. STOJANOVIC AND Z. ZVONAR, *Multisensor multiuser receivers for time-dispersive multipath fading channels*, Proc. 1995 IEEE Intl. Symp. on Info. Theory, (Sept. 1995).

[111] E. STROM AND S. MILLER, *Asynchronous DS-CDMA systems: Low-complexity near-far resistant receivers and parameter estimation*, tech. rep., Dept. of Elec. Eng. Univ. of Florida, Jan. 1994.

[112] B. SUARD, A. NAGUID, G. XU, AND A. PAULRAJ, *Performance of CDMA mobile communication systems using antenna arrays*, ICAASP Proceedings, IV (1993), pp. 153–156.

[113] TELECOMMUNICATIONS INDUSTRY ASSOCIATION, TIA/EIA, *Cellular System Dual-mode Mobile Station-Base Station Compatibility Standard IS-54B*, Washington, DC, 1992.

[114] ——, *Mobile Station-Base Station Compatibility Standard for Dual-Mode Wideband Spread Spectrum Cellular System IS-95A*, Washington, DC, 1995.

[115] U. TIMOR, *Multistage decoding of frequency-hopped FSK system*, The Bell System Technical Journal, 60 (April 1981), pp. 471–483.

[116] ——, *Improved decoding scheme for frequency-hopped multilevel FSK system*, The Bell System Technical Journal, 59 (December 1980), pp. 1839–1855.

[117] P. VAN ROOYEN AND F. SOLMS, *Maximum entropy investigation of the inter user interference distribution in a DS/SSMA system*, Proc. 1995 IEEE Personal, Indoor, Mobile Radio Communications Conference, (1995), pp. 1308–1312.

[118] M. VARANASI, *Noncoherent detection in asynchronous multiuser channels*, IEEE Trans. on Info. Theory, 39, No. 1 (January 1993), pp. 157–176.

[119] ——, *Group detection for synchronous Gaussian CDMA channels*, IEEE Trans. on Information Theory, 41 (July 1995), pp. 1083–1096.

[120] M. Varanasi and B. Aazhang, *Multistage detection in asynchronous code-division multiple-access communications*, IEEE Trans. Communications, 38 (Apr. 1990), pp. 509–519.

[121] ——, *Optimally near-far resistant multiuser detection in differentially coherent synchronous channels*, IEEE Trans. on Info. Theory, 37, No. 4 (July 1991), pp. 1006–1018.

[122] ——, *Near-optimum detection in synchronous code-division multiple-access systems*, IEEE Trans. Communications, (May 1991), pp. 725–736.

[123] M. Varanasi and S. Vasudevan, *Multiuser detectors for synchronous CDMA communications over non-selective rician fading channels*, IEEE Trans. on Communications, 42, No. 2/3/4 (1994), pp. 711–722.

[124] V. Veeravalli and C. Baum, *Hybrid acqisition of direct sequence CDMA signals*, Intl. Journal of Wireless Information Networks, 3 (Jan. 1996), pp. 55–65.

[125] S. Vembu and A. J. Viterbi, *Two different philosophies in CDMA-a comparison*, Proc. IEEE Vehicular Technology Conference, (Apr.-May 1996), pp. 869–873.

[126] J. Ventura-Traveset, G. Caire, E. Biglieri, and G. Taricco, *A multiuser approach to combating co-channel interference in narrowband mobile communications*, 7th Tyrrhenian Workshop on Digital Communication, (Sep. 1995).

[127] S. Verdú, *Optimum multi-user signal detection*, PhD thesis, University of Illinois at Urbana-Champaign, Aug. 1984.

[128] ——, *Computational complexity of optimum multiuser detection*, Algorithmica, 4 (1989), pp. 303–312.

[129] ——, *Multiuser detection*, in Advances in Statistical Signal Processing: Signal Detection, H. V. Poor and J. B. Thomas, eds., JAI Press, 1993, pp. 369–410.

[130] ——, *Adaptive multiuser detection*, in Code Division Multiple Access Communications, S. G. Glisic and P. A. Leppanen, eds., Kluwer Academic, 1995.

[131] ——, *Multiuser Detection*, in preparation, 1997.

[132] ——, *Minimum probability of error for asynchronous gaussian multiple-access channels*, IEEE Trans. on Information Theory, IT-32 (Jan. 1986), pp. 85–96.

[133] ——, *Minimum probability of error for asynchronous multiple access communication systems*, Proc. 1983 IEEE Military Communications Conference, 1 (Nov. 1983), pp. 213–219.

[134] ——, *Recent progress in multiuser detection*, Proc. 1988 Int. Conf. Advances in Communications and Control Systems, 1 (Oct. 1988), pp. 66–77.

[135] ——, *Optimum multiuser asymptotic efficiency*, IEEE Trans. Communications, COM-34 (Sept. 1986), pp. 890–897.

[136] ——, *Multiple-access channels with point-process observations: Optimum demodulation*, IEEE Trans. Information Theory, IT-32 (September 1986), pp. 642–651.

[137] S. Verdú and H. V. Poor, *Abstract dynamic programming models under commutativity conditions*, SIAM J. Control and Optimization, 24 (Jul. 1987), pp. 990–1006.

[138] A. Viterbi, *CDMA: Principles of Spread spectrum communications*, Addison-Wesley, 1995.

[139] ——, *Performance limits of error-correcting coding in multicellularCDMA systems with and without interference cancellation*, in Code Division Multiple Access Communications, Kluwer Academic, 1995, pp. 47–52.

[140] ——, *The orthogonal-random waveform dichotomy for digital mobile personal communication*, IEEE Personal Communications, (First Quarter, 1994), pp. 18–24.

[141] ——, *Very low rate convolutional codes for maximum theoretical performance of spread-spectrum multiple-access*, IEEE Journal on Selected Areas in Communications, (May 1990), pp. 641–649.

[142] S. Wijayasuriya, G. H. Norton, and J. McGeehan, *Sliding window decorrelating algorithm for DS-CDMA receivers*, Electronics Letters, 28 (Aug. 13th 1992), pp. 1596–1598.

[143] J. Winters, J. Salz, and R. Gitlin, *The impact of antenna diversity on the capacity of wireless communication systems*, IEEE Trans. Communications, (Feb. 1994), pp. 1740–1751.

[144] P. Wong and D. Cox, *Low-complexity diversity combining algorithms and circuit architectures for co-channel interference cancellation and frequency-selective fading mitigation*, IEEE Trans. Communications, (September 1996), pp. 1107–1116.

[145] H. WU AND A. DUEL-HALLEN, *Performance of multiuser decision-feedback detectors for flat fading synchronous CDMA channels*, Proc. 28th Annual Conf. on Information Sciences and Systems, (March 1994), pp. 133–138.

[146] Z. XIE, C. RUSHFORTH, R. SHORT, AND T. MOON, *Joint signal detection and parameter estimation in multi-user communications*, IEEE Transactions on Communications, (Aug. 1993), pp. 1208–1216.

[147] Z. XIE, R. SHORT, AND C. RUSHFORTH, *A family of suboptimum detectors for coherent multiuser communications*, IEEE J. Selected Areas in Communications, (May 1990), pp. 683–690.

[148] L. YANG AND R. A. SCHOLTZ, *δ-adjusted mth order multiuser detector*, in Multiaccess, Mobility and Teletraffic for Personal Communications, B. Jabbari, P. Godlewski, and X. Lagrange, eds., Kluwer Academic, 1996, pp. 249–263.

[149] K. YAO, *Error probability of asynchronous spread spectrum multiple access communication systems*, IEEE Trans. on Communications, COM-25 (Aug 1977), pp. 803–809.

[150] K. YOON, R. KOHNO, AND H. IMAI, *Cascaded co-channel interference cancelling and diversity combining for spread-spectrum multiaccess over multipath fading channels*, IEICE Trans. Commun., E76-B, No. 2 (Feb. 1993).

[151] Y. YOON, R. KOHNO, AND H. IMAI, *A spread-spectrum multi-access system with co-channel interference cancellation over multipath fading channels*, IEEE J. Select. Areas Commun., 11, No. 7 (Sept. 1993), pp. 1067–1075.

[152] Z. ZVONAR, *Combined multiuser detection and diversity reception for wireless CDMA systems*, to appear in IEEE Transactions on Vehicular Technology.

[153] ——, *Multiuser detection and diversity combining for wireless CDMA systems*, in Wireless and Mobile Communications, J. M. Holtzman and D. J. Goodman, eds., Kluwer Academic Publishers, 1994, pp. 51–65.

[154] Z. ZVONAR AND D. BRADY, *Differentially coherent multiuser detection in asynchronous CDMA flat rayleigh fading channels*, IEEE Trans. Communications, (Feb. 1995), pp. 1252–1259.

[155] ——, *Suboptimal multiuser detector for frequency-selective rayleigh fading synchronous CDMA channels*, IEEE Trans. Communications, (Feb. 1995), pp. 154–157.

[156] ——, *Linear multipath-decorrelating receivers for CDMA frequency-selective fading channels*, IEEE Trans. on Communications, (June 1996), pp. 650–653.

[157] Z. Zvonar and M. Stojanovic, *Performance of antenna diversity multiuser receivers in CDMA channels with imperfect fading estimation*, to appear in Wireless Personal Communications.

3

Intelligent Signal Detection

Simon Haykin

ABSTRACT
In this article we describe a novel system for the detection of a target signal in the presence of additive interference. The system operates under the premise that the statistics of the target signal and the interference are uknown and that both may be nonstationary. The design of the system is guided throughout by the information preservation rule.
The system consists of two independent channels fed from a common module that computes the Wigner-Ville distribution of the received signal.One channel, termed the interference channel, consists of a principal components analyzer followed by a multilayer perceptron classifier; these two components are tarined on different realizations of the received signal known to contain interference only. The other channel, termed the target channel, has a similar composition, except that this time the training data consist of different realizations of the received signal known to consist of a target signal plus interference.The two channels are linearly combined into a single output node where the final decision is made, whether a target is present or not.
Eperimental results, using real life ground-truthed data collected with an instrument-quality radar, are presented demonstrating the superior performance of the new detection system(receiver) over a conventional Doppler constant false-alarm receiver.

3.1 Introduction

In this article we focus on an important signal processing task, namely, that of detecting a weak target signal buried in a strong interference background that is nonstationary. This is a difficult signal processing task, particularly when the target-to-interference (noise) ratio is low.

The classical procedure of detecting a signal in the presence of "noise" is to use a matched filter/correlation receiver that represents a fundamental tool for the design of a conventional receiver for application in communications, radar, sonar, and so on. This approach, which goes back to the pioneering RCA report by North in 1943 and the 1946 classic paper by Middleton and Van Vleck, relies on the basic assumption that the noise is stationary. Yet in a typical operational environment, for example, a radar system operating in an ocean environment, it is known that the physical process responsible for generating the clutter is nonstationary. Moreover, radar clutter usually has non-Gaussian statistics, thereby complicating the theory of radar target detection that much more. We therefore find that there is a serious mismatch between (a) physical realities of the detection problem and (b) underlying mathematical assumptions of the classical approach. The net result is a violation of the information preservation rule.

The *information preservation rule* states that the information content of a received signal should be retained in its essential form as far as possible and used in a computationally efficient manner, until the system in question is ready to make its final decision. In violating this principle, we naturally end up with a suboptimal solution to the detection problem.

In this article we describe a novel neural-network-based strategy for the detection of a weak target signal of interest in an interference background that is strong enough to mask the signal. We have called the new strategy "Intelligent Signal Detection" in recognition of the intelligent way in which the information content of the received signal is preserved and then exploited for the task at hand. The idea of this new detection strategy is motivated by the echo-location (sonar) of a bat, which is known to detect, pursue and capture its target (e.g., an insect) with a facility and success rate that is the envy of every radar or sonar engineer [1]. We are not suggesting that our new detection strategy involves all the signal processing functions performed in the bat's echo-location system. What we are saying is that the three principal functions that characterize the new strategy, namely, time-frequency analysis, feature extraction, and pattern classification, are found in one form or another in the bat's echo-location system.

3.2 Three Basic Elements Of The Intelligent Detection System

Figure 1 shows a block diagram of the new detection strategy in its more basic form. It consists of three fundamental functional blocks that are designed to perform time-frequency analysis, feature extraction for dimensionality reduction, and pattern classification, in that order. In what follows, we shall attend to each one of these signal processing operations.

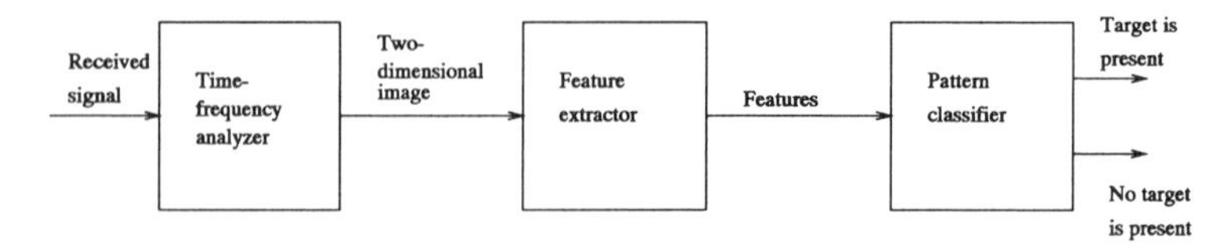

FIGURE 3.1. Block diagram showing the basic functional blocks of the signal detection system.

Time Frequency Analysis

Nonstationary signals have time-varying spectral properties, mandating the use of some form of joint time-frequency analysis. The technique employed for this purpose should do two things: (1) bring out the nonstationary behavior of the signal in a discernible fashion, and (2) allow the separation of multiple components contained in the signal. A possible approach is through the use of time-frequency distributions. Ideally, the aim is to approximate the elusive *time-frequency energy density function* $E_x(t,f)$, which is defined as the energy contained in a signal $x(t)$ within an infinitesimally small neighborhood around time t and frequency f. We say "approximate" because of the disjoint nature of time-frequency concentration. That is, a signal cannot be concentrated in both time and frequency simultaneously, by virtue of the *uncertainty principle* [2]. A thorough treatment of time-frequency distributions is beyond the scope of the present paper; a detailed exposition of the subject is presented in [2]. However, since this signal-processing step is highly critical to the whole detection scheme described herein, we present a brief overview of its relevant characteristics.

Time-frequency distributions are usually classified into linear and non-linear methods. The latter category includes the important subclass of *bilinear time-frequency representations* (BTFR), the formulation of which has been developed into a general framework by Cohen [2]. Specifically, the time-frequency distribution $C_x(t,f)$ for a signal $x(t)$ is said to exhibit the bilinear property if for every $(t,f) \in R^2$ there exists a linear operator $O(t,f)$ such that

$$C_x(t,f) = O(t,f)[x,x] \ for \ all \ x \in L^2(R) \tag{3.1}$$

At first glance, the importance of choosing this subclass as the basis of our detection strategy may not appear that obvious. But if we recall that, for our specific application, we are seeking a behavior akin to the time-frequency energy density function, though not necessarily in its exactly true form, then we must try to satisfy the time-frequency localized counterpart of the global energy conservation principle. That is, for any pair of signals $u(t)$ and $v(t)$ and an arbitrary pair of constants α and β, we should strive to satisfy the condition

$$E_{\alpha u+\beta v} = |\alpha|^2 E_u + |\beta|^2 E_v + \alpha\beta^* E_{u,v} + \beta\alpha^* E_{v,u} \tag{3.2}$$

where E is a measure of energy in the time-frequency domain, and the asterisk denotes complex conjugation. Correspondingly, in terms of the time-frequency distribution $C_x(t,f)$, we want

$$C_{\alpha u+\beta v}(t,f) = |\alpha|^2 C_u(t,f) + |\beta|^2 C_v(t,f) + \alpha\beta^* C_{u,v}(t,f) + \beta\alpha^* C_{v,u}(t,f) \tag{3.3}$$

which is just a statement of bilinearity imposed on $C_x(t, f)$. The first two terms on the right-hand side of (3) represent the auto-terms, and the remaining two terms represent the cross-terms.

For additional justification for the choice of a BTFR, we may refer to [3]. In this paper, Hlawatsch has considered the general class of BTFRs, and formulated the condition for a BTFR to be called *regular*[1]. The BTFR is said to be ***singular*** when it is not regular. It is only when the BTFR is regular that it is possible to recover the original signal from its BTFR to within a phase constant. This is an important property, since our primary design objective is to preserve the information content of the signal. It is now obvious that it is only by choosing a regular BTFR that the loss of information as a result of the transformation is minimized.

A particular BTFR that is regular is the *Wigner-Ville distribution* (WVD). Given a signal $x(t)$, its WVD is defined by [2]

$$W_x(t,f) = \int_{-\infty}^{\infty} x\left(t+\frac{\tau}{2}\right) x^*\left(t-\frac{\tau}{2}\right) e^{-2j\pi f\tau} d\tau \tag{3.4}$$

where the lag variable τ plays the role of a dummy variable. It is important to note that any other BTFR derived from the WVD by smoothing in the time-frequency plane is singular, in which case there is information loss due to the smoothing operation. This certainly points to the *optimum information-preserving property* of the standard WVD.

Although there are other BTFRs (e.g., the Rihaczek and Page distributions) that are also regular, the WVD amongst them all is the only one that satisfies two additional properties that are highly desirable from a signal detection viewpoint [4]:

- The WVD is real valued for any complex-valued input signal.
- It exhibits the least amount of spread in the time-frequency plane.

For these two reasons and, most importantly, because of the optimum information-preserving property of the WVD, we have chosen it as the tool

[1] Let $C_x(t, f)$ denote a bilinear time-frequency representation (BTFR) of a signal $x(t)$. According to Hlawatsch [3], $C_x(t, f)$ is defined in terms of a bilinear signal representation (BSR) operator $u_C(t, f; t_1, t_2)$ as follows:

$$C_x(t,f) = \int_{t_1}\int_{t_2} u_C(t,f;t_1,t_2)x(t_1)x^*(t_2)\,dt_1\,dt_2$$

The BTFT $C_x(t, f)$ is said to be ***regular*** if a bounded inverse operator $u_C^{-1}(t, f : t_1 t_2)$ exists such that:

$$\int_f\int_t u_C(t,f;t_1,t_2)u_C^{-1}(t,f;t_1',t_2')dtdf = \delta(t_1-t_1')\delta(t_2-t_2')$$

where $\delta(t)$ is the Dirac delta function.

for performing the time-frequency analysis that represents the first step in our modular detection strategy. A criticism that is often levelled against the WVD is the generation of cross-terms, or more precisely, *cross Wigner-Ville distributions*, due to the combined presence of two (or more) signal components. Recognizing that in an interference-dominated environment the cross-terms arise only when a target signal is present, it can be argued that the presence of cross-terms is in fact an asset. We say so because they provide another feature that can enhance the visibility of the target signal in the time-frequency image resulting from the application of the WVD. Indeed, cross-terms contribute to the optimal information-preserving property of the WVD in their own distinct way.

Feature Extraction

The amount of information contained in the WVD image is exactly the same as that contained in the original signal. Yet the dimension of the WVD image is N times that of the original signal, where N is the number of frequency bins in the WVD image. Accordingly, the use of WVD leads to a significant increase in the amount of redundant information contained in the time-frequency image of the input signal. To improve computational efficiency in accordance with the information preservation rule, it is necessary to follow up the WVD with some form of dimensionality reduction (i.e., data compression). A signal processing tool that is well suited for this task is *principal components analysis* (PCA) [5]. Basically, PCA performs a singular value decomposition (SVD) on the WVD image, and retains the singular vectors associated with the most significant singular values of the WVD image. The resulting output is represented by the combination of singular vectors retained by the PCA. Thus, the PCA is instrumental in extracting a finite set of *features* for the WVD image that is *optimum*, in that the original WVD image can be reconstructed from these features in a minimum mean-square (reconstruction) error sense. In other words, information loss brought on by the extraction of features by the PCA is kept to a minimum.

Figure 2 shows a two-channel receiver that expands on the basic scheme of Fig. 1. The two channels are fed from the WVD computer that performs time-frequency analysis on the input data. Both channels include a PCA network for feature extraction. In the interference channel of the receiver, the PCA network is tuned to input data known to contain interference only. In the target channel, on the other hand, the PCA network is tuned to input data known to contain a target signal plus interference.

The PCA is a linear method for dimensionality reduction. Alternatively, we may use a nonlinear method such as the *self-organizing feature map* (SOFM) devised by Kohonen [6]. The SOFM involves an array of neurons (e.g., two-dimensional in geometry), and the neurons undergo a form of competitive learning. For any one presentation of input data, a single neuron ends up winning the competition. In the system described in the next section, PCA was chosen for the feature extraction.

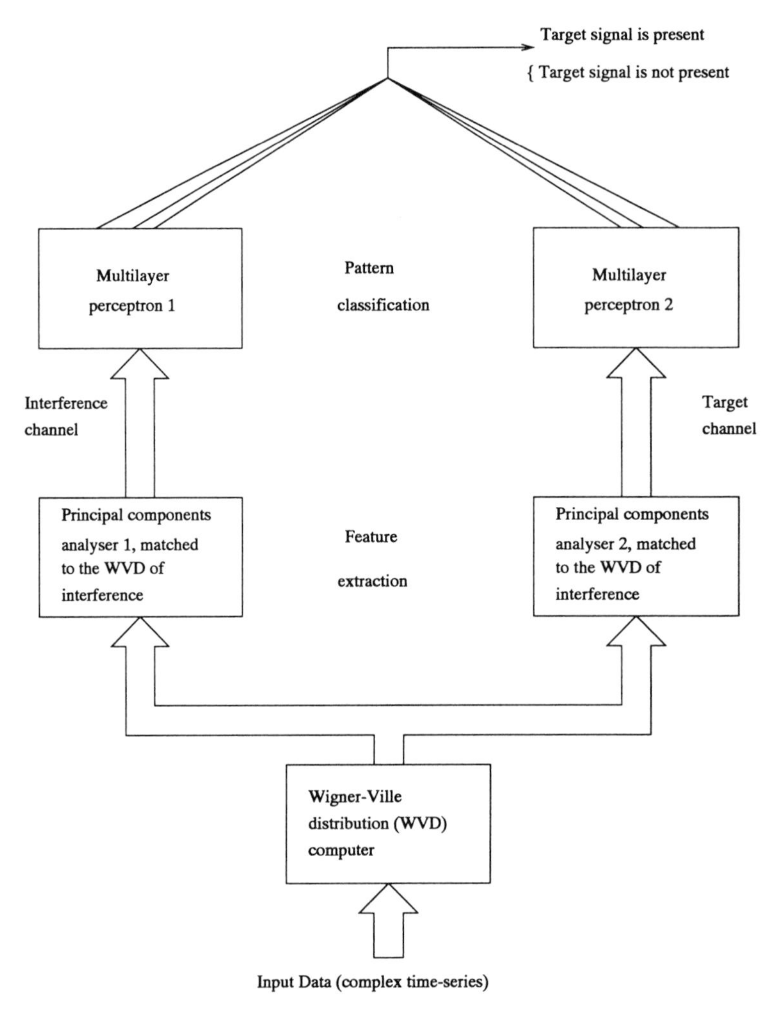

FIGURE 3.2. Two-channel signal detection system.

Pattern Classification

The final task is that of pattern classification, which is required to distinguish between the following two hypotheses (classes) in an optimum statistical sense:

- The null hypothesis H_0: the received signal consists of interference only.

- The other hypothesis H_1: the received signal consists of a target signal plus interference.

In other words, we have a binary hypothesis testing problem on our hands.

In the modular learning strategy of Fig. 2 this problem is tackled as follows. The two sets of features, extracted by PCA1 and PCA2 from the WVD image of a received signal, are applied to a corresponding pair of *multilayer perceptrons*, MLP1 and MLP2. Their use is motivated by the fact that an MLP is able to construct arbitrary decision boundaries between the two classes of interest, H_0 and H_1. The resulting "analog" outputs of the two MLPs are then *linearly combined* to produce an overall receiver output. A final decision is made by comparing this output against a preset threshold. If the *threshold* is exceeded, a decision is made that a target signal is present (i.e., hypothesis H_1 is true); otherwise, a decision is made that there is no target signal present (i.e., hypothesis H_0 is true).

In such a signal detection scenario, the following two decisions are of particular concern:

- *Missed detection*: the receiver decides in favor of hypothesis H_0 when H_1 is true.
- *False alarm*: the receiver decides in favor of hypotheses H_1 when H_0 is true.

Ordinarily, it is difficult to assign costs to these two wrong decisions. Accordingly, the customary practice is to follow the *Neyman-Pearson criterion*, in which the probability of detection (i.e., saying H_1 when H_1 is true) is maximized, subject to a constant probability of false alarm.

The design of the two MLPs and linear combiner is accomplished by using a *supervised learning* procedure, details of which are presented later in the article.

3.3 Neural Network-Based Two-Channel Receiver

The PCA network in the interference channel in Fig. 2 is trained by presenting it with WVD images known to represent interference only. Once the training is completed, the synaptic weights of the PCA network are fixed thereafter. The training procedure of the PCA network for the target channel follows a similar procedure, except for the fact that its training examples consist of WVD images known to contain target signal plus interference. The output of the PCA networks may be viewed as a specific number of dominant projections of the input WVD images on two subspaces; one subspace represents interference alone and the other subspace represents target plus interference. The problem is complicated by the fact that these two subspaces are typically unknown; we learn projections of the

WVD image onto them by way of real-life examples that are representative of the two scenarios. The training of the PCA networks in the two channels is performed in a self-organized manner. There are many ways in which this can be accomplished. In the work described in this article we have chosen the generalized Hebbian algorithm developed by Sanger [7, 8].

Each multilayer perceptron has two hidden layers, with the structure of the first hidden layer being constrained through the combined use of receptive fields and weight sharing. As for the output layer of each one of them, it terminates on three output nodes, the rationale for which is explained in the next section. The outputs of the two multilayer perceptron classifiers are *linearly combined* into a single output node, where the final decision is made. In other words, the decision as to whether a target is present or not is deferred to the overall output of the system, in accordance with the information preservation rule. The important point to note however is that the two multilayer perceptrons and linear combiner are *trained simultaneously* by presenting the whole network with WVD images that are known to contain (a) interference only, and (b) target signal plus interference; the presentations are made in an alternate manner. This training is performed in a supervised manner, using the backpropagation algorithm [8, 9]. The simultaneous training of the three networks performing the task of pattern recognition with labeled data ensures that the actions of the two multilayer perceptrons are constrained to reinforce each other.

3.4 Rationale For The Modular Detection Strategy

In mathematical terms, the two-channel receiver of Fig. 2 (i.e., the combination of WVD computer, PCAs and MLPs) *maps* the multidimensional *input (data) space* into a $2N$-dimensional *output (target) space*, where N is the number of output nodes per channel. In the input space, there is a precise separation between the two classes of received signal, H_0 and H_1, which is determined by monitoring the environment. However, in mapping the input space onto the output space, this precise separation between the classes H_0 and H_1 is learned in an imprecise fashion, with the result that decision (classification) errors are made at the final receiver output.

The details of the mapping from the input space to the output space depend, among other things, on the choice of N, the number of output nodes per channel, as explained next.

With binary hypothesis testing as the ultimate objective, an obvious choice for the number of output nodes per channel is $N = 2$. Under such a scheme, each channel of the receiver can only discern one of two possible outcomes. That is, given the received signal, each channel can only say that hypothesis H_0 or H_1 is true, depending on which particular output is higher. This "restricted" form of pattern classification results in a region of uncertainty, where the points representing classes H_0 and H_1 in the output space may overlap significantly.

In a real-life situation, however, we usually find that the target signal varies over a wide dynamic range. Specifically, a qualitative description of the received signal may fall under one of three likely categories:

- the received signal consists of a strong target signal plus interference,
- the received signal consists of a weak target signal plus interference, or
- the received signal consists solely of interference,

which suggests the use of $N = 3$ output nodes per channel. This alternative scheme permits a "finer" form of pattern classification with some beneficial effects: a more compact clustering of the points representing classes H_0 and H_1 in the output space, and a reduced overlap between them. This, in turn, means that the two-channel receiver of Fig. 2 with three output nodes per channel has the potential to outperform the same receiver with two output nodes per channel[2]. The experimental results presented in Section 5 bear out the validity of this statement. Indeed, it is for this reason that we have used $N = 3$ in Fig. 2.

Another question that needs to be addressed is why it is the receiver of Fig. 2 has two distinct channels in the first place. To answer this question, we note that in the traditional approach to radar target detection in a clutter-dominated environment, for example, we may use a "best" *mismatched filter* for clutter discrimination. In such an approach involving a single channel in the receiver, the requirement for best performance in additive noise is traded for an improvement in performance in clutter by purposely mismatching the filter. We may avoid the need for this tradeoff in performance by using two nonlinear matched filters, as depicted in Fig. 2, with each filter being *adaptively matched* to the received signal that arises under one of two hypotheses: H_0 or H_1. In addition, the use of two different channels as described herein provides two *independent assessments* of the

[2]In a loose sense, the arguments presented here, suggesting that the receiver of Fig. 2 with $N = 3$ can produce a better classification accuracy than the same receiver with $N = 2$, remind one of the issue of *soft decision coding* versus *hard decision coding* in digital communications. In hard-decision coding, binary quantization is applied to the demodulator output, resulting in an irreversible loss of information in the receiver. To reduce this loss, multi-level quantization (as an approximation to soft-decision coding) is used.

In Fig. 2 we may go on and increase the number of output nodes per channel beyond $N = 3$ to provide a finer description of the target strength and therefore better pattern classification. However, it is considered that for the task at hand $N = 3$ is the best compromise between improved classification performance and increased computational complexity

received signal, exploiting the fact that the WVD images of the received signal under hypotheses H_0 and H_1 look different even when the signal-to-interference ratio is relatively low. The simplest method of integrating the two channel outputs is through the use of *linear combining*, which is applicable to a wide class of optimization costs. This is precisely what has been done in designing the modular learning strategy of Fig.2.

The training of the two MLPs and linear combiner in Fig. 2 proceeds in a supervised manner. To do this we may use one of two methods:

1. Each of the two MLPs is trained *separately*, with *hard decisions* being made at their respective outputs. For example, in the case of three output nodes per channel, the MLP in the target channel is *constrained* to classify the received signal as containing a strong target signal, containing a weak target signal, or simply consisting of interference on its own. The "digital" outputs of the two MLPs are then linearly combined to produce an overall output, where the final decision is made whether a target is present or not.

2. The two MLPs and linear combiner are all trained *simultaneously*. The outputs of each MLP are now *free* to assume "analog" values within the limited range set by the activation functions of its output neurons (processing units), and in accordance with the training data. Under this second method, hard decision-making is deferred to the final output of the receiver.

The attractive feature of the first method is that the decision boundaries between the different classes of received signal are well defined at the outputs of the two channels. Nevertheless, in the study reported in this article we opted for the second method for two important reasons:

- Hard decisions are accompanied by an irreversible loss of information. In light of the information preservation rule saying that decision-making should be deferred to the very final output of the receiver, it may therefore be argued that the second method preserves the information content of the received signal better than the first method.

- The second method requires a single stage of supervised learning, which is perhaps computationally more efficient than the two different stages of supervised learning needed to implement the first method.

To summarize, the receiver of Fig. 2 "learns to learn" about its environment by proceeding as follows. First, the two PCA networks are individually trained on their respective examples of received signal, using a self-organized learning algorithm. Next, the two MLPs and linear combiner are simultaneously trained on a fresh set of examples, using a supervised learning algorithm and building on what has been learned by the PCA networks.

3.5 Case Study

To test the performance of the new receiver, we performed a case study involving the detection of a *growler* floating in an ocean environment. A growler is a small piece of ice that is broken off an iceberg. Typically, it is about the size of a grand piano, but recognizing that about 90% of the volume of ice lies below the water surface, a growler represents an object large enough to be hazardous to navigation in ice-infested waters, such as those found on the east coast of Canada. The radar task at hand is that of detecting a growler in the presence of sea clutter in the most reliable fashion possible.

For the collection of radar data representative of this environment, an instrument quality radar system called the IPIX radar was used. The IPIX radar is a fully coherent, polarimetric, X-band radar system, equipped with computer control, calibration and digital data acquisition capabilities. The present study is confined to the use of coherent data. That is, the received signal is complex-valued consisting of an in-phase and quadrature component.

A series of experiments using the IPIX radar were performed at two different sites located on the east coast of Canada. In each case the radar was mounted at a height above sea level that would be representative of a ship-mounted radar, with the radar antenna pointing toward a patch of the ocean surface. Ground-truthing of the data collected was maintained throughout the experiments, thereby providing knowledge of the prevalent ocean and weather conditions under which the various data sets were collected.

This case study was chosen for the application at hand because both the target of interest, a growler, and the background interference, sea clutter, are known to exhibit nonstationarity, which would require the use of adaptivity. Moreover, the generation of sea clutter is governed by a nonlinear dynamic process, which would therefore require the use of nonlinear processing. Thus, the detection of a growler in sea clutter provides a suitable medium for testing the capabilities of our new detection strategy.

Figure 3 presents WVD images pertaining to three different situations, with the horizontal axis representing time and the vertical axis representing frequency. Specifically:

- Figure 3(a) pertains to a strong target return. The target signal is clearly visible here as the bright horizontal line running about halfway across the image, at a frequency equal to the Doppler velocity of the target.

- Figure 3(b) pertains to a weak target return. In this case, the target signal is hardly visible. However, its presence is confirmed by the zebra-like pattern (alternating between black and white stripes) representing the cross Wigner-Ville terms; this pattern lies between the actual (but hardly visible) Doppler signature of the target and the signature representing the clutter. On the basis of Fig. 3(b), it may be justifiably said that "a barely visible target has been made visible in signal processing terms".

- Figure 3(c) pertains to the clutter acting alone.

These three figures are indeed distinctly different from each other, which should make the pattern classification task of the multilayer perceptron relatively straightforward.

Figure 4 presents the postdetection results for two different receivers: conventional Doppler CFAR receiver, and the modular learning strategy of Fig. 2. In both parts of Fig. 4, the vertical axis represents time, and the horizontal axis represents range (i.e., distance from the radar). The results presented in Fig. 4 pertain to a situation where the radar was dwelling on a patch of the ocean surface with a growler present in its field of view for the length of time (250 ms) indicated on the vertical axis. As such, for perfect performance, we should see a continuous black strip running vertically across the image. Comparing the results displayed in Fig. 4 for the two receivers, it is clear that the neural network-based receiver performs better than the conventional Doppler CFAR receiver. The reason for the large gap seen about half way in Fig. 4(b), pertaining to the neural network-based receiver, is that the growler was hidden behind an ocean wave for that particular time slot: insofar as the radar was concerned, there was no target to be seen.

Figure 5 presents the receiver operating characteristics (ROCs) for three different receivers using real-life radar data collected as part of this Case Study:

- Neural network-based receiver, based on the model shown in Fig. 2 using three output nodes per channel
- Neural network-based receiver, based on the model shown in Fig. 2, using two output nodes per channel
- Conventional constant false-alarm rate (CFAR) Doppler receiver.

The results shown in Fig. 5 clearly show that for a probability of false alarm smaller than 0.03, the neural-network-based receivers are superior to the conventional CFAR processor. Furthermore, they demonstrate that the neural-network-based receiver with three output nodes per channel always performs better than the one with two output nodes.

3.6 Summary And Discussion

In this paper we have described a modular receiver structure for the detection of a target signal buried in a nonstationary background. The primary design objective is to fully exploit the information content of the received signal in a computationally efficient manner. To achieve this objective, the receiver integrates the following tools in a principled fashion:

- The Wigner-Ville distribution (WVD), acting as the "carrier" of the full information content of the received signal.
- Two entirely different channels, for which the WVD image of the received signal provides a common input. Each channel is made up of a principal components analyser for dimensionality reduction on the WVD image, and a multilayer perceptron for pattern classification. One channel is adaptively matched to the interference acting alone, and the other channel is adaptively matched to the target signal plus interference.
- Linear combining, to combine the analog outputs of the two channels into a single overall output where the decision that a target signal is present or not is finally made.

Successful design of the new receiver rests on the premise that there is a sufficient number of real-life examples representative of the environment in which the receiver operates. Part of this database is used to train the receiver, and the remaining part is used to test it. In particular, the receiver undergoes a learning session that proceeds in two stages, one being unsupervised and the other supervised. Accordingly, the synaptic weights (free parameters) of the receiver are adjusted in a systematic fashion, whereby information contained in the examples about the environment is extracted and stored in those weights.

Operation of the new receiver was validated using a difficult case study. Specifically, experimental results were presented for the detection of a small piece of ice (growler) floating in an ocean environment, using a coherent radar. Highlights of the case study, based on the use of test data completely different from the training data, are:

- The performance of the adaptive modular receiver is superior compared to a conventional Doppler CFAR receiver.
- Implementation of the modular receiver with three output nodes per channel performs better than the same configuration using two output nodes per channel.

On the basis of other results not presented here, it can also be said that the adaptive behavior of the modular receiver permits a robust detection performance with respect to variations in the prevalent environmental conditions.

3.7 References

[1] Suga, N., Bisonar and Neural Computation in Bats, *Sci. Am.* vol. 262(6), pp. 60-68-, 1990.

[2] Cohen, L., *Time-Frequency Analysis*, Prentice-Hall, 1994.

[3] Hlawatsch, F., Regularity and unitarity of bilinear time-frequency signal representations", *IEEE Trans. Information Theory*, vol. 38, pp. 82-94, 1992.

[4] Flandrin, P., A time-frequency formulation of optimum detection, *IEEE Trans. Signal Process.*, vol. 36, pp. 1377-1384, 1988.

[5] Jolliffe, I.T., *Principal Component Analysis*, Springer-Verlag, New York, 1986.

[6] Kohonen, T., The self-organizing map, *Proceedings of the IEEE*, 78, pp. 1464-1480, 1990.

[7] Sanger, T.D., Optimal Unsupervised Learning in a Single-Layer Linear Feedforward Neural Network, *Neural Networks*, vol. 12, pp. 459-473, 1989.

[8] Haykin, S., *Neural Networks: A Comprehensive Foundation*, Macmillan, New York, 1994.

[9] Rumelhart, D.E., G.E. Hinton, and R.J. Williams, Learning Internal Representations by Error Propagation, in *Parallel Distributed Processing" Explorations in the Microstructure of Cognition* (D.E. Rumelhart and McClelland, eds.), vol. 1, chapter 8, MIT Press, Cambridge, MA, 1986.

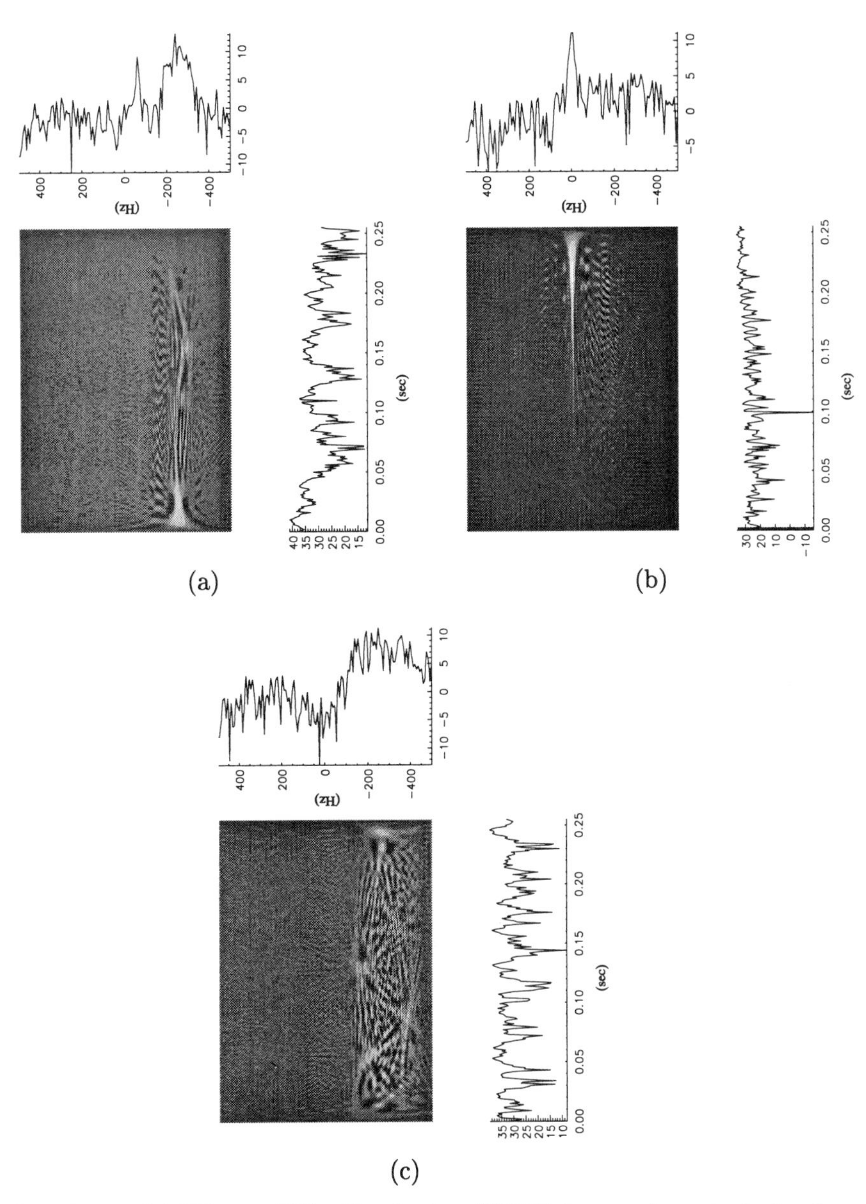

FIGURE 3.3. (a) Wignaer-Ville distribution (WVD) image of received signal containing a strong target signal. (b) WVD image of received signal containing a barely visible target signal. (c) WVD image of radar clutter (interference) acting alone.

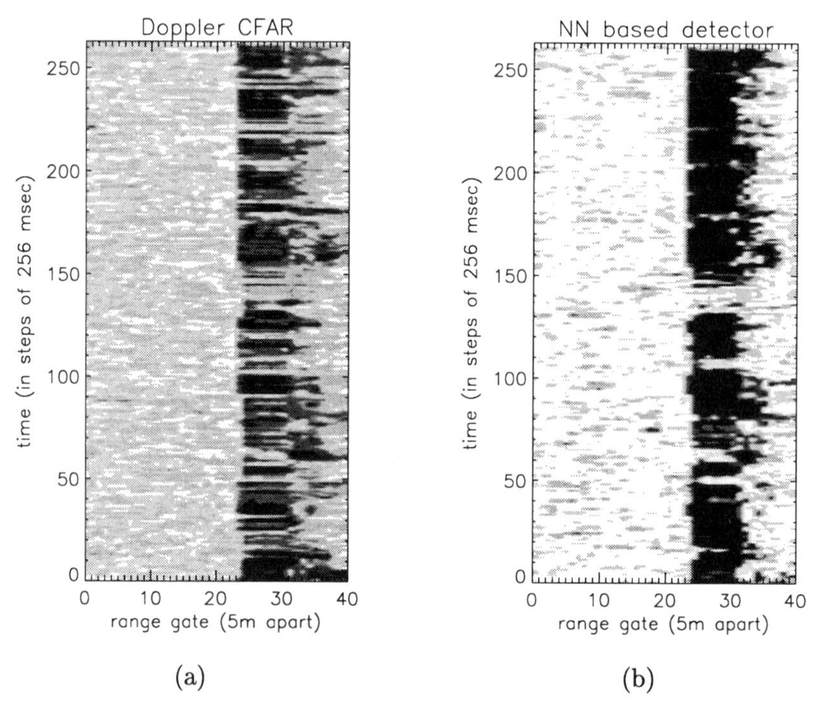

FIGURE 3.4. Postdetection results for (a) conventional constant false- alarm (CFAR) receiver and (b) neural network-based signal detection system.

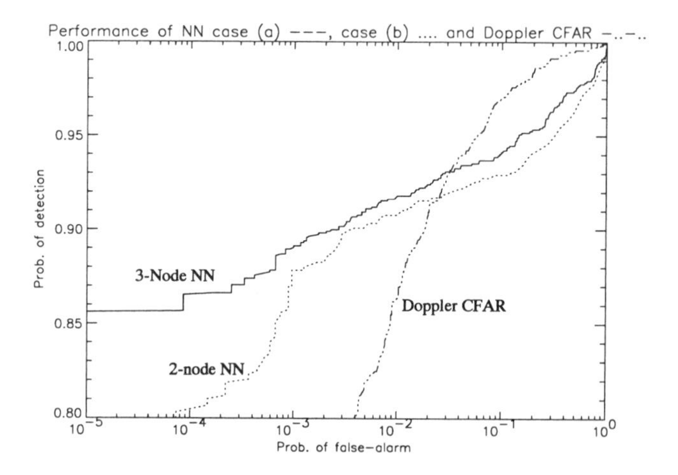

FIGURE 3.5. Receiver operating characteristics of conventional Doppler CFAR receiver and two different implementations of neural network-based receiver.

4

Biometric Identification for Access Control

Shang-Hung Lin
S. Y. Kung

ABSTRACT
Recently, with technological advance on microelectronic and vision system, true verification of individual identities has now become possible. This technology is based on a field called **biometrics**. Biometric systems are automated methods of verifying or recognizing the identity of a living person on the basis of some physiological characteristic, like a fingerprint or face pattern, or some aspect of behavior, like handwriting or keystroke patterns. The objectives of this chapter are to investigate various biometric identification methods and to develop useful techniques for implementing good biometric identification systems. In particular, we focus on two types of biometric methods, namely, **face recognition** and **palm print recognition**. Among all the biometric identification methods, face recognition has attracted much attention in recent years because it has potential to be most non-intrusive and user-friendly. In this chapter we propose an integrated face recognition system based on **probabilistic decision-based neural networks (PDBNN)**[33]. The face recognition system consists of three modules: First, a **face detector** finds the location of a human face in an image. Then an **eye localizer** determines the positions of both eyes in order to generate meaningful feature vectors. The facial region proposed contains eyebrows, eyes, and nose, but excluding mouth. (Eye-glasses will be allowed.) Lastly, the third module is a **face recognizer**. The PDBNN can be effectively applied to all the three modules. It adopts a hierarchical network structure with nonlinear basis functions and a competitive credit-assignment scheme. This chapter demonstrates a successful application of PDBNN to face recognition on the public ORL face database. Regarding the *performance*, experiments on three different databases all demonstrated high recognition accuracies as well as low false rejection and false acceptance rates. As to the *processing speed*, the whole recognition process (including PDBNN processing for eye localization, feature extraction, and classification) consumes approximately one second on a Sparc10, without using a hardware accelerator or co-processor. A new biometric identification scheme using human palm print information is proposed. This scheme extracts discriminant features from a grayscale palm image by edge filtering and Hough transform, and recognizes the input pattern by a novel structural matching algorithm.

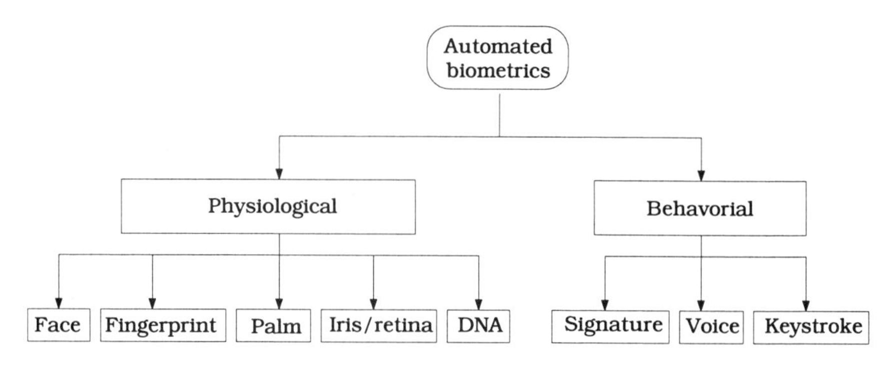

FIGURE 4.1. Different types of biometric ID methods.

4.1 Introduction

In today's complex world, the need to maintain the security of information or physical property is becoming both increasingly important and increasingly difficult. Conventional access control systems either rely on manual means of identity verification (e.g., a security guard recognizing each face), or they simplify the problem to verifying the identity of some physical object (e.g., an access card) or some numerical strings (e.g., password or PIN) rather than the actual individual. The first method is expensive. The second is based on a poor assumption that this object is always attached to its owner, or only the owner knows his/her password. Therefore, breaches in security, such as computer break-in's to sensitive databases or counterfeit bank cards, can still be found in newspapers from time to time.

Recently, technology has become available to allow true verification of individual identities. This technology is based on a field called "biometrics". Biometric systems are automated methods of verifying or recognizing the identity of a living person on the basis of some physiological characteristic, like a fingerprint or face pattern, or some aspect of behavior, like handwriting or keystroke patterns [38, 57]. Since biometric systems do not identify a person by what he/she knows (a code), or possesses (a card), but by a unique characteristic that is hard to reproduce, the possibility of forgery is greatly reduced.

The most commonly used biometric methods can be divided into two categories, as listed in Figure 4.1. The *physiological methods* verify a person's identity by means of his or her physiological characteristics such as fingerprint, iris pattern, palm geometry, or facial features. The *behavioral methods*, on the other hand, perform the identification task by recognizing people's behavioral patterns, such as signature, keyboard typing, and voice print.

Generally speaking, the traits used in the physiological category are relatively more stable than methods in behavioral category, because the physiological features are alterable only with severe damage to the individual. On

	Intrusiveness	conveniency	error rate
face	No	good	$10^{-1} \sim 10^{-3}$
palm	No?	middle	$< 10^{-3}$
fingerprint	Yes	middle	$10^{-2} \sim 10^{-6}$
iris/retina	Yes	very bad	$< 10^{-6}$
voice	No	middle	$10^{-1} \sim 10^{-2}$
signature	No	bad	$10^{-1} \sim 10^{-3}$

TABLE 4.1. Accuracy vs. intrusiveness in various biometric methods.

the other hand, although behavioral characteristics may be hard to measure due to influences such as stress, fatigue, or illness, they sometimes are more acceptable to users and generally cost less to implement in a system.

An important concern for a buyer to purchase a biometric system is its *intrusiveness*. If a security system makes the user feels discomfort, either psychologically or physically, then this system is an intrusive system. Generally speaking, a non-intrusive method is more user friendly and therefore more favorable. For example, in computer network security or access control for areas of middle or low security levels (e.g., apartments, hospitals, stores), an intrusive system will annoy users and therefore prevent them from using it. However, for high security areas, an intrusive system may sometimes turn out to be a benefit - it may appear to be a very serious recognition method and this seriousness may in itself discourage intruders.

In biometric methods, a tradeoff exists between recognition accuracy and intrusiveness. Table 4.1 depicts this tradeoff among various biometric methods. The "intrusiveness" in this figure is mainly for *psychological* intrusiveness (e.g., privacy being violated, or health being threatened), and "convenience" reflects *physical* intrusiveness (e.g., adjusting position to fit within the sensor range). We can observe the tradeoff among these physiological methods; iris identification is the most accurate biometric method in literature (error rate of 10^{-6} can be achieved). However, it is also the most intrusive method because users may feel that the device could hurt their eyes. On the other hand, face recognition is the most non-intrusive method, but unfortunately so far it is also the least accurate one among all the biometric methods. The system designer should carefully evaluate the customer's requirement on the identification accuracy and the intrusiveness concern in order to select a proper method to build the biometric identification system.

After the system designer decides the biometric method to build the security system, he or she needs to choose a proper biometric algorithm to implement the system. Every biometric identification algorithm can be modeled as a classification engine consisting of two consecutive processing components; a *feature extractor* reduces presumably relevant information (features) out of the raw data from sensors, and a *pattern classifier* categorizes the feature pattern into one of the pattern classes in the database. From a theoretical point of view, if there is a powerful feature extractor such that the resulting features are easily separable, then the job of the pattern classifier is trivial. On the other hand, if there is an omnipotent pattern classification algorithm which can always find a classifier which can solve all possible classification problems, then the feature extractor is unnecessary. Efforts on developing a theoretically omnipotent pattern classification algorithm can be seen in neural network literature [60, 9]. In real-world situations, however, since it is almost impossible to find either perfect feature extraction or omnipotent pattern classification algorithms under practical limitations, researchers put efforts on designing a good feature extraction algorithm for the given application, and on selecting a proper classification algorithm which is both powerful and compatible to the preceding feature extraction algorithm. In Section 4.2, we show several important feature extraction methods used in biometric identification applications. The commonly used biometric pattern classification algorithms are discussed in Section 4.3. In this chapter, we implement a face recognition system to demonstrate the importance of pattern classification algorithms. A neural network classifier called the probabilistic decision-based neural networks (PDBNN) is described in Section 4.4. The PDBNN-based face recognition system is discussed in Section 4.5. Also, in order to demonstrate the importance of feature extraction algorithms, we implement a palm print recognition system by applying edge filtering and Hough transform. The palm print system is shown in Section 4.6.

4.2 Feature Extraction for Biometric Identification

Table 4.2 shows several useful features in the image-based biometric identification problems. There are two main classes of feature extraction techniques applied to the recognition of digital images. The first technique is based on the computation of a set of *geometric features* of the object from the picture (e.g., the length of nose, or the distance between two minutiae). The second class of the technique is based on *template matching*, where a bidimensional array of intensity values (or more complicate features extracted from it) is involved. In Figure 4.2 we use a human face as an example to show to difference between these two feature extraction techniques. The non-image based biometric approaches such as thermography [48], pen pressure and movement [46], and motion [4] are not in the scope

	template					geometric features	auxiliary information
	low frequency component	high frequency component	principal component	texture	high level features		
	Gaussian Pyramid	Laplacian Pyramid, edge filtering	KL SVD APEX	Gabor filters, wavelet	NN		
face	O	O	O	O	O	O	color, motion, thermography
palm		O			O	O	
fingerprint	O	O		O	O	O	
iris/retina		O		O		O	
signature	O	O				O	pen movement, pen pressure

TABLE 4.2. Features used for image-based biometric identification. The circles indicate the features which have been used in different biometric identification problems. There are two categories of the features. The geometric features are obtained by measuring the geometric characteristics of several feature points from the sensory images. The template features are extracted by applying some global level processing to the subimage of interest. No prior knowledge about the object structural properties is involved in the feature extraction. Various template features and their corresponding feature extraction algorithms are listed in the table. Some other auxiliary features such as thermography or pen pressure have also been used for biometric identification, but they are not in the scope of this chapter.

of this chapter.

4.2.1 *Geometric Features*

Geometric feature extraction technique is widely used in many works of biometric identification. People either manually or automatically measure geometric features to verify tester's identity. In an early work done by Kelly [37], ten measurements are extracted manually for face recognition, including height, width of head, neck, shoulders, and hips. Kanade [22] uses an automatic approach to extract 16-dimensional vector of facial features. Cox et al. [19] manually extract 30 facial features for a mixed-distance face recognition approach. Kaufman and Breeding [23] measure the geometric features from face profile silhouettes. The ID3D Handkey system [1] automatically measures the hand geometric features by edge detection technique. Gunn and Nixon [15] use an active contour model (snake) to automatically extract human head boundary. Most automatic fingerprint identification systems (AFIS) [28, 8, 49, 47, 7] detect the locations of minutiae (anomalies on fingerprint map such as ridge ending and ridge bifurcation) and measure their relative distances.

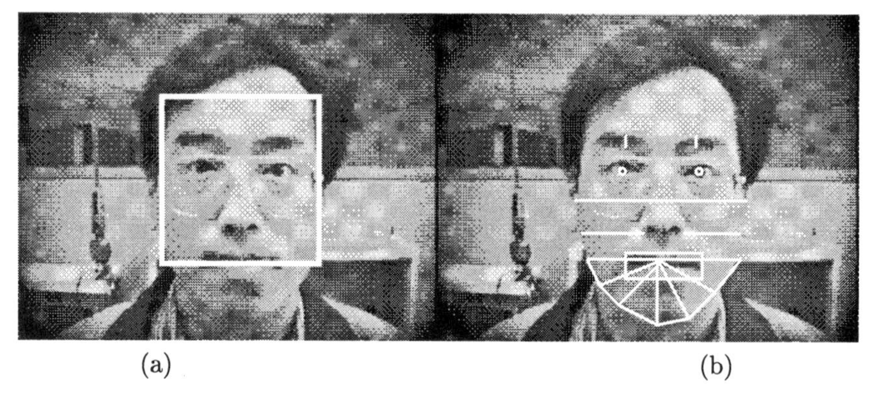

(a) (b)

FIGURE 4.2. Possible template ((a), image in the white box) and geometric measurements ((b), indicated in white lines) used for face recognition problem.

Feature point localization is an crucial step for geometric feature extraction. For example, two corners of a eye ("feature points") must be located before the width is measured. People have applied various techniques for automatic feature point localization, since manual localization is both tiresome and impractical. The automatic feature point localization techniques include Hough transform [42], spatial filters [13, 51, 66], deformable template [16, 67], morphological filters [49, 67], Gabor wavelets [6], knowledge-based approaches [65], and neural networks [29, 54, 58]. The difficulty of feature point localization varies from application to application. For fingerprint approaches, the localizer is easier to design since all the feature points are similar (most fingerprint approaches just detect no more than two types of minutiae [49]). For face recognition methods, on the other hand, designing localizers is more complex because much larger variation exists between various facial feature points (e.g., corners of eye, tip of nose, bottom of chin).

One disadvantage for geometric feature extraction is that a lot of "feature points" have to be correctly localized in order to make the correct measurements. Usually template-based approaches only need to localize two to four feature points (in order to locate the template position) [33, 41], but geometric methods may need to localize up to a hundred feature points in order to extract meaningful geometric features.

4.2.2 Template Features

Template-based methods crop a particular subimage (the template) from the original sensory image, and extract features from the template by applying some *global level* processing, without *a priori* knowledge about the structural properties of the object. Compared to geometric feature extraction algorithms, the image template approaches need much less points to localize in order to get a correct template. For example, in the PDBNN face recognition system (cf. Section 4.5), only two points (left and right eyes) need to be located in order to extract facial template for recognition. Brunelli and Poggio [50] compare these two types of feature extraction methods on face recognition problem. They find out that the template approach is faster and is able to generate more accurate recognition results than the geometric approach.

The template which is directly cropped out of the original image is not suitable for recognition without further processing. The reasons are the following: (1) The feature dimension of the original template is usually too high. A template of 320 x 240 pixels means a feature vector of dimension 76800. It is impractical to use such a high dimensional feature vector for recognition. (2) The original template may be easily corrupted by many types of variation. Table 4.2 lists five types of commonly used feature extraction methods for template-based approaches.

Low frequency components (or Gaussian pyramid) are good for removing high frequency noise and for dimension reduction. Sung and Poggio [58] use low-resolution images for face detection. Baldi [3] uses low pass filtering techniques as the first preprocessing step of his neural network fingerprint recognizer. Golomb and Sejnowski [2], Turk and Pentland [61] also use low-pass filtering and down-sampling as the preprocessing steps for face recognition.

The most commonly used features are *high frequency components* (or edge features). Although edge information is more prone to be affected by noise, it reflects more structural properties of the object. Many fingerprint approaches use directional edge filtering and thinning process to extract fingerprint structures [28, 49]. Rice analyzes the vein structures on human hands by extracting edge information from infra red images [52]. Wood et al. use a modified Laplacian of Gaussian ($\nabla^2 G$) operator to detect retina vessel structures [63].

If the diversity of input patterns is sufficient to show statistical properties, *principal component analysis* can also be used for feature extraction. Pentland et al. [61, 40, 45] assume that the set of all possible face patterns occupies a small and parameterized subspace, derived from the original high-dimensional input image space, i.e., the "eigenface space". Therefore they apply the principal component analysis to obtain the "eigenfaces" (the principal components in the eigenspace) for face recognition.

Another useful feature for biometric identification is the *texture information*. Wavelet methods are often used to extract texture information. FBI has used wavelet transform to be the compression standard for fingerprint images [5]. Daugman [10] uses Gabor wavelets to transform an iris pattern into a 256-byte "iris code". He claims that the theoretical error rate could be as low as 10^{-6}. Lades et al. [39] also use Gabor wavelets as the features for their dynamic link architecture (DLA, an elastic matching type of neural network) for face recognition system.

Recently, more and more people are using *neural networks* to extract information for classification. One advantage of neural network feature extractor is its learning ability. The weighting parameters in a neural network feature extractor can be trained so that the network can extract features which contain the most discriminant information for the following classifier. In fact, most neural network feature extractors are connected with the neural network classifiers, so that both modules can be trained together. Examples of neural network feature extractors are Baldi and Yves' convolutional neural network for fingerprint recognition [3], Weng, Huang, and Ahuja's Cresceptron for face detection [21], Rowley's convolutional neural network for face detection [54], and Lades' dynamic link architecture for face recognition. Lawrence et al. [27] combine self-organizing map [24] and convolutional neural network for face recognition. Over 96% recognition rate has been reported on the face database from Olivetti Research Laboratory [55].

From Figure 4.2 we can see that high frequency component features are widely used for various biometric methods. In our two biometric systems (face and palm), we will apply this type of features. In the face recognition system, we will also use low frequency component features to assist the recognition. Moreover, since our face recognition system has two neural network components (face detector and eye localizer) which are used to locate and normalize the facial template, the neural network feature extraction approach is also applied to our system.

4.3 Pattern Classification for Biometric Identification

From a theoretical point of view, pattern classification algorithms are usually independent of application domain. Several classification algorithms have been proven to be universal approximators [53, 60, 9] (given unlimited training patterns). However, by considering the compatibility with the corresponding feature extraction algorithms, different biometric methods (e.g., face, palm, iris, fingerprint, ...) may have different preference on selecting pattern classification algorithms. In this section, we discuss several pattern classification algorithms which are commonly used for biometric identification applications.

4.3.1 Statistical Pattern Recognition

The conventional pattern classification uses statistical recognition methods. If patterns of each person are available, one can build a statistical model for each person class. A common approach models each class as a normal density, so the system estimates for each person the corresponding mean feature vector and covariance matrix. Using a prior distribution on the individuals in the database, the classification task is completed by computing the Bayesian posterior probability of each person, conditioned on observation of the query. If the computation is performed using log probability, the classification process can be considered as a nearest neighbor search using the *Mahalanobis* distance metric.

There are other statistical approaches for pattern classification. The *k-nearest neighbors algorithm* [12] determines the class of a test pattern by comparing it with the most likely class of the k nearest training patterns. The likelihood of each class is estimated by its relative frequency among the k nearest training patterns. As the training pattern size grows, these relative frequencies converge to the true posterior class probability. The *Parzen windows* [44] attempt to estimate the class-conditional densities via a linear superposition of window functions - one for each training pattern. The window function is required to be unimodal and has a unit area under its curve. As the training pattern size grows, the linear superposition of window functions for a given class converges to the true class-conditional density. Daugman [10] evaluates the recognition performance of his IriScan system by estimating the distribution of Hamming distances between two iriscodes.

Topological Matching Algorithms

Topological matching algorithms are often used to estimate the class-conditional probability when the input pattern is a *undirected graph.* They are widely used in fingerprint identification problems.

Following the method developed by Sir E. R. Henry nearly 100 years ago [17], almost all the existing fingerprint systems use minutiae as features for identification [28]. For each fingerprint image, they detect the presence of minutiae and use them to construct a undirected graph. Each vertex in the graph corresponds to a minutia. Its attributes could include the minutia position, orientation, minutia type, or the ridge counts between this particular minutia and its neighboring ones. After the minutiae graph for the query pattern is constructed, its class-conditional probability (the "similarity") is computed by a matching process between the query graph and the reference graph of that class. Chen and Kuo [7] build tree-structural minutiae graphs and match two graphs via a breadth-first search scheme. Costello et al. [8] match the topological minutiae map by a coincident sequencing scheme. Ratha et al. [49] implement a parallel topological matching algorithm on a specialized hardware called Splash 2.

Notice that in fingerprint applications, most topological matching algorithms can achieve high recognition rate by just using *one* example for each person (which means there is no need to use several training examples to construct the reference graph). Therefore they can be used in the applications where few training examples are available (e.g., criminal identification).

4.3.2 Neural Networks

Neural networks are powerful pattern classifiers. Although the development of neural networks is motivated by the functions of human brain, neural networks encompass lots of similarities with statistical pattern recognition approaches. In the following we discuss five types of neural networks which are commonly used for image-based physiological biometric systems. They are vector quantization, multi-layer perceptron, radial basis function network, hierarchical mixture of expert, and probabilistic decision based neural network.

Vector Quantization

Vector quantization (VQ) classifies input patterns based on the *nearest-neighbor rule.* Let $\mathbf{X} = \{\mathbf{x}_1, \mathbf{x}_2, \cdots, \mathbf{x}_l\}$ be a set of l classified patterns, and let $\mathbf{x}'_l \in \mathbf{X}$ be the pattern nearest to the test pattern $\mathbf{x}$. Then the nearest-neighbor rule for classifying $\mathbf{x}$ is to assign it the class associated with $\mathbf{x}'_l$.

VQ does not have complicated learning rules or network structures. The classification scheme is also straightforward. Therefore it is often used by the systems that have powerful feature extraction algorithms. For example, Daugman [10] and Rice [52] use minimum hamming distance rule for classifying iris patterns and hand vein structural patterns, respectively. Pentland et al. [61, 40, 45] project the face images onto the eigenface subspace and determine the classes of the input patterns by applying nearest neighbor search using Euclidean distance metric. Cox et al. [19] propose an algorithm for face recognition called *mixture-distance* algorithm, which uses VQ to estimate both the true pattern generative process (the "platonic" process) and the process which generates the vectors we ultimately observed (the "observation" process).

The nearest-neighbor rule is a sub-optimal procedure; its use will usually lead to an error rate greater than the minimum possible, the Bayes rate. However, people have proven that with an unlimited number of patterns the error rate is never worse than twice the Bayes rate [12].

Multi-Layer Perceptron

Multi-layer perceptron (MLP) is one of the most popular neural network models. Usually the basis function of each neuron (the perceptron) is the linear basis function (LBF), and the activation function is either the step function or the sigmoid function. The most commonly used learning scheme for MLP is the *back-propagation* algorithm. Sung and Poggio [58] use MLP for face detection. Golomb et al. [2] present a cascade of two MLPs for gender classification.

Radial Basis Function Network

Another type of feed-forward network is the radial basis function (RBF) network. Each neuron in the hidden layer employs a radial basis function, and uses a Gaussian kernel to serve as the activation function. The weighting parameters in the RBF network are the centers and the widths of these neurons. The output functions are the linear combination of these radial basis functions. It has been shown that the RBF network has the same asymptotic approximation power as a multi-layer perceptron[60]. Brunelli and Poggio [50] use a special type of RBF network called the "HyperBF" network for face recognition, and they report a 100% recognition rate on a 47 people database.

Hierarchical Mixture of Experts

The hierarchical mixture of experts (HME)[20] is a modular architecture in which the outputs of a number of "experts", each performing classification task in a particular portion of the input space, are combined in a probabilistic way by a "gating" network which models the probability that each portion of the input space generates the final network output. Each local expert network performs multiway classification over K classes by using either K independent binomial model, each modelling only one class, or one multinomial model for all classes. The HME is based on the principle of "divide and conquer" in which a large, difficult problem is broken into many smaller, more tractable problems. Waterhouse and Robinson [62] have reported a satisfactory result by using HME for large vocabulary speech recognition.

HME has an explicit relationship with statistical pattern classification methods. Given a pattern, each expert network estimates the pattern's conditional *a posteriori* probability on local areas, and the outputs of the gating network represent the probabilities that its corresponding expert subnet produces the correct answer. The final output is the weighted sum of the estimated probabilities from all the expert networks.

Probabilistic Decision-Based Neural Network

Probabilistic decision-based neural network (PDBNN) [33, 34, 31, 30, 32] is a probabilistic variant of its predecessor, Decision Based Neural Network (DBNN)[26]. DBNN is an efficient neural network for many pattern classification problems (for example, optical character recognition [25]). Similar to DBNN, PDBNN has a modular network structure. One subnet is designated to represent one object class (this "One-Class-One-Network" (OCON) property is discussed in Section 4.4.3). For multi-class classification problems, the outputs of the subnets (the *discriminant functions*) will compete with each other, and the subnet with the largest output values will claim the identity of the input pattern. PDBNN has been used for face and palm recognition problems [35, 34].

PDBNN also has a explicit relationship with statistical pattern classification methods. Unlike the HME, each neuron in PDBNN estimates the pattern's class-conditional ***likelihood density*** on local areas, not the conditional posterior probability. The output of a class subnet is the weighted sum of the estimated densities from all its neurons. A detailed discussion of PDBNN is given in the next section.

4.4 Probabilistic Decision-Based Neural Network

PDBNN is the core technology of our face recognition system. In this section we will describe the network structure and learning rules of the PDBNN for both binary classification and multi-class pattern classification problems.

4.4.1 Discriminant Functions of PDBNN

One major difference between the PDBNN and prototypical DBNN (or other RBF networks) is that PDBNN follows probabilistic constraint. That is, the subnet discriminant functions of PDBNN are designed to model the log-likelihood functions. The reinforced and antireinforced learning is applied to **all** the clusters of the global winner and the supposed (i.e. the correct) winner, with a weighting distribution proportional to the degree of possible involvement (measured by the likelihood) by each cluster.

Given a set of iid patterns $\mathbf{X}^+=\{\mathbf{x}(t);t = 1,2,\ldots,N\}$, we assume that the class likelihood function $p(\mathbf{x}(t)|\omega)$ for class ω (i.e., face class or eye class) is a mixture of Gaussian distributions. Define $p(\mathbf{x}(t)|\omega,\Theta_r)$ to be one of the Gaussian distributions which comprise $p(\mathbf{x}(t)|\omega)$ ($p(\mathbf{x}(t)|\omega,\Theta_r) = N(\mu_r, \Sigma_r)$),

$$p(\mathbf{x}(t)|\omega) = \sum_{r=1}^{R} P(\Theta_r|\omega)p(\mathbf{x}(t)|\omega,\Theta_r)$$

where Θ_r represents the r-th cluster in the subnet, and $P(\Theta_r|\omega)$ denotes the prior probability of cluster r. By definition $\sum_{r=1}^{R} P(\Theta_r|\omega) = 1$.

The discriminant function of one-subnet PDBNN models the log-likelihood function:

$$\phi(\mathbf{x}(t),\mathbf{w}) = \log p(\mathbf{x}(t)|\omega) = \log[\sum_r P(\Theta_r|\omega)p(\mathbf{x}(t)|\Theta_r,\omega)] \qquad (4.1)$$

where

$$\mathbf{w} \equiv \{\mu_r, \Sigma_r, P(\Theta_r|\omega), T\} \qquad (4.2)$$

T is the threshold of the subnet. It is trained in the GS learning phase. Again, an explicit teacher value would not be required, just like the original DBNN.

Elliptic Basis Function(EBF)

In most general formulation, the basis function of a cluster should be able to approximate the Gaussian distribution with full-rank covariance matrix. A hyper-basis function (HyperBF) is meant for this[60]. However, for those applications which deal with high dimensional data but finite number of training patterns, the training performance and storage space discourage such matrix modelling. A natural simplifying assumption is to assume uncorrelated features of unequal importance. That is, suppose that $p(x|\omega, \Theta_r)$ is a D-dimensional Gaussian distribution with uncorrelated features,

$$p(\mathbf{x}(t)|\omega, \Theta_r) = \frac{1}{(2\pi)^{D/2}\prod_d^D \sigma_{rd}} exp(-\frac{1}{2}\sum_{d=1}^{D}\frac{(x_d(t) - w_{rd})^2}{\sigma_{rd}^2}) \sim N(\mu_r, \Sigma_r) \tag{4.3}$$

where $\mathbf{x}(t) = [x_1(t), x_2(t), \ldots, x_D(t)]^T$ is the input pattern, $\mu_r = [w_{r1}, w_{r2}, \ldots, w_{rD}]^T$ is the mean vector, and diagonal matrix $\mathbf{\Sigma}_r = diag[\sigma_{r1}^2, \sigma_{r2}^2, \ldots, \sigma_{rD}^2]$ is the covariance matrix.

To approximate the density function in Eq. 4.3, we apply the elliptic basis functions (EBF) to serve as the basis function for each cluster:

$$\psi(\mathbf{x}(t), \omega, \Theta_r) = -\frac{1}{2}\sum_{d=1}^{D}\beta_{rd}(x_d(t) - w_{rd})^2 + \theta_r \tag{4.4}$$

where $\theta_r = -\frac{D}{2}\ln 2\pi - \sum_{d=1}^{D}\ln\sigma_{rd}$. After passing an exponential activation function, $exp\{\psi(\mathbf{x}(t), \omega, \Theta_r)\}$ can be viewed the same Gaussian distribution as described in Eq. 4.3, except a minor notational change: $\frac{1}{\beta_{rd}} = \sigma_{rd}^2$.

4.4.2 Learning Rules for PDBNN

There are two properties of the PDBNN learning scheme. The first one is **decision dependent learning rules**. Unlike the approximation neural networks, where exact target values are required, the teacher in DBNN only tells the correctness of the classification for each training pattern. Based on the teacher information, DBNN performs a distributed and localized updating rule. There are three main aspects of this training rule:

(1) *When to update?* A selective training scheme can be adopted, e.g. weight updating only when misclassification.

(2) *What to update?* The learning rule is distributive and localized. It applies *reinforced learning* to the subnet corresponding to the correct class and *antireinforced learning* to the (unduly) winning subnet.

(3) *How to update?* Adjust the boundary by updating the weight vector $\mathbf{w}$ either in the direction of the gradient of the discriminant function (i.e., reinforced learning) or opposite to that direction (i.e., antireinforced learning).

The second property of the decision-based learning scheme is **hybrid locally unsupervised and globally supervised learning**. It is based on the so-called **LUGS** (Locally Unsupervised Globally Supervised) learning. There are two phases in this scheme: during the locally-unsupervised (LU) phase, each subnet is trained individually, and no mutual information across the classes may be utilized. After the LU phase is completed, the training enters the Globally-Supervised (GS) phase. In GS phase teacher information is introduced to reinforce or anti-reinforce the decision boundaries obtained during LU phase. The discriminant functions in all clusters will be trained by the two-phase learning.

Unsupervised Training for LU learning

The values of the parameters in the network are initialized in the LU learning phase. In the LU learning phase of PDBNN, we usually use k-mean or VQ to determine initial positions of the cluster centroids, and then use Expectation Maximization algorithm [11] to update the cluster parameters of each class. The EM algorithm is a special kind of quasi-Newton algorithm with a searching direction having a positive projection on the gradient of the log likelihood. In each EM iteration, there are two steps: Estimation (E) step and Maximization (M) step. The M step maximizes a likelihood function which is further refined in each iteration by the E step. The EM algorithm in the LU phase is as follows[64]: Use the data set $\mathbf{X}^+=\{\mathbf{x}(t);\mathbf{x}(t) \in \omega,\ t = 1, 2, \ldots, N\}$. The goal of the EM learning is to maximize the log likelihood of data set $\mathbf{X}^+$

$$l(\mathbf{w}; \mathbf{X}^+) = \sum_{t=1}^{N} \log p(\mathbf{x}(t)|\omega) = \sum_{t=1}^{N} \log[\sum_{r} P(\Theta_r|\omega) p(\mathbf{x}(t)|\Theta_r, \omega)] \quad (4.5)$$

The followings are the operations taken in an iteration of the EM algorithm. At iteration j, (1) **E-step:** we first compute the conditional posterior probabilities $h_r^{(j)}(t), \forall r$:

$$h_r^{(j)}(t) \quad = \quad \frac{P^{(j)}(\Theta_r|\omega) p^{(j)}(\mathbf{x}(t)|\omega, \Theta_r)}{\sum_k P^{(j)}(\Theta_k|\omega) p^{(j)}(\mathbf{x}(t)|\omega, \Theta_k)} \quad (4.6)$$

(2) **M-step:** maximizing $Q(\mathbf{w}, \mathbf{w}^{(j)})$ with respect to $\mathbf{w}$ (cf. Eq. 4.2), we have:

$$P^{(j+1)}(\Theta_r|\omega) \quad = \quad (1/N) \sum_{t=1}^{N} h_r^{(j)}(t),$$

$$\mu_r^{(j+1)} \quad = \quad (1/\sum_{t=1}^{N} h_r^{(j)}(t)) \sum_{t=1}^{N} h_r^{(j)}(t)\mathbf{x}(t)$$

$$\mathbf{\Sigma}_r^{(j+1)} = (1/\sum_{t=1}^{N} h_r^{(j)}(t)) \sum_{t=1}^{N} h_r^{(j)}(t)[\mathbf{x}(t) - \mu_r^{(j)}][\mathbf{x}(t) - \mu_r^{(j)}]^T \quad (4.7)$$

Notice that since the threshold T will not affect the likelihood value, it is not updated here. When the EM iteration converges, it should ideally obtain Maximum Likelihood Estimation (MLE) of the data distribution. EM has been reported to deliver excellent performance in several data clustering problems[64].

Supervised Training for GS learning

One major difference between PDBNN and traditional statistical approaches is that after maximum likelihood estimation, the PDBNN has one more learning phase, the GS learning, for minimizing the classification error. The GS learning algorithm is used after the PDBNN finishes the LU training. In the Global Supervised (GS) training phase, teacher information is utilized to fine-tune decision boundaries. When a training pattern is misclassified, the LVQ-type reinforced or anti-reinforced learning technique is applied[24]. The reinforced and anti-reinforced learning rules for the network are the following:

$$\begin{array}{ll} \text{Reinforced Learning:} & \mathbf{w}^{(j+1)} = \mathbf{w}^{(j)} + \eta l'(d(t))\nabla\phi(\mathbf{x}(t),\mathbf{w}) \\ \text{Antireinforced Learning:} & \mathbf{w}^{(j+1)} = \mathbf{w}^{(j)} - \eta l'(d(t))\nabla\phi(\mathbf{x}(t),\mathbf{w}) \end{array} \tag{4.8}$$

where

$$d(t) = \begin{cases} T - \phi(\mathbf{x}(t),\mathbf{w}) & \text{if } \mathbf{x}(t) \in \mathbf{X}^+ \\ \phi(\mathbf{x}(t),\mathbf{w}) - T & \text{if } \mathbf{x}(t) \in \mathbf{X}^- \end{cases} \tag{4.9}$$

The *penalty function* $l(d(t))$ can be either a piecewise linear function ($l(d) = \zeta$ if $d > 0, l(d) = 0$ if $d \leq 0$) or a sigmoid function. If the misclassified training pattern is from positive training (i.e., face or eye) set, reinforced learning will be applied. If the training pattern belongs to the so-called negative training (i.e. "non-face(or eye)") set, then only the anti-reinforced learning rule will be executed - since there is no "correct" class to be reinforced.

The gradient vectors in Eq. 4.8 are computed as follows:

$$\begin{aligned} \frac{\partial\phi(\mathbf{x}(t),\mathbf{w})}{\partial w_{rd}}\Big|_{\mathbf{w}=\mathbf{w}^{(j)}} &= h_r^{(j)}(t)\cdot\beta_{rd}^{(j)}(x_d(t) - w_{rd}^{(j)}) \\ \frac{\partial\phi(\mathbf{x}(t),\mathbf{w})}{\partial \beta_{rd}}\Big|_{\mathbf{w}=\mathbf{w}^{(j)}} &= h_r^{(j)}(t)\cdot\frac{1}{2}(\frac{1}{\beta_{rd}^{(j)}} - (x_d(t) - w_{rd}^{(j)})^2) \end{aligned} \tag{4.10}$$

where $h_r^{(j)}(t)$ is the conditional posterior probability as shown in Eq. 4.6, and $w_{rd}^{(j)}$ and $\beta_{rd}^{(j)}$ are defined in Eq. 4.3 and Eq. 4.4 respectively. As to the conditional prior probability $P(\Theta_r|\omega)$, since the EM algorithm can automatically satisfy the probabilistic constraints $\sum_r P(\Theta_r|\omega) = 1$ and $P(\Theta_r|\omega) \geq 0$, it is applied to update the $P(\Theta_r|\omega)$ values in the GS phase so that the influences of different clusters are regulated: At the end of the epoch j,

$$P^{(j+1)}(\Theta_r|\omega) = (1/N)\sum_{t=1}^{N} h_r^{(j)}(t) \tag{4.11}$$

Threshold Updating

The threshold value of PDBNN detector can also be learned by reinforced/antireinforced learning rules. Since the increment of the discriminant function $\phi(\mathbf{x}(t), \mathbf{w})$ and the decrement of the threshold T have the same effect on the decision making process, the direction of the reinforced and anti-reinforced learning for the threshold is the opposite of the one for the discriminant function. For example, given an input $\mathbf{x}(t)$, if $\mathbf{x}(t) \in \omega$ but $\phi(\mathbf{x}(t), \mathbf{w}) < T$, then T should reduce its value. On the other hand, if $\mathbf{x}(t) \notin \omega$ but $\phi(\mathbf{x}(t), \mathbf{w}) > T$, then T should increase.

$$\begin{aligned} T^{(j+1)} &= T^{(j)} - \eta l'(d(t)) \quad \text{if } \mathbf{x}(t) \in \omega \quad \text{(reinforced learning)} \\ T^{(j+1)} &= T^{(j)} + \eta l'(d(t)) \quad \text{if } \mathbf{x}(t) \notin \omega \quad \text{(antireinforced learning)} \end{aligned} \tag{4.12}$$

4.4.3 Extension of PDBNN to Multiple-Class Pattern Recognition

Similar to the one-subnet PDBNN, we use mixture of Gaussian as the class likelihood function $p(\mathbf{x}(t)|\omega_i)$ for the multi-class PDBNN. The discriminant function of each subnet in PDBNN is as follows:

$$\phi(\mathbf{x}(t), \mathbf{w}_i) = \log p(\mathbf{x}(t)|\omega_i) = \log[\sum_r P(\Theta_{r|i}|\omega_i)p(\mathbf{x}(t)|\Theta_{r|i}, \omega_i)] \tag{4.13}$$

where $\mathbf{w_i} \equiv \{\mu_{r|i}, \Sigma_{r|i}, P(\Theta_{r|i}|\omega_i), T_i\}$ is the parameter set for subnet i. The EBF can also be applied to serve as cluster basis function $p(\mathbf{x}(t)|\Theta_r, \omega_i)$. The overall diagram of such discriminant function is depicted in Figure 4.3. Notice that since the output of PDBNN is of probability form, it can be used as an indicator of the confidence score of the recognition result.

There are two important characteristics in the multi-subnet PDBNN. First, it is a OCON structure. Second, it generates low false acceptance and false rejection rates. The details are discussed as follows.

- **OCON Structure** Inherited from the prototypical DBNN, the PDBNN adopts a One-Class-in-One-Network (OCON) structure, where one subnet is designated to one class only. The structure of PDBNN recognizer is depicted in Figure 4.3. Each subnet specializes in distinguishing its own class from the others, so the number of hidden units

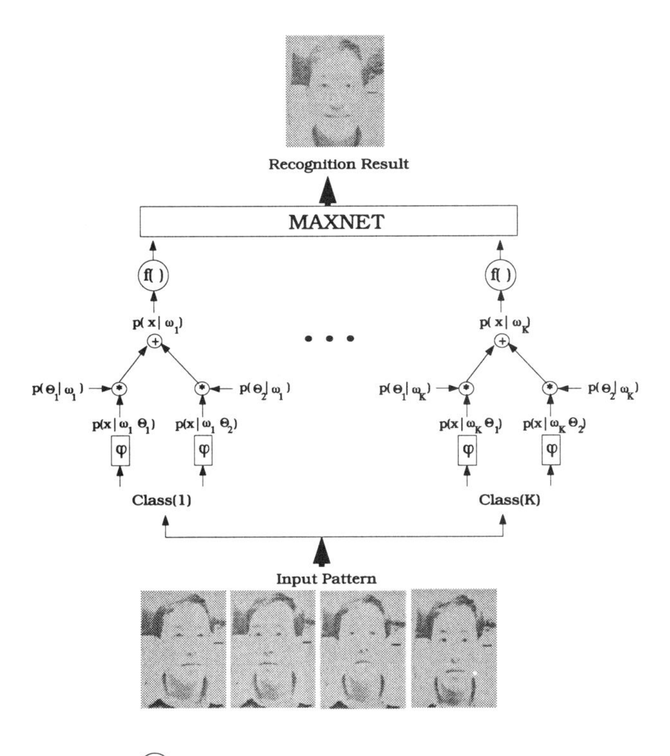

FIGURE 4.3. Structure of PDBNN face recognizer. Each class subnet is designated to recognize one person. All the network weightings are in probabilistic format.

is usually small. Compared to the All-Class-One-Network (ACON) structure (e.g., multi-layer perceptron), the OCON structure is superior in many aspects. Pandya and Macy [43] observe that OCON can achieve better generalization and training accuracy, and yet requires only one fourth of the training time of the ACON structure. Moreover, the OCON structure is suitable for **incremental training**, i.e. *network upgrading upon adding/removing memberships.* For our face recognition system, we found that the trained PDBNN face recognizer can be easily adapted to a face verification system due to its OCON structure. Any individual person's database can be individually stored, either in computer or in user's magnet card, and individually retrieved for verification of his/her identity as proclaimed.

- **False Acceptance and False Rejection**

The PDBNN is very suitable for tackling false rejection/acceptance problem. The reason is that PDBNN focuses only on density distributions in individual classes (rather than global partitioning), no additional intruder class subnet is needed. Also, since density value usually drops as the distance to the centers increases, the decision regions tend to be more locally conserved, and thus a lower false acceptance rate will be expected. Numerous experiments indicate this characteristic.

Learning Rules for Multi-Subnet PDBNN

The multi-subnet PDBNN follows the same learning rules as the one-subnet PDBNN in the LU learning phase. Each subnet performs k-mean, VQ, or EM to adjust its parameters. In the GS learning phase, only the misclassified patterns are used. The GS learning rules for the multi-class PDBNN is the straightforward extension of the one-class version of the GS learning mentioned in Section 4.4.2. Notice that there are two types of misclassified patterns. The first type is from the positive dataset. A pattern of this type will be used for (1) reinforced learning of the threshold and discriminant function of the class which it actually belongs to, and (2) anti-reinforced learning of the thresholds and discriminant functions of the classes which have higher discriminant function values than the true class has. The second type of the misclassified patterns is from the negative dataset. A negative training pattern is either a intruder face pattern or a non-face pattern. Since there is no subnet representing the negative data, a pattern of this type is used only for anti-reinforced learning of the threshold and discriminant function of the class which it is misclassified to. The thresholds T_i in the multi-subnet PDBNN are also trained by the reinforced and anti-reinforced rules, just like the one-subnet case.

4.5 Biometric Identification by Human Faces

A PDBNN-based face recognition system[35, 34, 59, 31] is being developed under a collaboration between Siemens Corporate Research, Princeton, and Princeton University. The total system diagram is depicted in Figure 4.4. All the four main modules, face detector, eye localizer, feature extractor, and face recognizer are implemented on a SUN Sparc10 workstation. An RS-170 format camera with 16 mm, F1.6 lens is used to acquire image sequences. The S1V digitizer board digitizes the incoming image stream into 640x480 8-bit grayscale images and stores them into the frame buffer. The image acquisition rate is on the order of 4 to 6 frames per second. The acquired images are then down sized to 320x240 for the following processing[30].

As shown in Figure 4.4, the processing modules are executed sequentially. A module will be activated only when the incoming pattern passes the preceding module (with an agreeable confidence). After a scene is obtained by

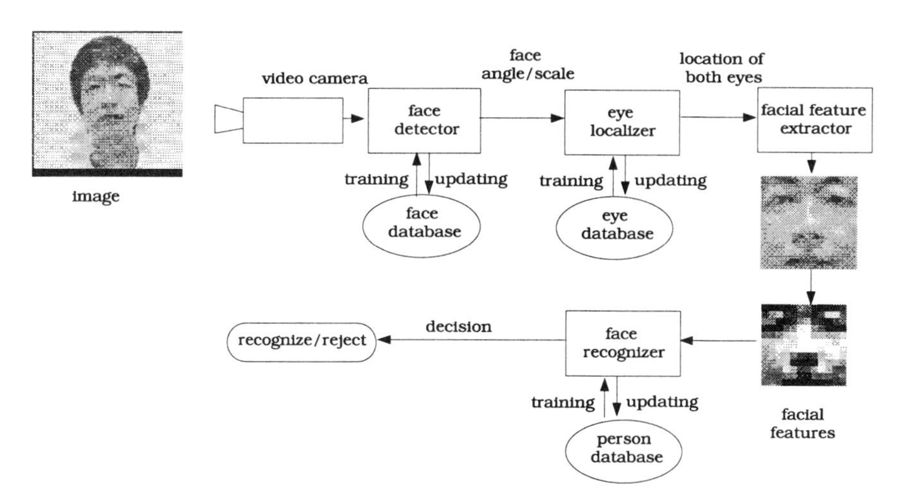

FIGURE 4.4. System configuration of the face recognition system. Face recognition system acquires images from video camera. Face detector determines if there are faces inside images. Eye localizer indicates the exact positions of both eyes. It then passes their coordinates to facial feature extractor to extract low-resolution facial features as input of face recognizer.

the image acquisition system, a quick detection algorithm based on binary template matching is applied to detect the presence of a proper sized moving object. A PDBNN face detector is then activated to determine whether there is a human face. If positive, a PDBNN eye localizer is activated to locate both eyes. A subimage (approx. 140 × 100) corresponding to the face region will then be extracted [32]. Finally, the feature vector is fed into a PDBNN face recognizer for recognition and subsequent verification.

In order to test the capability of the PDBNN classifier, the feature extraction method used in this system simply extracts the low frequency (the "intensity") and high frequency components (the "edges") from the input image template (cf. Figure 4.5).

The system built upon the proposed has been demonstrated to be applicable under reasonable variations of orientation and/or lighting, and with possibility of eye glasses. This method has been shown to be very robust against large variation of face features, eye shapes and cluttered background[29]. The algorithm takes only 200 ms to find human faces in an image with 320x240 pixels on a SUN Sparc10 workstation. For a facial image with 320x240 pixels, the algorithm takes 500 ms to locate two eyes. In the face recognition stage, the computation time is linearly proportional to the number of persons in the database. For a 200 people database, it takes less than 100 ms to recognize a face. Furthermore, because of the inherent parallel and distributed processing nature of DBNN, the technique can be easily implemented via specialized hardware for real time performance.

The experimental results of the PDBNN face detector, eye localizer, and

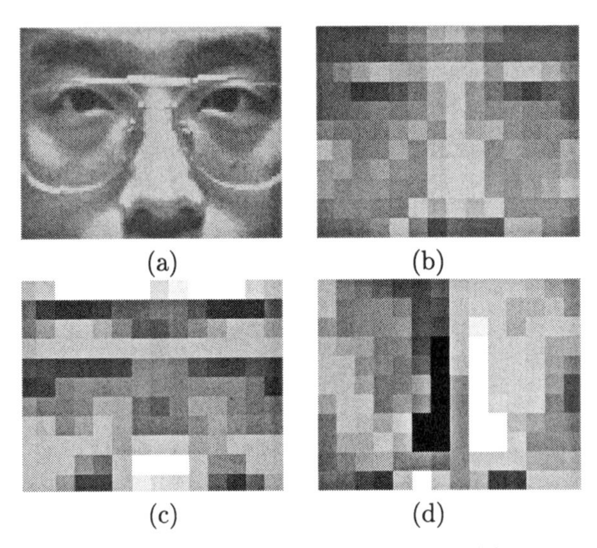

FIGURE 4.5. (a)Facial region used for face recognition. (b)Intensity feature extracted from (a). (c)X-directional gradient feature. (d)Y-directional gradient feature.

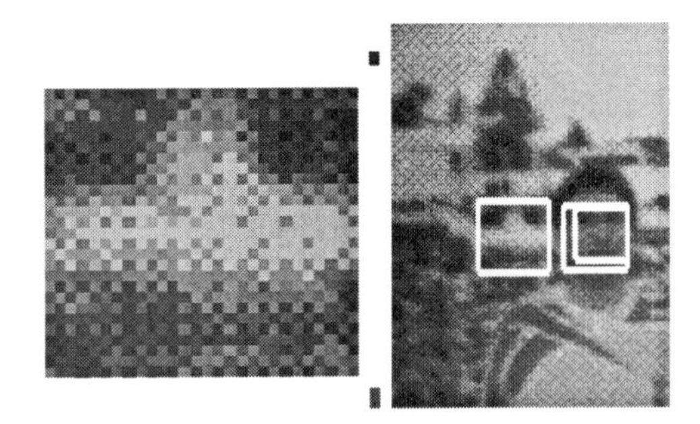

FIGURE 4.6. A pattern looks much like a face.

face recognizer are shown in the following sections.

4.5.1 Face Detection

Face detection is in general a very hard computer vision problem. One example is shown in Figure 4.6. The picture on the left may looks like a smiling face, but actually it's just an image of stadium crowd. For our automatic face recognition system, we have developed a face detection module that is mainly for security and access control; the faces are usually vertically oriented and unoccluded, and they are often in the dominant portion of an image. Different lighting conditions are allowed, but it has to perform very well in indoor lighting conditions.

A one-class PDBNN model is used to build a face detector. There are

FIGURE 4.7. Some faces detected by DBNN face detector.

displ. from true face location	0 to 5 pixels	5 to 10 pixels	> 10 pixels
% of test patterns	98.5%	1.5%	0 %

TABLE 4.3. Performance of face detection. We can see that most faces are detected with accuracy of less than 5-pixel error.

four clusters(neurons) in the network. The number of clusters are determined based on empirical result. The input dimension of the network is $12 \times 12 \times 2 = 288$. (12x12 x-directional and y-directional gradient feature vector. Readers can refer to Figure 4.5(c) and (d).) The network weighting parameters and thresholds are trained following the procedures described in Section 4.4.2. For an input pattern, if the output of the network is larger than the threshold, it is recognized as a "face". Otherwise it is classified as a "non-face".

The examples in Figure 4.7 shows the performance of the PDBNN face detector. We can see that this algorithm performs pretty well on the image with faces of different sizes (from the anchorwoman's face to a pedestrian's face), and lighting conditions. To further evaluate the performance of the face detector module in the PDBNN face recognition system, 92 annotated images were used for training (each image generates approximately 25 *virtual training patterns*), and 473 images for testing. Since our face detector is used mainly for surveillance purpose, all the images are taken in the normal indoor lighting condition with cluttered background. The image size is 320x240 pixels, and the face size is about 140x100 pixels. The variation of head orientation is about 15 degrees toward the four directions (up, down, right, left). The performance was measured by the error (in terms of pixels) between the detected face location and the true location. To be fair to dif-

displ. from true eye location	0 to 3 pixels	3 to 5 pixels	> 5 pixels
% of test patterns	96.4%	2.5%	1.1 %

TABLE 4.4. Performance of eye localization. We can see that most eyes are detected with accuracy of less than 3-pixel error.

ferent sizes of faces, the errors were normalized with the assumption that the distance between the two eyes is 40 pixels, which is the average distance in our annotated images. In order to reduce the searching time, the images (320x240 pixels) are normally down-sized by a factor of 7. Working with the low resolution images (search range approximately 46x35 pixels, search step 1 pixel, and block size 12x12 pixels), our pattern detection system detects a face within 200 ms. For our face recognition application, a 10-pixel error is acceptable, since the main purpose of face detection is merely to restrict the searching areas for eye localization. Among all the testing images, 98.5% of the errors are within 5 pixels and 100% are within 10 pixels in the original high-resolution image (which is less than 1 pixel error in the low resolution images). Table 4.3 lists the detection accuracy of the PDBNN face detector.

4.5.2 Eye Localization

Eye detection is a much harder problem than face detection, because (1) the eye areas are usually very small, (2) eyes may be open, closed, or semi-closed, and (3) eye glasses often blur the eye areas. Therefore, it requires more training images than face detection. Besides, since an eye occupies much smaller area than a face does, it is much harder to detect the location of an eye. In our experiments, we used 250 annotated images for training and 323 images for testing. The conditions on the images are the same as the database for face detection experiment. Similar to the face detector, the eye localizer is also implemented by a one-class PDBNN. There are also four clusters in the network. The input feature dimension is $14 \times 14 \times 2 = 392$. We only train a left eye detector. To detect the right eye, we generate the mirror image of the original one, and try to detect a left eye in the mirror image. Like the face detector experiment, the errors are normalized with the assumption that the eye-to-eye distance is 40 pixels. For our face recognition system, 3-pixel errors (or less) are well within the tolerance, and even 5-pixel errors are often still acceptable. Table 4.4 shows the experimental result. 96.4% of the errors are within 3 pixels, and 98.9% are within 5 pixels.

Figure 4.8 shows several eyes that are detected by the PDBNN eye localizer. We can see that the PDBNN eye localizer can detect both open and closed eyes, eyes looking in different directions, eyes with different sizes, and even eyes under eyeglass reflection. Figure 4.9 illustrates two examples where the eye localizer fails. The effect of specular reflection on the eye-

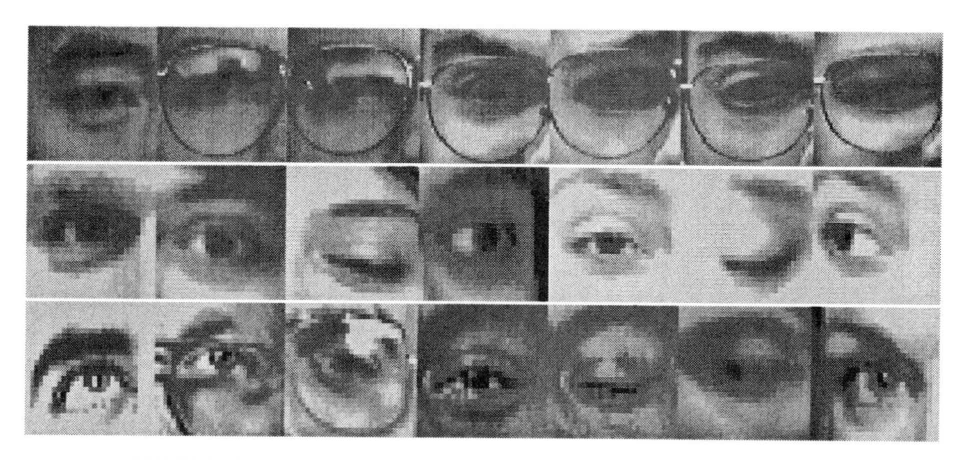

FIGURE 4.8. Eyes that are detected by the PDBNN eye localizer.

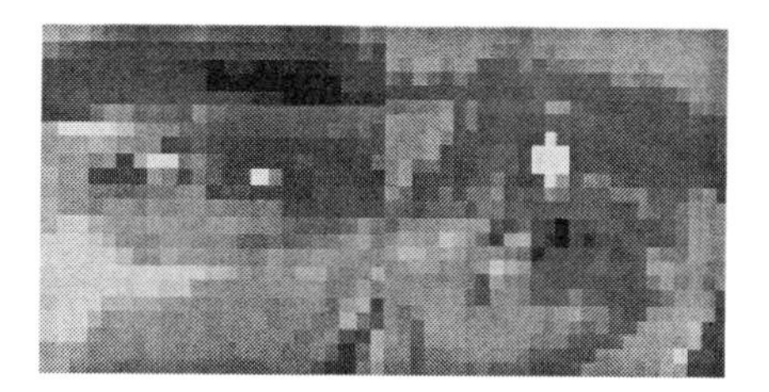

FIGURE 4.9. Eyes that cannot be detected by the PDBNN eye localizer. The specular reflection on the eye glasses blurs the eye in the left picture, and creates a "fake eye" on the eye glasses in the right picture.

System	Error rate	Classification time	Training Time
PDBNN	4%	< 0.1 seconds	20 minutes
SOM + CN	3.8%	< 0.5 seconds	4 hours
Pseudo 2D-HMM	5%	240 seconds	n/a
Eigenface	10%	n/a	n/a
HMM	13%	n/a	n/a

TABLE 4.5. Performance of different face recognizers on the ORL database. Part of this table is adapted from S. Lawrence et al., "face recognition: a convolutional neural network approach", technical report, NEC research institute, 1995.

glasses are so severe that it either greatly blur the eyes, or creates a "fake eye" on the glasses (see the right picture in Figure 4.9.

4.5.3 Face Recognition

The PDBNN face recognizer can be considered as an extension of the PDBNN face detector. As mentioned in Section 4.4.3, PDBNN is the OCON network, that is, there is one subnet in the PDBNN designated for a person to be recognized. For a K-person recognition problem, a PDBNN face recognizer consists of K different subnets. Analogous to the PDBNN detector, a subnet i in the PDBNN recognizer models the distribution of person i only, and treats those patterns which do not belong to person i as "non-i" patterns.

Various experimental results of the PDBNN face recognizer can be seen in [33]. In this chapter we show one front-view experiment on the face database from the Olivetti Research Laboratory in Cambridge, UK (the ORL database). There are 10 different images of 40 different persons. There are variations in facial expression (open/close eyes, smiling/non-smiling), facial details (glasses/no glasses), scale (up to 10%), and orientation (up to 20 degree). A HMM-based approach is applied to this database and achieves 13% error rate[55]. The popular eigenface algorithm[61] reports an error rate around 10% [55, 27]. In [56], a pseudo 2D HMM method is used and achieves 5% at the expense of long computation time (4 minutes/pattern on Sun Sparc II). In [27] Lawrence et al. use the same training and test set size as Samaria did and a combined neural network (self organizing map and convolutional neural network) to do the recognition. This scheme spent four hours to train the network and less than one second for recognizing one facial image. The error rate for ORL database is 3.8%. Our PDBNN-based system reaches similar performance (4%) but has much faster training and recognition speed (20 minutes for training and less than 0.1 seconds for recognition). Both approaches run on SGI Indy. Table 4.5 summarizes the performance numbers on ORL database.

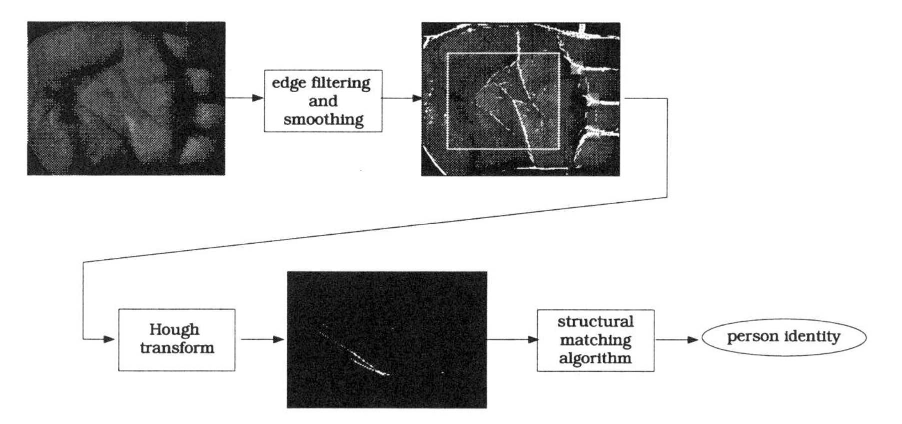

FIGURE 4.10. Palm print recognition system. Edge filtering and smoothing algorithms are applied to a 640x480 greyscale image to extract palm prints . The extracted edge points (the white dots in the bounding box) are then transformed to Hough domain. After thresholding, the feature points in Hough domain are considered as a feature pattern for the palm in the original image. Structural matching algorithm are then applied to calculate the similarity between the incoming feature pattern and the patterns stored in the database.

4.6 Biometric Identification by Palm Prints

In this section, we introduce a palm print recognition scheme. Compared to the face recognition system in Section 4.5, whose emphasis is on the PDBNN pattern classifier, the focus of the palm print system in this section is on its feature extraction part. The palm print system diagram is shown in Figure 4.10. This scheme is a template-based recognition approach. A CCD camera hooked to an SGI Indy workstation is placed under a transparent plate. On the transparent plate there are some marks guiding the users to place their hands in the right position. A 640 x 480 greyscale image that containing the palm area (see the leftmost picture in Figure 4.10) is taken after the user correctly places his or her hand on the plate. An image template containing the central palm region is determined based on the positions of the roots of middle finger and ring finger (the white bounding box in Figure 4.10). The size of the cropped image is proportional to the width of the two fingers, therefore the size varies from person to person.

4.6.1 Feature Extraction for Palm Print Recognition

A Sobel edge filtering is applied on the template to highlight the palm prints and the thresholding and median filtering are applied afterwards in order to reduce noise. Furthermore, since palm prints are often in the shape of straight lines or curves with low curvature, and since Hough transform has the effect of line linking [18, 14], the edge points in the palm template are transformed to Hough domain. After thresholding, several clusters in Hough domain indicating the major palm prints will be formed. Figure 4.11 shows those clusters in Hough domain and their corresponding edge points on the image plane.

4.6.2 Pattern Classification for Palm Print Recognition

After Hough transform, the feature pattern of the palm are formed by the coordinates of the remaining points on the Hough domain. A *structural matching algorithm* is applied to calculate the similarity between this feature pattern and the reference patterns stored in the database.

Figure 4.12 shows the flow chart of the structural matching algorithm. This algorithm finds out how similar two point patterns are by first finding how many pairs of points (a pair of points consists of one point from test pattern and the other point from reference pattern) are "good matches". If there are many good match pairs, the two patterns are similar, otherwise they are not. Notice that if two points are close to each other, they do not necessarily make a good match; a pair of points is a good match when their "relative locations" in their own pattern structures are similar. The details of the structural matching algorithm is described in the following.

Structural Matching Algorithm

Given two sets of planar point patterns

$$\begin{aligned} A &= \{a_1, a_2, \ldots, a_m\} \\ B &= \{b_1, b_2, \ldots, b_n\} \end{aligned}$$

the structural matching algorithm performs the following procedures:

1. **Structural Matching Scores**

   For each pair $\{a_i, b_j\}$, compute the "structural matching score" C_{ij} by the following:

   $$C_{ij} = \frac{\sum_{h \neq i}\{\min_{k \neq j} \Delta_{hk}\}}{m-1} \tag{4.14}$$

   where Δ_{hk} is defined as

   $$\begin{aligned} \Delta_{hk} &= |D_{ih} - D_{jk}| \\ D_{ih} &= \|a_i - a_h\|^2, \quad D_{jk} = \|b_j - b_k\|^2 \end{aligned} \tag{4.15}$$

   where D_{ih} is the distance between point a_i and point a_h. If, for each D_{ih}, the point b_j has an associated point in the set B such that

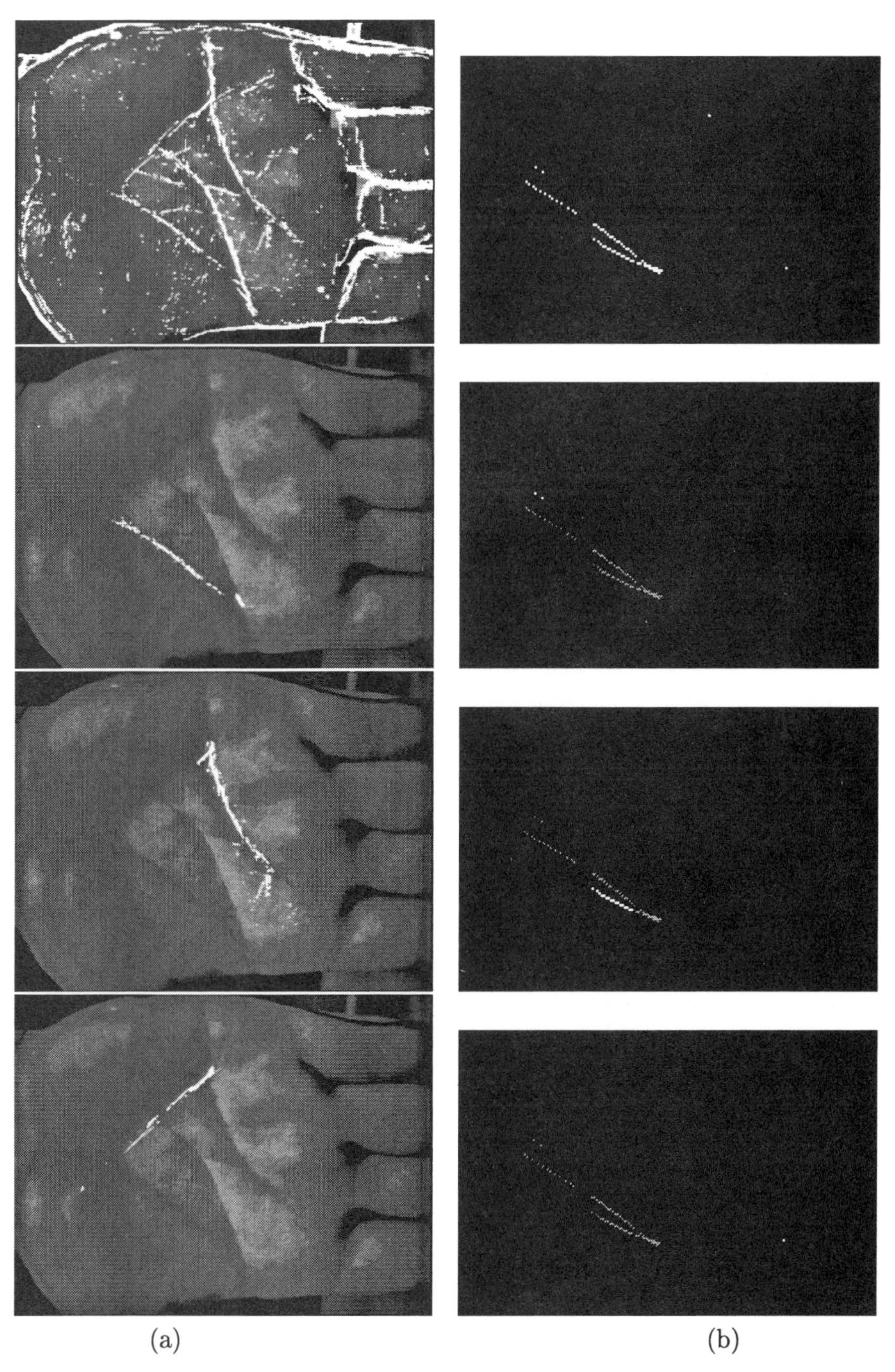

FIGURE 4.11. (a) A palm print edge map (the top left) and some palm print line segments on the map. (b) The corresponding Hough domain features (the highlighted points). We can see that the Hough transform of the edge points on major palm lines form line-shaped clusters in Hough domain.

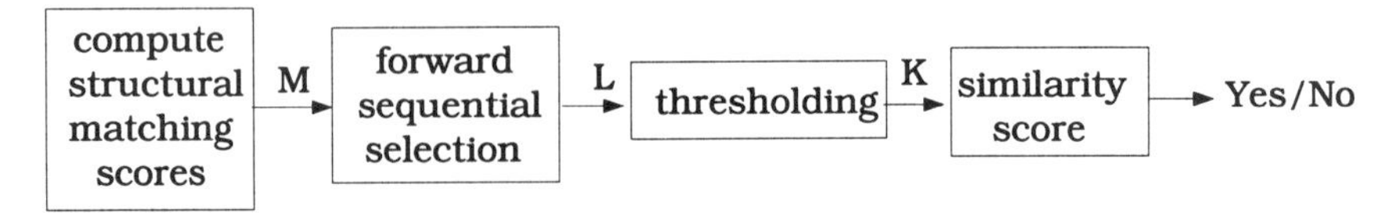

FIGURE 4.12. Structural Matching Algorithm. Given two sets of points, the structural matching algorithm first computes the structural matching score for each possible pair of points and put this score into a $n \times m$ matrix M . A forward sequential search is then applied to select the pairs from the one with the highest matching score in M. These pairs forms a queue L in the order of matching scores. After Thresholding, only K pairs with large scores remain in the queue. The number K is then used to calculate the final similarity score.

$D_{ih} = D_{jk}$, then C_{ij} will be zero (which means "good match"). Notice that since no inter-class distance is measured, the matching algorithm is *translation invariant.* Also, since no directional information is included, the algorithm is *rotation invariant.*

When all the structural matching scores are available, form a $m \times n$ "matching matrix" M:

$$M = \begin{pmatrix} C_{11} \cdots C_{1n} \\ \vdots \quad \vdots \quad \vdots \\ C_{m1} \cdots C_{mn} \end{pmatrix}$$

2. **Forward Sequential Selection** Once the matching matrix M is available, we select the pairs which have low structural matching scores (i.e., good matching pairs) by the following *forward sequential selection* procedure:

   (a) Select the smallest C_{ij} from matrix M, and put it in the queue L.

   (b) Remove all the elements that are in the same row or column with C_{ij} (C_{ix} and C_{xj}) from the matrix M.

   (c) Continue step 1 and step 2, until M is empty.

   Although this selection procedure may not achieve the solution in which the summation of the structural matching scores of all the selected pairs is minimum, it is one way to efficiently find a close-to-optimum solution, especially when the number of points in the two patterns are different.

3. **Thresholding**

   Set a threshold T. Remove all the elements in the queue L which are larger than T.

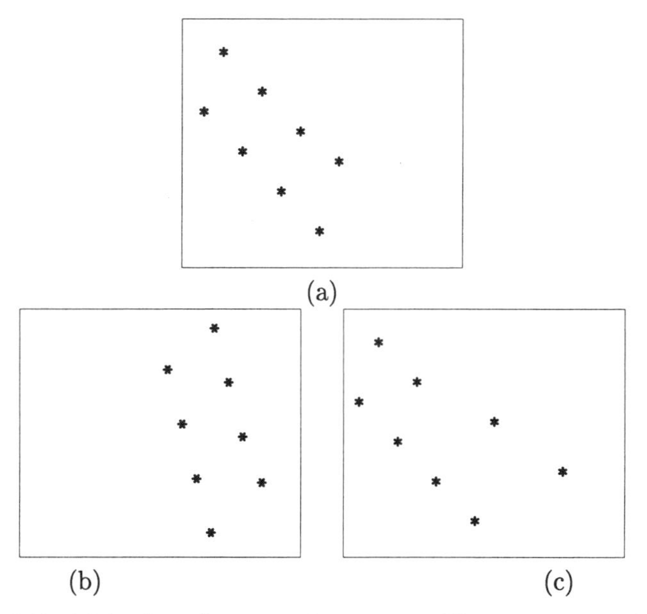

FIGURE 4.13. (a) is the reference point pattern. The test pattern in (b) has higher similarity score than the test pattern in (c) does.

4. **Similarity Score**

   Suppose there are K pairs in the queue after thresholding. The similarity score is defined as follows:

$$S = \frac{K^2}{m \times n}$$

Figure 4.13 illustrates the performance of the structural matching algorithm. Suppose the point pattern in Figure 4.13(a) is the reference point. Since the structural matching algorithm is rotational and translational invariant, the similarity score of the pattern in Figure 4.13(b) is high. On the other hand, since the relative positions between the points in the pattern (c) are different from those in the reference pattern, the similarity score for pattern (c) is low.

4.6.3 Experimental Results

We have conducted two experiments using this palm print recognition scheme. The first one is a "recognition" experiment. That is, the user does not claim his or her identity. The system has to recognize him or her based on the palm print image. The second experiment is the "verification" experiment. In this experiment the user claimed his or her identity and the task of the system was to verify the claim. The database we used consists

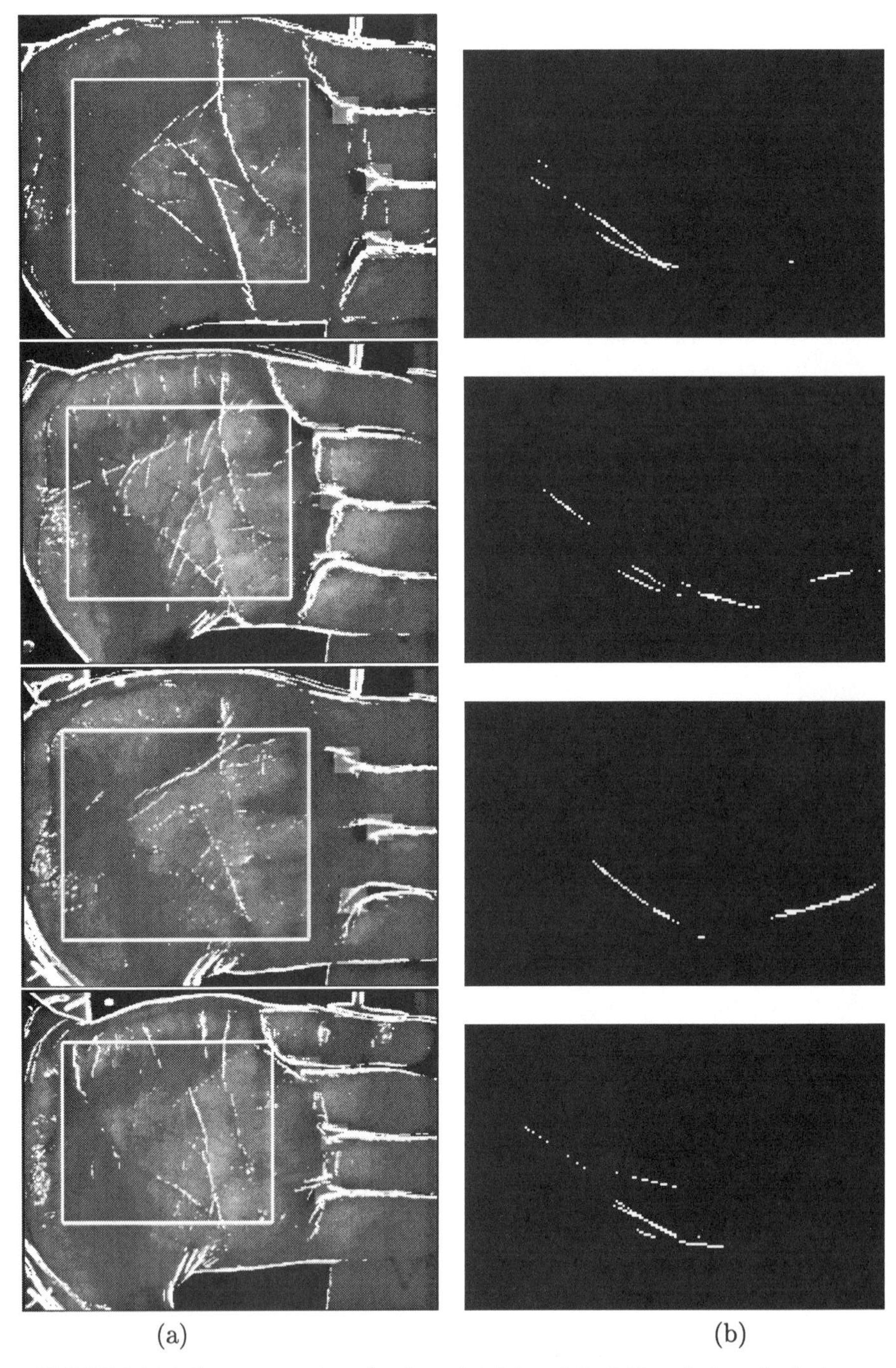

FIGURE 4.14. Some examples of palm print (a) and their Hough domain features (b).

of palm print pictures of 32 people. There are three images for each person. One of them was used for extracting the reference pattern, and the other two were used for testing. Some image examples are included in Figure 4.14(a). The number of test images in the recognition experiment is 2 image/person x 32 persons = 64 images. In the verification experiment, the number of images used for generating false rejection rate was also 64, while the number of images used for false acceptance rate was (3x31) images/person x 32 persons = 2976 images; each image in the database can be used to attack the system 32 times (one of them is for false rejection test) by changing its identity claim, including the reference images.

The experimental results are as follows:

1. Palm recognition:
   - Recognition rate = 100% (64/64).
2. Palm Verification:
   - False rejection rate = 0% (0/64).
   - False acceptance rate = 0.5% (15/2976).

The experimantal results tell us that, like fingerprints, palm prints also contains rich discriminative information. Although the palm print features are not as unique as fingerprints and may change over years, encouraging results of 0.5% FAR and 0% FRR can be achieved in a 32-people database. The major drawback of the current system is the processing speed. It takes about 1 minute to identify a person. The worst case of median filter computation complexity is $O(p^2)$ (p = number of pixels in a picture) and the computation complexity of Hough transform is $O(m^3)$ (m = number of edge points). A much faster feature extraction scheme needs to be developed in order for the system to be in practical use. Therefore, a more thorough investigation of palm print recognition is considered as an important future work.

4.7 Concluding Remarks

With the vast advance of the computer technology, the need to maintain secure access control and financial transactions, etc., is becoming both increasingly important and increasingly difficult. A new trend of security technology called "biometric", which performs the true verification of individual identity, is thus attracting more and more attention from researchers and companies. In this chapter, we investigated several biometric identification systems and compared different feature extraction and classification algorithms. Two system implementations, the face and palm print recognition systems, were developed and experimental results are provided. The face recognition system demonstrated the power of the PDBNN pattern classifier, and the palm print recognition system indicates the importance of a good feature extractor. Due to their non-intrusiveness, both systems are very suitable for gateway access control and computer security management.

4.8 References

[1] Julian Ashboum. Practical implementation of biometrics based on hand geometry. In *IEE Colloquium on Image Processing for Biometric Measurement*, pages 5/1 – 5/6, April 1994.

[2] B.A.Golomb and T.J.Sejnowski. SEXNET: a neural network identifies sex from human faces. In D.S.Touretzky and R.Lipmann, editors, *Advances in Neural Information Proceedings Systems 3*, San Mateo, CA, 1991. Morgan Kaufmann.

[3] Pierre Baldi and Yves Chauvin. Neural networks for fingerprint recognition. *Neural Computation*, 5:402–418, 1993.

[4] A Bobick and A. Wilson. A state-based technique for the summarization and recognition of gesture. In *International Conference on Computer Vision*, 1995.

[5] J. Bradley, C. Brislawn, and T. Hopper. The FBI wavelet/scalar quantization standard for gray-scale fingerprint image compression. Technical Report LA-UR-93-1659, Los Alamos National Lab, 1993.

[6] B.S.Manjunath, R.Chellapa, and C.v.d.Malsburg. A feature based approach to face recognition. In *Proc. IEEE Computer Soc. Conf. on Computer Vision and Patt. Recog.*, pages 373–378, 1992.

[7] Z. Chen and C. H. Kuo. A topology-based matching algorithm for fingerprint authentication. In *Proc. IEEE Intl. Carnahan Conference on Security Technology*, 1991.

[8] B. D. Costello, C. A. Gunawardena, and Y. M. Nadiadi. Automated coincident sequencing for fingerprint verification. In *IEE Colloquium on Image Processing for Biometric Measurement*, pages 3/1 – 3/5, April 1994.

[9] G. Cybenko. Approximation by superpositions of a sigmoidal function. *Mathematics of Control, Signals, and Systems*, 2:303–314, 1989.

[10] John G. Daugman. High confidence visual recognition of persons by a test of statistical independence. *IEEE Transactions on Pattern Analysis and Machine Intelligence*, 15(11):1148–1161, November 1993.

[11] A.P. Dempster, N.M. Laird, and D.B. Rubin. Maximum likelihood from incomplete data via the EM algorithm. In *J. of Royal Statistical Society, B39*, pages 1–38, 1976.

[12] Richard. O. Duda and Peter. E. Hart. *Pattern Classification and Scene Analysis*. Wiley Interscience, 1973.

[13] M. Fang, A. Singh, and M-Y. Chiu. A fast method for eye localization. Technical Report SCR-94-TR-488, Siemens Corporate Research, Inc., 1994.

[14] Rafael C. Gonzalez and Richard E. Woods. *Digital Image Processing.* Addison Wesley, 1993.

[15] S. R. Gunn and M. S. Nixon. A dual active contour for head boundary extraction. In *IEE Colloquium on Image Processing for Biometric Measurement,* pages 6/1 – 6/4, April 1994.

[16] P. W. Hallinan. Recognizing human eyes. In *SPIE Proc.: Geometric Methods in Computer Vision,* volume 1570, pages 214–226, 1991.

[17] Sir W. J. Herschel. *The Origin of Fingerprinting.* AMS Press, New York, 1974.

[18] P.V.C. Hough. *Methods and means for recognizing complex patterns.* U.S. Patent 3,069,654.

[19] I.J.Cox, J.Ghosn, and P.N. Yianilos. Feature-based face recognition using mixture distance. Technical Report 95-09, NEC Research Institute, 1995.

[20] M. I. Jordan and R. A. Jacobs. Hierarchies of adaptive experts. *Neural Information Systems,* 4, 1992.

[21] J.Weng, T.S.Huang, and N.Ahuja. Learning recognition and segmentation of 3D objects from 2D images. In *Proc. IEEE Int. Conf. on Computer Vision,* pages 121–128, 1993.

[22] T. Kanade. *Computer Recognition of Human Faces.* Birkhauser Verlag, Stuttgart, Germany, 1977.

[23] G. J. Kaufman and K. J. Breeding. Automatic recognition of human faces from profile silhouettes. *IEEE Trans. on Systems, Man and Cybernetics,* SMC-6(2):113–121, 1976.

[24] T. Kohonen. *Self-organization and Associative Memory.* Springer-Verlag, 1984.

[25] S. Y. Kung. *Digital Neural Networks.* Prentice Hall, 1993.

[26] S.Y. Kung and J.S. Taur. Decision-based neural networks with signal/image classification applications . *IEEE Trans. on Neural Networks,* 6(1):170–181, Jan 1995.

[27] S. Lawrence, C.L. Giles, A.C. Tsoi, and A.D. Back. Face recognition: a convolutional neural network approach. Technical report, NEC Research Institute, 1995.

[28] H. C. Lee and R. E. Gaensslen. *Advances in Fingerprint Technology.* Elsevier, New York, 1991.

[29] Shang-Hung Lin, Yin Chan, and S. Y. Kung. A probabilistic decision-based neural network for locating deformable objects and its applications to surveillance system and video browsing. In *to appear in International Conference on Acoustics, Speech and Signal Processing,* Atlanta, USA, May 1996.

[30] Shang-Hung Lin and S. Y. Kung. *Neural network for locating and recognizing a deformable object.* patent no. 95P7504.

[31] Shang-Hung Lin and S. Y. Kung. Probabilistic DBNN via expectation-maximization with multi-sensor classification applications. In *Proc. of 1995 IEEE International Conference on Image Processing,* volume III, pages 236–239, Washington, D.C., USA, Oct 1995.

[32] Shang-Hung Lin, S. Y. Kung, and L. J. Lin. *An adaptive pattern detection technique based on neural networks with applications to face detection and eye localization.* patent no. 95P7504.

[33] Shang-Hung Lin, S.Y. Kung, and L.J. Lin. Face recognition/detection by probabilistic decision-based neural network. *IEEE Trans. on Neural Networks, special issue on Artificial Neural Networks and Pattern Recognition*, 1996.

[34] Shang-Hung Lin, S.Y. Kung, and Long-Ji Lin. A probabilistic DBNN with applications to sensor fusion and object recognition. In *Proc. of 5th IEEE workshop on Neural Networks for Signal Processing*, pages 333–342, Cambridge,MA,USA, Aug 1995.

[35] Shang-Hung Lin, S.Y.Kung, and M.Fang. A neural network approach for face/palm recognition. In *Proc. of 5th IEEE workshop on Neural Networks for Signal Processing*, pages 323–332, Cambridge,MA,USA, Aug 1995.

[36] R. P. Lippmann. An introduction to computing with neural nets. *IEEE ASSP Magazine*, 4(4-22):153, 1987.

[37] M.D.Kelly. Visual identification of people by computer. Technical Report AI-130, Stanford AI Proj., 1970.

[38] B. Miller. Vital signs of identity. *IEEE Spectrum*, pages 22–30, February 1994.

[39] M.Lades, J.Vorbruggen, J.Buhmann, J.Lange, and C.v.d.Malsburg. Distortion invariant object recognition in dynamic link architecture. *IEEE Trans. Computers*, 42:300–311, 1993.

[40] B. Moghaddam and A. Pentland. Face recognition using view-based and modular eigenspaces. *Automatic Systems for the Identification and Inspection of Humans, SPIE*, 2257, 1994.

[41] B. Moghaddam and A. Pentland. Probabilistic visual learning for object detection. In *The 5th Intl. Conf. on Computer Vision*, Cambridge, MA, June 1995.

[42] M. Nixon. Eye spacing measurement for facial recognition. In *SPIE Proceedings*, volume 575, pages 279–285, 1985.

[43] A. S. Pandya and R. B. Macy. *Pattern Recognition with Neural Networks in C++*. CRC Press and IEEE Press, 1996.

[44] E. Parzen. On estimation of a probability density function and mode. *Annals of Mathematical Statistics*, 33:1065–1076, 1962.

[45] A. Pentland, B. Moghaddam, and T. Starner. View-based and modular eigenspaces for face recognition. *Proceedings IEEE Conf. on Computer Vision and Pattern Recognition*, pages 84–91, June 1994.

[46] R. Plamondon, editor. *Progress in automatic signature verification.* World Scientific, 1994.

[47] Fred Preston. Automatic fingerprint matching. In *ICCST*, pages 199–202, 1989.

[48] F. J. Prokoski, R. B. Riedel, and J. S. Coffin. Identification of individuals by facial thermography. In *Proc. IEEE Intl. Carnahan Conference on Security Technology*, 1992.

[49] N. K. Ratha, A. K. Jain, and D. T. Rover. Fingerprint matching on Splash 2. Technical Report March, Dept. of Computer Science, Michigan State University, 1994.

[50] R.Brunelli and T.Poggio. Face recognition: features versus templates. *IEEE Trans. Patt. Anal. and Mach. Intell.*, 15:1042–1052, 1993.

[51] D. Reisfeld and Y. Yeshuran. Robust detection of facial features by generalized symmetry. In *Proc. 11th International Conference on Pattern Recognition*, pages 117–120, 1992.

[52] J. Rice. A quality approach to biometric imaging. In *IEE Colloquium on Image Processing for Biometric Measurement*, pages 4/1 – 4/5, April 1994.

[53] M.D. Richard and R.P. Lippmann. Neural network classifiers estimate bayesian *a posteriori* probabilities. *Neural Computation*, 3:461–483, 1991.

[54] H.A. Rowley, S. Baluja, and T. Kanade. Human face detection in visual scenes. Technical Report CMU-CS-95-158R, School of Computer Science, Carnegie Mellon University, 1995.

[55] F.S. Samaria and A.C. Harter. Parameterization of a stochastic model for human face identification. In *Proc. IEEE workshop on Applications of Computer Vision*, Sarasota, FL, 1994.

[56] F.S. Saramia. *Face Recognition Using Hidden Markov Model.* PhD thesis, University of Cambridge, 1994.

[57] Dave Sims. Biometric recognition: our hands, eyes, and faces give us away. *IEEE Computer Graphics and Applications*, pages 14–15, September 1994.

[58] K.K. Sung and T. Poggio. Learning human face detection in cluttered scenes. In *Computer Analysis of Image and Patterns*, pages 432–439, 1995.

[59] S.Y.Kung, M.Fang, S.P.Liou, and J.S.Taur. Decision-based neural network for face recognition system. In *Proc. of 1995 IEEE International Conference on Image Processing*, volume I, pages 430–433, Washington, D.C., USA, Oct 1995.

[60] T.Poggio and F.Girosi. Networks for approximation and learning. *Proc. IEEE*, 78:1481–1497, Sept. 1990.

[61] M. Turk and A. Pentland. Eigenfaces for recognition. *J. of Cognitive Neuroscience*, 3:71–86, 1991.

[62] S. R. Waterhouse and A. J. Robinson. Classification using hierarchical mixtures of experts. Technical Report Feb, Engineering Department, Cambridge University, 1994.

[63] S. L. Wood, Gongyuan Qu, and L. W. Roloff. Detection and labeling of retinal vessels for longitidunal studies. In *Proc. of 1995 IEEE International Conference on Image Processing*, volume III, pages 164–167, Washington, D.C., USA, Oct 1995.

[64] L. Xu, M.I. Jordan, and G.E. Hinton. A modified gating network for the mixture of experts architecture. In *World Congress on Neural Networks,San Diego*, pages II405–II410, 1994.

[65] Y.Cheng, K.Liu, J.Yang, and H.Wang. A robust algebraic method for face recognition. In *Proc. 11th Int. Conf. on Patt. Recog.*, pages 221–224, 1992.

[66] K. C. Yow and R. Cipolla. Finding initial estimates of human face locations. In *Proc. 2nd Asian Conf. on Computer Vision*, volume 3, pages 514–518, Singapore, 1995.

[67] A. Yuille, D. Cohen, and P. Hallinan. Feature extraction from faces using deformable templates. In *Porc. IEEE Computer Soc. Conf. on Computer Vision and Pattern Recognition*, pages 104–109, 1989.

5

Multidimensional Nonlinear Myopic Maps, Volterra Series, and Uniform Neural-Network Approximations

Irwin W. Sandberg

ABSTRACT Our main result is a theorem which gives necessary and sufficient conditions under which discrete-space multidimensional myopic input-output maps with vector-valued inputs drawn from a certain large set can be uniformly approximated arbitrarily well using a structure consisting of a linear preprocessing stage followed by a memoryless nonlinear network. Noncausal as well as causal maps are considered. Approximations for noncausal maps for which inputs and outputs are functions of more than one variable are of current interest in connection with, for example, image processing.

5.1 Introduction

The most important single result in the theory of linear systems is that under conditions that are ordinarily satisfied, a linear system's output is represented by a convolution operator acting on the system's input. This result provides understanding concerning the behavior of linear systems as well as an analytical basis for their design. There is no such simple result for nonlinear systems, but certain classes of nonlinear systems possess Volterra series (or Volterra-like series) representations in which the output is given by a familiar infinite sum of iterated integral operators operating on the system's input. The pressing need to have available representations for nonlinear systems, as well as to understand the limitations and advantages of such representations, is illustrated by fact that Volterra series representations have been discussed and studied in applications prior to the appearance of any results that established their existence, and even recent publications often cite [1] in seeking to justify the use of a Volterra series approach even though no justification is given (or intended) there.[1]

---

[1]The most pertinent material in Volterra's path-breaking [1] is Fréchet's Weierstrass-like result in [1, p. 20]. While interesting and important, it has significant limitations

A Volterra series representation, when it exists, is a very special representation in that, under weak assumptions, the terms in the infinite sum are unique. This uniqueness, at least in principle, often facilitates the determination of the terms in the series. But just as there are many real-valued functions on the interval $[0, 1]$ that do not have a power series expansion, there are many systems whose input-output maps do not have a Volterra series representation. However, the failure of a system to have an exact representation of a certain form does not rule out the possibility that a useful approximate representation may exist. This is the main reason that approximations to input-output maps are of interest.

The general problem of approximating input-output maps arises also in the neural-networks field where much progress has been made in recent years. Initially attention was focused on the problem of approximating real-valued maps of a finite number of real variables. For example, Cybenko [4] proved that any real-valued continuous map defined on a compact subset of $\mathbb{R}^n$ (n an arbitrary positive integer) can be uniformly approximated arbitrarily well using a neural network with a single hidden layer using any continuous sigmoidal activation function. And subsequently Mhaskar and Micchelli [5], and also Chen and Chen [6], showed that even any non-polynomial continuous function would do. There are also general studies of approximation in $L_p(\mathbb{R}^n)$, $1 \leq p < \infty$ (and on compact subsets of $\mathbb{R}^n$) using radial basis functions as well as elliptical basis functions. It is known [7], for example, that arbitrarily good approximation of a general $f \in L_1(\mathbb{R}^n)$ is possible using uniform smoothing factors and radial basis functions generated in a certain natural way from a single g in $L_1(\mathbb{R}^n)$ if and only if g has a nonzero integral.

The material just described is concerned with the approximation of systems without dynamics. At about the same time there began [8] a corresponding study of the network (e.g., neural network) approximation of functionals and approximately-finite-memory maps. It was shown that large classes of approximately-finite-memory maps can be uniformly approximated arbitrarily well by the maps of certain simple nonlinear structures

---

with regard to Volterra series representations of input-output maps in that, for example, it concerns approximations rather than series expansions in the usual sense. An analogy is that while any continuous real function f on the interval $[0, 1]$ can be approximated uniformly arbitrarily well by a polynomial, this does not mean that every such f has a power series representation. Another limitation is that Fréchet's result directly concerns functionals rather than mappings from one function space to another, and the domain of Fréchet's functionals is a set of functions defined on an interval that is finite. In contrast, in most circuits and systems studies involving Volterra series, inputs are defined on $\mathbb{R}$ or $\mathbb{R}_+$ (or on the discrete-time analogs of these sets). For influential early material concerning Volterra series and applications, see [2]. Additional background material, references, and a description of relatively recent results concerning the existence and convergence of Volterra series can be found in [3].

using, for example, sigmoidal nonlinearities or radial basis functions.[2] This is of interest in connection with, for example, the general problem of establishing a comprehensive analytical basis for the identification of dynamic systems.[3] The approximately-finite-memory approach in [8] is different from, but is related to, the fading-memory approach in [13] where it is proved that certain scalar single-variable causal fading-memory systems with inputs and outputs defined on $\mathbb{R}$ can be approximated by a finite Volterra series.

The study in [8] addresses noncausal as well as causal systems, and also systems in which inputs and outputs are functions of several variables. In a recent paper [14] we give strong corresponding results within the framework of an extension of the fading-memory approach. A key idea in [14] is a substantial generalization of the proposition in [13] to the effect that a certain set of continuous functions defined on $\mathbb{R}$ is compact in a weighted-norm space. In [14] attention is restricted to the case of inputs and outputs defined on the infinite m-dimensional interval $(-\infty, \infty)^m$ where m is an arbitrary positive integer. In this paper we consider the important "discrete-space" case in which inputs and outputs are defined on $(\ldots, -1, 0, 1, \ldots)^m$. We use the term "myopic" to describe the maps we study because the term "fading-memory" is a misnomer when applied to noncausal systems, in that noncausal systems may anticipate as well as remember. In our setting, a shift-invariant causal myopic map has what we call "uniform fading memory."

Our main result is a theorem which gives necessary and sufficient conditions under which multidimensional myopic input-output maps with vector-valued inputs drawn from a certain large set can be uniformly approximated arbitrarily well using a structure consisting of a linear preprocessing stage followed by a memoryless nonlinear network.[4] Such structures were first considered in an important but very special context by Wiener [2, pp. 380-382]. We consider causal as well as noncausal maps. Approximations for noncausal maps for which inputs and outputs are functions of more than one variable, with each variable taking values in $(\ldots, -1, 0, 1, \ldots)$, are of

---

[2] It was later found [9] (see also [10]) that the approximately-finite-memory condition is met by the members of a certain familiar class of stable continuous-time systems.

[3] It was also observed that any continuous real functional on a compact subset of a real normed linear space can be uniformly approximated arbitrarily well using only a feedforward neural network with a linear-functional input layer and one hidden nonlinear (e.g., sigmoidal) layer. This has applications concerning, for instance, the theory of classification of signals, and is a kind of general extension of an idea due to Wiener concerning the approximation of input-output maps using a structure consisting of a bank of linear maps followed by a memoryless map with several inputs and a single output (see, for instance, [2, pp. 380-382]). For further results along these lines, see [11]. And for related work, see [12].

[4] This theorem was obtained in joint work with the writer's graduate student Lilian Xu.

current interest in connection with, for example, image processing.

Section 5.2 begins with a description of notation and key definitions. Our main result is given in Section 5.2.2. This is followed by a section on comments and one on the specialization of our result to generalized discrete-space finite Volterra series approximations.

5.2 Approximation of Myopic Maps

5.2.1 Preliminaries

Throughout the paper, $\mathbb{R}$ is the set of reals, $\mathcal{Z}$ is the set of all integers, and $\mathbb{N}$ is the set of positive integers. Let n and m in $\mathbb{N}$ be arbitrary. $\|\cdot\|$ and $\langle\cdot,\cdot\rangle$ are the Euclidean norm and inner product on $\mathbb{R}^n$, respectively, and for each $\alpha := (\alpha_1, \ldots, \alpha_m) \in \mathcal{Z}^m$, $|\alpha|$ stands for $\max_j |\alpha_j|$. With $\mathcal{Z}_- = \{\ldots, -1, 0\}$, $\mathcal{Z}_-^m$ denotes $(\mathcal{Z}_-)^m$.

For any positive integer n_0, let $C(\mathbb{R}^{n_0}, \mathbb{R})$ denote the set of continuous maps from $\mathbb{R}^{n_0}$ to $\mathbb{R}$, and let D_{n_0} stand for any subset of $C(\mathbb{R}^{n_0}, \mathbb{R})$ that is dense on compact sets, in the usual sense that given $\epsilon > 0$ and $f \in C(\mathbb{R}^{n_0}, \mathbb{R})$, as well as a compact $V \subset \mathbb{R}^{n_0}$, there is a $q \in D_{n_0}$ such that $|f(v) - q(v)| < \epsilon$ for $v \in V$. The D_{n_0} can be chosen in many different ways, and may involve, for example, radial basis functions, polynomial functions, piecewise linear functions, sigmoids, or combinations of these functions.

Let w be an $\mathbb{R}$-valued function defined on $\mathcal{Z}^m$ such that $w(\alpha) \neq 0$ for all α and $\lim_{|\alpha|\to\infty} w(\alpha) = 0$. With $X(\mathcal{Z}^m, \mathbb{R}^n)$ the set of all $\mathbb{R}^n$-valued maps on $\mathcal{Z}^m$, denote by X_w the normed linear space given by

$$X_w = \{x \in X(\mathcal{Z}^m, \mathbb{R}^n) : \sup_{\alpha \in \mathcal{Z}^m} \|w(\alpha)x(\alpha)\| < \infty\}$$

with the norm

$$\|x\|_w = \sup_{\alpha \in \mathcal{Z}^m} \|w(\alpha)x(\alpha)\| .$$

Later we will use the fact that X_w is complete (see Appendix A).

Now let $\mathcal{X}(\mathcal{Z}^m, \mathbb{R}^n)$ be the set of all bounded functions contained in $X(\mathcal{Z}^m, \mathbb{R}^n)$, and let S be a nonempty subset of $\mathcal{X}(\mathcal{Z}^m, \mathbb{R}^n)$. For each $\beta \in \mathcal{Z}^m$, define T_β on S by

$$(T_\beta x)(\alpha) = x(\alpha - \beta), \quad \alpha \in \mathcal{Z}^m.$$

The set S is said to be ***closed under translation*** if $T_\beta S = S$ for each $\beta \in \mathcal{Z}^m$.[5] Let G map S to the set of $\mathbb{R}$-valued functions on $\mathcal{Z}^m$. Such a G is

[5]With regard to the choice of words ***closed under translation***, an equivalent definition is $T_\beta S \subseteq S$ for all β. To see this, observe that if $T_\beta S \subseteq S$ for all β and $T_{\beta_0} S \subset S$ for some β_0, then $T_{\beta_0} A \subset S$ for any $A \subseteq S$, which leads to the contradiction that $T_{\beta_0} T_{-\beta_0} S \subset S$.

shift-invariant if S is closed under translation and

$$(Gx)(\alpha - \beta) = (GT_\beta x)(\alpha), \quad \alpha \in \mathcal{Z}^m$$

for each $\beta \in \mathcal{Z}^m$ and $x \in S$. The map G is *causal* if

$$x(\alpha) = y(\alpha) \text{ whenever } \alpha_j \le \beta_j \ \forall j \quad \Longrightarrow \quad (Gx)(\beta) = (Gy)(\beta)$$

for each $\beta \in \mathcal{Z}^m$ and every x and y in S.

We assume throughout the paper that G is shift-invariant. We say that G is *myopic* on S with respect to w if given an $\epsilon > 0$ there is a $\delta > 0$ with the property that x and y in S and

$$\sup_{\alpha \in \mathcal{Z}^m} \|w(\alpha)[x(\alpha) - y(\alpha)]\| < \delta \quad \Longrightarrow \quad |(Gx)(0) - (Gy)(0)| < \epsilon. \tag{5.1}$$

Thus, and roughly speaking, G is myopic if the value of $(Gx)(\alpha)$ is always relatively independent of the values of x at points remote from α.

In our theorem in the next section we refer to certain sets $\mathcal{G}(w)$ and $\mathcal{G}_-(w)$. These sets concern sums of the form

$$\sum_{\beta \in D} \langle g(\beta), x(\beta) \rangle \tag{5.2}$$

in which $x \in \mathcal{X}(\mathcal{Z}^m, \mathbb{R}^n)$ and $D = \mathcal{Z}^m$ or $\mathcal{Z}^m_-$. Such sums are well defined in the sense of absolute summability (see Appendix B) for any g from D to $\mathbb{R}^n$ such that

$$\sum_{\beta \in D} \left\| w(\beta)^{-1} g(\beta) \right\| < \infty \tag{5.3}$$

(because (5.3) implies the absolute summability of g). Let $D = \mathcal{Z}^m$. By $\mathcal{G}(w)$ we mean any set of functions g from D to $\mathbb{R}^n$ such that (5.3) is met for each g, and for each nonzero $x \in \mathcal{X}(\mathcal{Z}^m, \mathbb{R}^n)$ there corresponds a g for which (5.2) is nonzero. Similarly, with $D = \mathcal{Z}^m_-$, the set $\mathcal{G}_-(w)$ is any set of functions g from D to $\mathbb{R}^n$ such that (5.3) is met for each g, and for each $x \in \mathcal{X}(\mathcal{Z}^m, \mathbb{R}^n)$ whose restriction to D is nonzero there is a g for which (5.2) is nonzero. The sets $\mathcal{G}(w)$ and $\mathcal{G}_-(w)$ can be chosen in many ways (see Appendix C). Finally, let Q be the map from S to $X(\mathcal{Z}^m, \mathbb{R}^n)$ defined by $(Qs)(\alpha) = s(\beta)$ for each s and α, where $\beta_j = -|\alpha_j|$ for all j.

5.2.2 *Our Main Result*

Our main result below gives a necessary and sufficient condition for the uniform approximation of myopic maps with vector-valued inputs of a finite number of variables. In stating this result, we use the fact that sums of the form

$$\sum_{\beta \in \mathcal{Z}^m} \langle p(\beta - \alpha), x(\beta) \rangle$$

and

$$\sum_{\beta \in (-\infty,\alpha]} \langle q(\beta - \alpha), x(\beta) \rangle$$

are well defined and finite for each $\alpha \in \mathcal{Z}^m$ when x is an element of $\mathcal{X}(\mathcal{Z}^m, I\!R^n)$, $p \in \mathcal{G}(w)$, and $q \in \mathcal{G}_-(w)$. This follows from the observation that by (5.3) both p and q are absolutely summable on their respective domains (see Appendix B).

Theorem 1: Assume that S is uniformly bounded.[6] (Recall that G is a shift-invariant map from S to the set of $I\!R$-valued functions defined on $\mathcal{Z}^m$.) Then the following two statements are equivalent.

(i) G is myopic on S with respect to w.

(ii) For each $\epsilon > 0$, there are an $n_0 \in I\!N$, elements $g_1, \ldots, g_{n_0}$ of $\mathcal{G}(w)$, and an $N \in D_{n_0}$ such that

$$|(Gx)(\alpha) - N[(Lx)(\alpha)]| < \epsilon, \quad \alpha \in \mathcal{Z}^m \tag{5.4}$$

for all $x \in S$, where L is given by

$$(Lx)_j(\alpha) = \sum_{\beta \in \mathcal{Z}^m} \langle h_j(\alpha - \beta), x(\beta) \rangle \tag{5.5}$$

with $h_j(\beta) = g_j(-\beta)$ for all β and j.

Moreover, if G is causal and $QS \subseteq S$ then (ii) can be replaced with:

(ii′) For each $\epsilon > 0$, there are an $n_0 \in I\!N$, elements $g_1, \ldots, g_{n_0}$ of $\mathcal{G}_-(w)$, and an $N \in D_{n_0}$ such that $|(Gx)(\alpha) - N[(Lx)(\alpha)]| < \epsilon$ for all $\alpha \in \mathcal{Z}^m$ and $x \in S$, where L is given by

$$(Lx)_j(\alpha) = \sum_{\beta \in (-\infty,\alpha]} \langle h_j(\alpha - \beta), x(\beta) \rangle \tag{5.6}$$

with $h_j(\beta) = g_j(-\beta)$ for all β and j, and where $(-\infty, \alpha]$ means $\{\ldots, (\alpha_1 - 1), \alpha_1\} \times \cdots \times \{\ldots, (\alpha_m - 1), \alpha_m\}$.

Proof

(i) $\Rightarrow$ (ii): Assume that (i) holds. Recall that $w(\alpha) \to 0$ as $|\alpha| \to \infty$ and that S, which is uniformly bounded, is a subset of the complete space X_w. By the characterization in [19, p. 136] of relative compactness of subsets of a general Banach space with a basis, it is not difficult to show that S is

[6] This means that there is a positive constant c for which $\|x(\alpha)\| \leq c$ for all $x \in S$ and all α.

a relatively compact subset of X_w.[7] Since G is myopic with respect to w, the functional $(G\cdot)(0)$ is uniformly continuous on S with respect to $\|\cdot\|_w$. We will use the following lemma proved in Appendix D.

Lemma 1: Assume that U is a nonempty uniformly bounded subset of $X(\mathbb{Z}^m, \mathbb{R}^n)$ that is relatively compact in X_w. Let $F: U \to \mathbb{R}$ be uniformly continuous with respect to $\|\cdot\|_w$. Then given $\epsilon > 0$, there are an $n_0 \in \mathbb{N}$, elements $g_1, \ldots, g_{n_0}$ of $\mathcal{G}(w)$, and an $N \in D_{n_0}$ such that

$$|F(u) - N[y_1(u), \ldots, y_{n_0}(u)]| < \epsilon, \quad u \in U$$

where each y_j is defined on U by

$$y_j(u) = \sum_{\beta \in \mathbb{Z}^m} \langle g_j(\beta), u(\beta) \rangle .$$

Continuing with the proof, let $\epsilon > 0$, $\alpha \in \mathbb{Z}^m$, and $x \in S$ be given. By Lemma 1 with $U = S$, $F = (G\cdot)(0)$, and $u = T_{-\alpha}x$, we have

$$|G(T_{-\alpha}x)(0) - N[y_1(T_{-\alpha}x), \ldots, y_{n_0}(T_{-\alpha}x)]| < \epsilon$$

with N, n_0, and the g_j as described in Lemma 1. By the shift-invariance of G, $(Gx)(\alpha) = (GT_{-\alpha}x)(0)$. And since $T_{-\alpha}x(\beta) = x(\alpha + \beta)$ for all β, we have

$$y_j(T_{-\alpha}x) = \sum_{\beta \in \mathbb{Z}^m} \langle g_j(\beta), x(\alpha + \beta) \rangle = \sum_{\beta \in \mathbb{Z}^m} \langle g_j(\beta - \alpha), x(\beta) \rangle$$

for each j. Thus we have (ii).

(i) $\Rightarrow$ (ii$'$): Assume that $QS \subseteq S$ and that G is causal as well as myopic on S with respect to w. Here we use the following variant of Lemma 1 proved in Appendix E.

Lemma 2: Let U be a nonempty uniformly bounded subset of $X(\mathbb{Z}^m, \mathbb{R}^n)$ that contains QU. Assume also that U is relatively compact in X_w, and let $F: U \to \mathbb{R}$ be uniformly continuous with respect to $\|\cdot\|_w$. Then given $\epsilon > 0$, there are an $n_0 \in \mathbb{N}$, elements $g_1, \ldots, g_{n_0}$ of $\mathcal{G}_-(w)$, and an $N \in D_{n_0}$ such that

$$|F(u) - N[y_1(u), \ldots, y_{n_0}(u)]| < \epsilon, \quad u \in QU$$

[7] Let $e^j_\alpha : \mathbb{Z}^m \to \mathbb{R}^n$ with $\alpha \in \mathbb{Z}^m$ and $j \in \{1, \ldots, n\}$ be defined by the condition that $e^j_\alpha(\beta) = 0$ for $\beta \neq \alpha$ and $e^j_\alpha(\beta)$ for $\beta = \alpha$ is the element of $\mathbb{R}^n$ with jth component unity and all other components zero. Then $\{e^j_\alpha\}$ is clearly a basis for X_w. The result in [19, p. 136] is applicable because $\left\|e^j_\alpha\right\|_w \to 0$ as $|\alpha| \to \infty$ uniformly in j.

where each y_j is defined on QU by

$$y_j(u) = \sum_{\beta \in \mathcal{Z}_-^m} \langle g_j(\beta), u(\beta) \rangle .$$

Now let $x \in S$ and $\alpha \in \mathcal{Z}^m$ be given and, referring to Lemma 2, take $U = S$, $F = (G \cdot)(0)$, and $u = QT_{-\alpha}x$. Here for each j, $y_j(QT_{-\alpha}x) = y_j(T_{-\alpha}x)$ and

$$y_j(T_{-\alpha}x) = \sum_{\beta \in \mathcal{Z}_-^m} \langle g_j(\beta), x(\alpha + \beta) \rangle = \sum_{\beta \in (-\infty, \alpha]} \langle g_j(\beta - \alpha), x(\beta) \rangle .$$

In addition, by the causality and shift-invariance of G,

$$(GQT_{-\alpha}x)(0) = (GT_{-\alpha}x)(0) = (Gx)(\alpha).$$

This gives (ii′).

(ii) $\Rightarrow$ (i): Assume that (ii) holds. Fix $\epsilon > 0$, $u_1 \in S$, and $u_2 \in S$. By (ii) choose n_0, elements $g_1, \ldots, g_{n_0}$ of $\mathcal{G}(w)$, and an $N \in D_{n_0}$, all independent of u_1 and u_2, such that $|(Gu_1)(0) - N[(Lu_1)(0)]| < \epsilon/3$ and $|(Gu_2)(0) - N[(Lu_2)(0)]| < \epsilon/3$.

By the triangle inequality in $\mathbb{R}$,

$$|(Gu_1)(0) - (Gu_2)(0)| < 2\epsilon/3 + |N[(Lu_1)(0)] - N[(Lu_2)(0)]| .$$

Since N is continuous and (by the absolute summability of the g_j) $(LS)(0)$ is bounded, N is uniformly continuous on $(LS)(0)$. Thus, there is a $\delta_0 > 0$, independent of u_1 and u_2, such that

$$\|(Lu_1)(0) - (Lu_2)(0)\| < \delta_0 \quad \Longrightarrow \quad |N[(Lu_1)(0)] - N[(Lu_2)(0)]| < \epsilon/3.$$

It thus suffices to show that there exists a $\delta > 0$, independent of u_1 and u_2, such that

$$\sup_{\alpha \in \mathcal{Z}^m} \|w(\alpha)[u_1(\alpha) - u_2(\alpha)]\| < \delta \quad \Longrightarrow \quad \|(Lu_1)(0) - (Lu_2)(0)\| < \delta_0.$$

For $j = 1, \ldots, n_0$ we have

$$\begin{aligned} |(Lu_1)_j(0) - (Lu_2)_j(0)| &\leq \sum_{\alpha \in \mathcal{Z}^m} |\langle g_j(\alpha), u_1(\alpha) - u_2(\alpha) \rangle| \\ &\leq \sum_{\alpha \in \mathcal{Z}^m} \left\| w(\alpha)^{-1} g_j(\alpha) \right\| \cdot \|w(\alpha)[u_1(\alpha) - u_2(\alpha)]\| \\ &\leq \sup_{\alpha \in \mathcal{Z}^m} \|w(\alpha)[u_1(\alpha) - u_2(\alpha)]\| \cdot \sum_{\alpha \in \mathcal{Z}^m} \left\| w(\alpha)^{-1} g_j(\alpha) \right\| , \end{aligned}$$

from which it is clear that we can choose δ as required.[8] (ii$'$) $\Rightarrow$ (i) can be established similarly. This completes the proof of the theorem.

5.2.3 Comments

The approximating structure of Theorem 1 is illustrated in Figure 5.1. Our proof shows that (i) implies (ii) and (ii$'$) even without the hypothesis that the elements of D_{n_0} are continuous. Also, (ii) implies (i), and (ii$'$) implies (i), without the assumption that S is closed under translation.

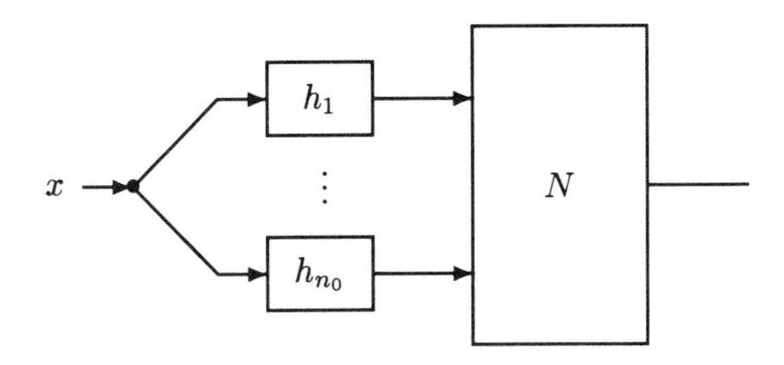

FIGURE 5.1. Approximation structure.

The map Q in the theorem can be replaced with any map from S to $X(\mathcal{Z}^m, I\!R^n)$ such that: $(Qx)(\alpha) = x(\alpha)$ for α in $\mathcal{Z}^m_-$ and $x \in S$, and for distinct elements x and y of QS the restriction $(x-y)|_{\mathcal{Z}^m_-}$ of $(x-y)$ to $\mathcal{Z}^m_-$ is not zero. For example, Q can be taken to be defined by

$$(Qx)(\alpha) = x(\hat{\alpha}), \quad \alpha \in \mathcal{Z}^m \tag{5.7}$$

where $\hat{\alpha}$ is the unique point in $\mathcal{Z}^m_-$ that minimizes $|\alpha - \hat{\alpha}|$.[9]

The following condition is closely related to the concept of myopic maps: given an $\epsilon > 0$ there is a $\delta > 0$ such that x and y in S and

$$\sup_{\alpha \in \mathcal{Z}^m_-} \|\mu(\alpha)[x(\alpha) - y(\alpha)]\| < \delta \quad \Longrightarrow \quad |(Gx)(0) - (Gy)(0)| < \epsilon \tag{5.8}$$

(note the $\mathcal{Z}^m_-$ instead of $\mathcal{Z}^m$), where μ is an $I\!R$-valued function on $\mathcal{Z}^m_-$ that satisfies $\lim_{|\alpha|\to\infty} \mu(\alpha) = 0$ and never vanishes. Appendix F is a brief study

[8] A related but different way to argue that (ii) $\Rightarrow$ (i) is to observe that under (ii) the real-valued function $(G\cdot)(0)$ is uniformly continuous on S with respect to the norm in X_w, since $(G\cdot)(0)$ can be uniformly approximated arbitrarily well over S by uniformly continuous functions.

[9] $\hat{\alpha}$ is defined by the simple condition that each of its elements belongs to $\{\ldots, -1, 0\}$ and is as close as possible to the corresponding element of α.

of implications concerning myopic maps, causality, and condition (5.8). In particular, it is shown there that, under the hypotheses of Theorem 1 and assuming $QS \subseteq S$, condition (5.8) is met for some μ if and only if G is causal and myopic on S with respect to some w. The uniform boundedness hypothesis of Theorem 1 concerning S, as well as the requirement that $QS \subseteq S$, are often satisfied. For example, they are satisfied for S the subset of $X(\mathcal{Z}^m, \mathbb{R}^n)$ consisting of all functions that are uniformly bounded with some bound c. This case with $m = n = 1$ is considered in [13]. And condition (5.8) is closely related to the *fading memory* concept described there for causal G's. In this context condition (5.8) is what might be called a *uniform fading memory* condition.

With S as described in Theorem 1, the condition that G is myopic on S with respect to w is equivalent to the seemingly weaker condition that given $\epsilon > 0$ and $x \in S$ there is a $\delta > 0$ such that $|(Gx)(0) - (Gy)(0)| < \epsilon$ for $y \in S$ with $\sup_{\alpha \in \mathcal{Z}^m} \|w(\alpha)[x(\alpha) - y(\alpha)]\| < \delta$. This is true because any continuous $\mathbb{R}$-valued function defined on a compact set is uniformly continuous, and the extension described in Appendix D of the functional F of Lemma 1 is defined on a compact set.[10]

A simple useful observation concerning (5.3) is this: If G is myopic with respect to to some w, and q is any positive map from $\mathcal{Z}^m$ to $\mathbb{R}$ such that $q(\beta) \to \infty$ as $|\beta| \to \infty$, then G is myopic with respect to some w for which $|w(\beta)^{-1}| \leq q(\beta)$ for all β.[11] As an application, we note that this observation together with an example described in a footnote in Appendix C can be used to show that a network structure related to one studied in [15] can approximate arbitrarily well any discrete-time time-invariant causal uniformly fading memory input-output map for which the conditions on inputs of Theorem 1 are met and $m = n = 1$. For related results see, for example, [11] and [16].

The implication (ii$'$) $\Rightarrow$ (i), which concerns causal maps, is new even for the $m = n = 1$ case.

The class of maps G addressed by our theorem that are myopic with respect to some w is very large. In fact, a natural question to ask is whether there are G's that are not myopic. It turns out that one answer to this question is closely related to the theory of representations of linear G's and

---

[10] Similarly it can be shown that, under the uniform boundedness hypothesis of Theorem 1, condition (5.8) is equivalent to the condition that given $\epsilon > 0$ and $x \in S$ there is a $\delta > 0$ such that $|(Gx)(0) - (Gy)(0)| < \epsilon$ for $y \in S$ with $\sup_{\alpha \in \mathcal{Z}^m_-} \|\mu(\alpha)[x(\alpha) - y(\alpha)]\| < \delta$. This for $m = n = 1$ is (essentially) the discrete-time fading memory condition that plays a role in [13]. It can also be shown that, under the hypotheses of Theorem 1, (ii$'$) holds if G is causal and condition (5.8) is met with μ the restriction to $\mathbb{R}^m_-$ of w. (The condition that $QS \subseteq S$ is not needed.)

[11] This is true because if G is myopic with respect to w, then it is myopic with respect to $|w|$, and thus with respect to ω defined by $\omega(\beta) = \max\{|w(\beta)|, q(\beta)^{-1}\}$. A similar observation is made in [13].

to a pertinent oversight in the literature. This is discussed in Appendix G, where an example is given of a linear G that is myopic for no w.

We do not consider in this paper the problem of actually determining the elements of the structure in Figure 5.1. What we have done is to show that in an important setting it is possible to choose acceptable elements. And of course it is ordinarily useful to know that certain elements exist before one attempts to determine them by analytical, adaptive, or experimental techniques. Another illustration of our viewpoint is the following. In the area of signal processing there is considerable current interest in exploring nonlinear filtering techniques to overcome some of the basic limitations of linear filtering. Our theorem shows that the members of a certain class of nonlinear filters can be represented (as accurately as one wishes) by the special structure of Figure 5.1. These are what might be called the "myopic-map filters," and they form an interesting very large class of filters. In the next section we observe that the structure can provide Volterra-series approximations.

5.2.4 *Finite Generalized Volterra-Series Approximations*

If the D_{n_0} are chosen to be certain sets of polynomial functions, the approximating maps $N[L(\cdot)]$ of (ii) and (ii′) of Theorem 1 are generalizations of discrete-time finite Volterra series. More specifically, suppose that the maps q of each D_{n_0} are such that $q(v)$ is a real polynomial in the components of v, and consider the $n = 1$ case. By writing products of sums as iterated sums, it is not difficult to see that $N[L(\alpha)]$ of (ii) has the form

$$k_0 + \sum_{j=1}^{p} \sum_{\beta_j \in \mathcal{Z}^m} \cdots \sum_{\beta_1 \in \mathcal{Z}^m} k_j(\alpha - \beta_1, \ldots, \alpha - \beta_j) x(\beta_1) \cdots x(\beta_j) \qquad (5.9)$$

in which k_0 is a constant, p is a positive integer, and each $k_j(\alpha-\beta_1,\ldots,\alpha-\beta_j)$ is a finite linear combination of products of the form $h_{i_j(1)}(\alpha - \beta_1)$ $h_{i_j(2)}(\alpha-\beta_2) \cdots h_{i_j(j)}(\alpha\mathrm{i}-\beta_j)$, where the indices $i_j(1),\ldots,i_j(j)$ are drawn from $\{1,\ldots,n_0\}$ and the h's are as described in (ii). Similarly, in the case of (ii′), $N[L(\alpha)]$ has the form

$$k_0 + \sum_{j=1}^{p} \sum_{\beta_j \in (-\infty,\alpha]} \cdots \sum_{\beta_1 \in (-\infty,\alpha]} k_j(\alpha - \beta_1, \ldots$$

$$, \alpha - \beta_j) x(\beta_1) \cdots x(\beta_j) \qquad (5.10)$$

where the k_j are formed from the h_j of (ii′) in the way described above in connection with (ii). For each j, each of the two j-fold iterated sums in the summations over j of sums in (5.9) and (5.10) is a sum of products of sums over $\mathcal{Z}^m$ and $(-\infty,\alpha]$, respectively. Since in both cases the h_j are absolutely summable, by a result in Appendix B (and the boundedness of x) each sum in each product can be written as an m-fold iterated sum. Thus, each of the j-fold iterated sums can be simplified in that each can be written as a (jm)-fold iterated sum in which each summation is with respect to a scalar variable. For example, for $m = j = 2$,

$$\sum_{\beta_j \in \mathcal{Z}^m} \cdots \sum_{\beta_1 \in \mathcal{Z}^m} k_j(\alpha - \beta_1, \ldots, \alpha - \beta_j) x(\beta_1) \cdots x(\beta_j)$$

can be written in the form

$$\sum_{\gamma_4 \in \mathcal{Z}} \cdots \sum_{\gamma_1 \in \mathcal{Z}} \tilde{k}_2(\alpha_1 - \gamma_1, \alpha_2 - \gamma_2, \alpha_1 - \gamma_3, \alpha_2 - \gamma_4) x(\gamma_1, \gamma_2) x(\gamma_3, \gamma_4)$$

where $x(\gamma_1, \gamma_2) = x[(\gamma_1, \gamma_2)]$, $x(\gamma_3, \gamma_4) = x[(\gamma_3, \gamma_4)]$, and $\tilde{k}_2$ is related to k_2 in the obvious way.

Theorem 1 addresses only input-output maps that are shift-invariant. Some thoughts on Volterra series approximations for shift-varying maps are described in Appendix H.

5.3 Appendices

APPENDIX A: The Completeness of X_w

To see that X_w is complete, let $\{x_j\}$ be a Cauchy sequence in X_w, and notice that for each positive ϵ there is a positive N such that

$$\|wx_p - wx_q\|_0 < \epsilon, \quad p, q > N$$

where $\|\cdot\|_0$ is the norm in $\mathcal{X}(\mathcal{Z}^m, \mathbb{R}^n)$ given by $\|x\|_0 = \sup_{\alpha \in \mathcal{Z}^m} \|x(\alpha)\|$. By the completeness of $\mathcal{X}(\mathcal{Z}^m, \mathbb{R}^n)$, there is a $y \in \mathcal{X}(\mathcal{Z}^m, \mathbb{R}^n)$ such that $\{wx_j\}$ converges to y in $\mathcal{X}(\mathcal{Z}^m, \mathbb{R}^n)$. Thus, for an arbitrary positive ϵ there exists a positive $\tilde{N}$ such that

$$\|y - wx_j\|_0 < \epsilon, \quad j > \tilde{N}.$$

Since w never vanishes, the function y/w can be seen to be the limit of $\{x_j\}$ in X_w.

APPENDIX B: Absolute Summability

Here, for the reader's convenience, we describe certain facts that we use concerning absolutely summable families of functions. The material is essentially standard.[12] As in Section 5.2.1, let $D = \mathcal{Z}^m$ or $\mathcal{Z}^m_-$. Let f be a map from D to E, where E is either $\mathbb{R}$ or $\mathbb{R}^n$. Let $\|\cdot\|_*$ denote the absolute value operator if $E = \mathbb{R}$, and let $\|\cdot\|_* = \|\cdot\|$ if $E = \mathbb{R}^n$.

We say that f belongs to $S_1(D)$ if there is a positive constant c such that

$$\sum_{\beta \in J} \|f(\beta)\|_* \leq c$$

for all finite subsets J of D. Such an f is said to be ***absolutely summable*** on D, and another way to denote this is to write

$$\sum_{\beta \in D} \|f(\beta)\|_* < \infty.$$

If $f \in S_1(D)$, then $\sum_{|\beta| \leq k} f(\beta)$ converges in E to a limit l as $k \to \infty$, and we denote this l by $\sum_{\beta \in D} f(\beta)$.

Write $\beta = (\beta_1, \ldots, \beta_m)$, and for each j let $\sum_{\beta_j}$ denote a sum in which β_j ranges over $\mathcal{Z}$ if $D = \mathcal{Z}^m$, or over $\mathcal{Z}_-$ if $D = \mathcal{Z}^m_-$. If $f \in S_1(D)$, the iterated sum $\sum_{\beta_m} \cdots \sum_{\beta_1} f(\beta)$ is well defined in the sense that each sum converges absolutely, and we have

$$\sum_{\beta \in D} f(\beta) = \sum_{\beta_m} \cdots \sum_{\beta_1} f(\beta). \tag{5.11}$$

[12]It can be viewed as a direct modification of material in [20, pp. 97-100] or as a special case of the general theory of integration with respect to a measure.

It is also true that (5.11) holds if $\beta_m, \ldots, \beta_1$ is replaced with any permutation $\beta_{q(m)}, \ldots, \beta_{q(1)}$ of $\beta_m, \ldots, \beta_1$.

APPENDIX C: On $\mathcal{G}(w)$ and $\mathcal{G}_-(w)$

Let $D = \mathcal{Z}^m$ or $\mathcal{Z}_-^m$. Denote by A the set of all functions g from D to $\mathbb{R}^n$ satisfying (5.3), and recall that these functions are absolutely summable. Let B be any dense subset of A, in the sense that for each $\epsilon > 0$ and each $g \in A$ there is an $f \in B$ such that

$$\sum_{\beta \in D} \|g(\beta) - f(\beta)\| < \epsilon.$$

It is clear that any dense subset of $l_1(D)$[13] whose elements satisfy (5.3) is a dense subset of A.[14] We now show that $\mathcal{G}(w)$ and $\mathcal{G}_-(w)$ can be taken to be B when $D = \mathcal{Z}^m$ and $D = \mathcal{Z}_-^m$, respectively.

Let $x \in \mathcal{X}(\mathcal{Z}^m, \mathbb{R}^n)$ be such that the restriction of x to D is nonzero. We are going to show that there is an $f \in B$ such that the sum $\sum_{\beta \in D} \langle f(\beta), x(\beta) \rangle$ is not zero. Choose any $r \in l_1(D)$ that never vanishes, and let g be the element of A given by

$$g(\beta) = \|r(\beta)\| \, w^2(\beta) x(\beta), \quad \beta \in D.$$

Using

$$\sum_{\beta \in D} \langle g(\beta), x(\beta) \rangle = \sum_{\beta \in D} \|r(\beta)\| \cdot \|w(\beta)x(\beta)\|^2 > 0,$$

select $f \in B$ such that

$$2 \sup_{\beta \in \mathcal{Z}^m} \|x(\beta)\| \cdot \sum_{\beta \in D} \|f(\beta) - g(\beta)\| < \sum_{\beta \in D} \langle g(\beta), x(\beta) \rangle .$$

---

[13] For $D = \mathcal{Z}_-^m$ or $\mathcal{Z}^m$, $l_1(D)$ denotes the normed space of all $r : D \to \mathbb{R}^n$ such that r is absolutely summable on D, with the norm $\|\cdot\|_1$ in $l_1(D)$ given by $\|r\|_1 = \sum_{\beta \in D} \|r(\beta)\|$. In other words, $l_1(D)$ denotes the set $S_1(D)$ of Appendix B with $E = \mathbb{R}^n$ and with the norm indicated here.

[14] For example, suppose that $D = \mathcal{Z}_-^m$, that $n = m = 1$, and that σ and λ are positive constants such that $\lambda > \sigma$ and $w(\beta)^{-1} = O(e^{-\sigma\beta})$. Then B can be taken to be the set of all finite linear combinations of the functions $k_1, k_2, \ldots$ where $k_j(\beta) = (\beta)^{(j-1)} e^{\lambda\beta}$ for each j. (A proof of the denseness of this B in $l_1(\mathcal{Z}_-)$ can be obtained by modifying material in [17, pp. 75–76] concerning a related continuous-time case.) As another example, assume that that $D = \mathcal{Z}^m$, that $n = m = 1$, and that σ and λ are positive constants such that $\lambda > \sigma$ and $w(\beta)^{-1} = O(e^{\sigma\beta^2})$. In this case B can be taken to be the set of all finite linear combinations of the functions $k_1, k_2, \ldots$ where $k_j(\beta) = (\beta)^{(j-1)} e^{-\lambda\beta^2}$ for each j (see [17, p. 80] for a corresponding continuous-time denseness result). As a third example, for $D = \mathcal{Z}^m$ we can take B to be the family of all finite linear combinations of the elements of the set $\{e_\alpha^j\}$ described in an earlier footnote.

Thus,

$$\left| \sum_{\beta \in D} \langle f(\beta), x(\beta) \rangle \right| \geq \sum_{\beta \in D} \langle g(\beta), x(\beta) \rangle - \sum_{\beta \in D} \|f(\beta) - g(\beta)\| \cdot \|x(\beta)\|,$$

showing that

$$\left| \sum_{\beta \in D} \langle f(\beta), x(\beta) \rangle \right| > \frac{1}{2} \sum_{\beta \in D} \langle g(\beta), x(\beta) \rangle > 0.$$

APPENDIX D: Proof of Lemma 1

Let $cl(U)$ denote the closure of U in X_w. We shall use the proposition that $cl(U) \subset \mathcal{X}(\mathcal{Z}^m, \mathbb{R}^n)$, which follows from the uniform boundedness of U and the fact that convergence in X_w implies uniform convergence on bounded subsets of $\mathcal{Z}^m$.

Since $cl(U)$ is closed, F has a continuous extension F_e to $cl(U)$ [18, p. 99, Problem 13]. Let A be the set of all real-valued functions a on $cl(U)$ given by

$$a(u) = \rho + c \sum_{\beta \in \mathcal{Z}^m} \langle g(\beta), u(\beta) \rangle \tag{5.12}$$

with $g \in \mathcal{G}(w)$ and ρ and c real numbers. By the Schwarz inequality in $\mathbb{R}^n$, and the observation that $\langle g(\beta), u(\beta) \rangle = \langle w(\beta)^{-1} g(\beta), w(\beta) u(\beta) \rangle$, each a is continuous. Let u_1 and u_2 be elements of $cl(U)$, and let r_1 and r_2 be real numbers such that $r_1 = r_2$ if $u_1 = u_2$. If $u_1 = u_2$, choose $c = 0$. If u_1 and u_2 are distinct, pick $g \in \mathcal{G}(w)$ and c such that

$$c \sum_{\beta \in \mathcal{Z}^m} \langle g(\beta), u_1(\beta) - u_2(\beta) \rangle = r_1 - r_2.$$

Observe that with $\rho = r_1 - c \sum_{\beta \in \mathcal{Z}^m} \langle g(\beta), u_1(\beta) \rangle$, we have $a(u_1) = r_1$ and $a(u_2) = r_2$, where a is given by (5.12). We will use the following lemma which is a direct consequence of [17, p. 35, Theorem 1].

Lemma : Let $\mathcal{X}$ be the family of continuous real-valued functions on $cl(U)$, let $\mathcal{X}_0$ be an arbitrary subfamily of $\mathcal{X}$, and let $\mathcal{L}_0$ be the family of all functions generated from $\mathcal{X}_0$ by the lattice operations on real-valued functions and uniform passage to the limit. Suppose that $x \in \mathcal{X}$, and that for any u_1 and u_2 in $cl(U)$ and any real numbers r_1 and r_2 such that $r_1 = r_2$ if $u_1 = u_2$ there exists an $a \in \mathcal{X}_0$ such that $a(u_1) = r_1$ and $a(u_2) = r_2$. Then $x \in \mathcal{L}_0$.

Let $\epsilon > 0$ be given. By the lemma, with $\mathcal{X}_0 = A$ and $x = F_e$, there are n_0, elements $a_1, \ldots, a_{n_0}$ of A, and a lattice map[15] $M : \mathbb{R}^{n_0} \to \mathbb{R}$ such that

$$|F_e(u) - M[a_1(u), \ldots, a_{n_0}(u)]| < \epsilon/2.$$

for $u \in cl(U)$. For $j = 1, \ldots, n_0$ let ρ_j, c_j, and g_j be the ρ, c, and g associated with a_j, respectively. For each j define y_j on $cl(U)$ by

$$y_j(u) = \sum_{\beta \in \mathcal{Z}^m} \langle g_j(\beta), u(\beta) \rangle ,$$

and notice that each $K_j := y_j(cl(U))$ is a compact set in $\mathbb{R}$. Let $N \in D_{n_0}$ satisfy

$$|M(c_1 v_1 + \rho_1, \ldots, c_{n_0} v_{n_0} + \rho_{n_0}) - N(v_1, \ldots, v_{n_0})| < \epsilon/2$$

for $v := (v_1, \ldots, v_{n_0}) \in (K_1 \times \cdots \times K_{n_0})$. Thus, for any $u \in U$

$$\begin{aligned} & |F(u) - N[y_1(u), \ldots, y_{n_0}(u)]| \\ \leq \; & |F(u) - M[a_1(u), \ldots, a_{n_0}(u)]| \\ & + |M[a_1(u), \ldots, a_{n_0}(u)] - N[y_1(u), \ldots, y_{n_0}(u)]| \\ < \; & \epsilon/2 + \epsilon/2, \end{aligned}$$

as required.

Another way to prove Lemma 1 is to make use of the Stone-Weierstrass theorem instead of the lemma described above. The details are almost the same. The relationship in [17] between [17, p. 35, Theorem 1] and the Stone-Weierstrass theorem is that [17, p. 35, Theorem 1] is used in the proof of a version of the Stone-Weierstrass theorem. Of course the proof given above shows that for our purposes it is not necessary to use the Stone-Weierstrass theorem, and it focuses attention on the important concept of a lattice map.

APPENDIX E: Proof of Lemma 2

Let $cl(QU)$ denote the closure of QU in X_w. Since $QU \subseteq U$, QU is relatively compact in X_w and (by the comments at the beginning of Appendix D) $cl(QU) \subseteq cl(U) \subset \mathcal{X}(\mathcal{Z}^m, \mathbb{R}^n)$. In particular, the map F_e described in Appendix D is continuous on $cl(QU)$ with respect to $\|\cdot\|_w$. Here let A be the set of all real-valued functions a on $cl(QU)$ defined by

$$a(u) = \rho + c \sum_{\beta \in \mathcal{Z}^m_-} \langle g(\beta), u(\beta) \rangle$$

[15] The *lattice operations* $a \vee b$ and $a \wedge b$ on pairs of real numbers a and b are defined by $a \vee b = \max(a, b)$ and $a \wedge b = \min(a, b)$. We say that a map $M : \mathbb{R}^{n_0} \to \mathbb{R}$ is a *lattice map* if Mv is generated from the components $v_1, \ldots, v_{n_0}$ of v by a finite number of lattice operations that do not depend on v.

with $g \in \mathcal{G}_-(w)$ and ρ and c real numbers. Here too each a is continuous. Let u_1 and u_2 be distinct elements of $cl(QU)$, and let r_1 and r_2 be arbitrary real numbers. Then $(u_1 - u_2) \in \mathcal{X}(\mathcal{Z}^m, \mathbb{R}^n)$ is nonzero and, by the definition of Q, the restriction $(u_1 - u_2)|_{\mathcal{Z}_-^m}$ of $(u_1 - u_2)$ to $\mathcal{Z}_-^m$ is nonzero. Hence there is an $a \in A$ such that $a(u_1) = r_1$ and $a(u_2) = r_2$, and obviously the same conclusion holds even if u_1 and u_2 are not distinct provided that $r_1 = r_2$. This establishes the applicability of the lemma in Appendix D, with $\mathcal{X}_0 = A$ and $x = F_e$ restricted to $cl(QU)$. The remainder of the proof is essentially the same as that of Lemma 1. Here the y_j are defined on $cl(QU)$, and

$$y_j(u) = \sum_{\beta \in \mathcal{Z}_-^m} \langle g_j(\beta), u(\beta) \rangle .$$

APPENDIX F: Myopic Maps and Causality

Let R be any map from S to $X(\mathcal{Z}^m, \mathbb{R}^n)$ such that $(Rs)(\alpha) = s[\gamma(\alpha)]$, where $\gamma : \mathcal{Z}^m \to \mathcal{Z}_-^m$ satisfies $\gamma(\alpha) = \alpha$ for $\alpha \in \mathcal{Z}_-^m$. For example, we can take R to be the Q of Section 5.2 or the Q defined by (5.7).

Proposition 1: Let S be a uniformly bounded subset of $X(\mathcal{Z}^m, \mathbb{R}^n)$. If $RS \subseteq S$ and G is causal and myopic on S with respect to w, then G meets condition (5.8) with $\mu = w|_{\mathcal{Z}_-^m}$.

Proof: Let $\epsilon > 0$ be given, and let $\delta > 0$ be such that (5.1) is met. Let c be the uniform bound for S, and let $b_k = \{\alpha \in \mathcal{Z}^m : |\alpha| \leq k\}$ for each $k \in \mathbb{N}$. Using $w(\alpha) \to 0$ as $|\alpha| \to \infty$, choose k such that $2c\,|w(\alpha)| < \delta$ for $\alpha \notin b_k$, and hence so that

$$\sup_{\alpha \notin b_k} \|w(\alpha)[(Rx)(\alpha) - (Ry)(\alpha)]\| < \delta$$

for any x and y in S. Let x and y satisfy

$$\max_{\alpha \in b_k} |w(\alpha)| \cdot \sup_{\alpha \in \mathcal{Z}_-^m} \|w(\alpha)[x(\alpha) - y(\alpha)]\| < \delta \min_{\beta \in d_k} |w(\beta)| ,$$

where $d_k = \{\gamma(\alpha) : \alpha \in b_k\}$. Then

$$\begin{aligned}
& \sup_{\alpha \in b_k} \|w(\alpha)[(Rx)(\alpha) - (Ry)(\alpha)]\| \\
\leq\ & \max_{\alpha \in b_k} |w(\alpha)| \cdot \max_{\beta \in d_k} \left|w(\beta)^{-1}\right| \cdot \sup_{\beta \in d_k} \|w(\beta)[x(\beta) - y(\beta)]\| \\
\leq\ & \max_{\alpha \in b_k} |w(\alpha)| \cdot (\min_{\beta \in d_k} |w(\beta)|)^{-1} \cdot \sup_{\alpha \in \mathcal{Z}_-^m} \|w(\alpha)[x(\alpha) - y(\alpha)]\| \\
<\ & \delta,
\end{aligned}$$

Since

$$\sup_{\alpha \in \mathcal{Z}^m} \|w(\alpha)[(Rx)(\alpha) - (Ry)(\alpha)]\| < \delta$$

and $RS \subseteq S$ and G is myopic, we have

$$|(GRx)(0) - (GRy)(0)| < \epsilon.$$

Using $(Rs)(\alpha) = s(\alpha)$ for all $s \in S$ and $\alpha \in \mathcal{Z}^m_-$, together with the causality of G, we have

$$|(Gx)(0) - (Gy)(0)| < \epsilon,$$

This completes the proof.

Proposition 2: Suppose that condition (5.8) is met. Then G is causal.[16]

Proof: Let $\alpha \in \mathcal{Z}^m$ be arbitrary, and suppose that x and y are two elements of S with $x(\beta) = y(\beta)$ for all $\beta_j \leq \alpha_j$. Clearly, $(T_{-\alpha}x)(\beta) = (T_{-\alpha}y)(\beta)$ for $\beta \in \mathcal{Z}^m_-$. By condition (5.8), $|(GT_{-\alpha}x)(0) - (GT_{-\alpha}y)(0)| < \epsilon$ for all $\epsilon > 0$, which gives $(GT_{-\alpha}x)(0) = (GT_{-\alpha}y)(0)$. Thus, using the shift-invariance of G, we have $(Gx)(\alpha) = (Gy)(\alpha)$, which proves the proposition.

Finally, suppose that μ is as described in connection with condition (5.8), and let ν be any extension of μ to $\mathcal{Z}^m$ that never vanishes and satisfies $\lim_{|\alpha| \to \infty} \nu(\alpha) = 0$. (For example, we can take $\nu(\alpha) = \mu(\beta)$ for all α, where $\beta_j = -|\alpha_j|$ for each j.) Since

$$\sup_{\alpha \in \mathcal{Z}^m_-} \|\mu(\alpha)[x(\alpha) - y(\alpha)]\| = \sup_{\alpha \in \mathcal{Z}^m_-} \|\nu(\alpha)[x(\alpha) - y(\alpha)]\| \leq$$

$$\sup_{\alpha \in \mathcal{Z}^m} \|\nu(\alpha)[x(\alpha) - y(\alpha)]\|,$$

it follows directly that G is myopic with respect to ν if condition (5.8) is met.

APPENDIX G: A Representation Theorem for Linear Discrete-Space Systems

The material is this appendix, while related to the topic of myopic maps, is a self-contained section with its own notation and definitions.

The cornerstone of the theory of discrete-time single-input single-output linear systems is the idea that every such system has an input-output map H that can be represented by an expression of the form

$$(Hx)(n) = \sum_{p=-\infty}^{\infty} h(n,p)x(p) \tag{5.13}$$

[16] This is observed in [13] for $m = n = 1$.

in which x is the input and h is the system function associated with H in a certain familiar way. It is widely known that this, and a corresponding representation for time-invariant systems in which $h(n,p)$ is replaced with $h(n-p)$, are discussed in many books (see, for example, [21, p. 70], [22, p. 96]). Almost always it is emphasized that these representations hold *for all* linear input-output maps H. On the other hand, in [13, p. 1159] attention is directed to material in [23, p. 58] which shows that certain time-invariant H's in fact do not have convolution representations.[17] This writer does not claim that these H's are necessarily of importance in applications, but he does feel that their existence shows that the analytical ideas in the books are flawed.[18]

One of the main purposes of this appendix is show that, under some mild conditions concerning the set of inputs and H, (5.13) becomes correct if an additional term is added to the right side. More specifically, we show that

$$(Hx)(n) = \sum_{p=-\infty}^{\infty} h(n,p)x(p) + \lim_{k\to\infty} (HE_kx)(n)$$

for each n, in which h has the same meaning as in (5.13), and E_kx denotes the function given by $(E_kx)(p) = x(p)$ for $|p| > k$ and $(E_kx)(p) = 0$ otherwise. This holds whenever the input set is the set of bounded functions, the outputs are bounded, and H is continuous. In particular, we see that in this important setting, an H has a representation of the form given by (5.13) if and only if

$$\lim_{k\to\infty} (HE_kx)(n) = 0$$

for all x and n. Since this is typically a very reasonable condition for a system map H to satisfy, it is clear that the H's that cannot be represented using just (5.13) are rather special.

Our results concerning H are given in the following section, in which the setting is more general in that we address H's for which inputs and outputs depend on an arbitrary finite number of variables. We also consider H's for which inputs and outputs are defined on just the nonnegative integers. In that setting the situation with regard to the need for an additional term

[17]The claim in [13] that there exists a time-invariant causal H that has no convolution representation is correct, but it may not be clear that the argument given there actually shows this. Specifically, it may not be clear from what is said in [13] that the pertinent map constructed there is causal and time-invariant. However, it is not difficult to modify what is said so that it establishes the claim (see [23, p. 58] or the proof of the proposition in Section G.2).

[18]The oversight in the books is due to the lack of validity of the interchange of the order of performing a certain infinite sum and then applying $(H\cdot)(n)$. The infinite sum at issue clearly converges pointwise, but that is not enough to justify the interchange. A special case in which the interchange is justified is that in which the inputs are elements of ℓ_p $(1 \le p < \infty)$ and each $(H\cdot)(n)$. is a bounded linear functional on ℓ_p.

in the representation is different: no additional term is needed for causal maps H.

G.1. Linear-System Representation Result

G.1.1. Preliminaries

Let m be a positive integer, let $\mathcal{Z}$ be the set of all integers, and let $\mathcal{Z}_+$ denote the set of nonnegative integers. Let D stand for either $\mathcal{Z}^m$ or $\mathcal{Z}_+^m$. Let $\ell_\infty(D)$ denote the normed linear space of bounded $\mathbb{R}$-valued functions x defined on D, with the norm $\|\cdot\|$ given by $\|x\| = \sup_{\alpha \in D} |x(\alpha)|$.

For each positive integer k, let c_k stand for the discrete hypercube $\{\alpha \in D : |\alpha_j| \leq k \ \forall j\}$ (α_j is the jth component of α), and let $\ell_1(D)$ denote the set of $\mathbb{R}$-valued maps g on D such that

$$\sup_k \sum_{\beta \in c_k} |g(\beta)| < \infty.$$

For each $g \in \ell_1(D)$ the sum $\sum_{\beta \in c_k} g(\beta)$ converges to a finite limit as $k \to \infty$, and we denote this limit by $\sum_{\beta \in D} g(\beta)$.

Define maps Q_k and E_k from $\ell_\infty(D)$ into itself by $(Q_k x)(\alpha) = x(\alpha)$, $\alpha \in c_k$ and $(Q_k x)(\alpha) = 0$ otherwise, and $(E_k x)(\alpha) = x(\alpha)$, $\alpha \notin c_k$ and $(E_k x)(\alpha) = 0$ otherwise.

In the next section H stands for any linear map from $\ell_\infty(D)$ into itself that satisfies the condition that

$$\sup_k (HQ_k u)(\alpha) < \infty \tag{5.14}$$

for each $u \in \ell_\infty(D)$ and each $\alpha \in D$. This condition is met whenever H is continuous because then $(HQ_k u)(\alpha) \leq |(HQ_k u)(\alpha)| \leq \|HQ_k u\| \leq \|H\| \cdot \|Q_k u\| \leq \|H\| \cdot \|u\|$.

G.1.2. Our Theorem

In the following theorem, $h(\cdot, \beta)$ for each $\beta \in D$ is defined by $h(\cdot, \beta) = H\delta_\beta$, where $(\delta_\beta)(\alpha) = 1$ for $\alpha = \beta$ and $(\delta_\beta)(\alpha)$ is zero otherwise. Of course $h(\cdot, \beta)$ is the response of H to a unit "impulse" occurring at $\alpha = \beta$.

Theorem: For any H as described, and for each $\alpha \in D$ and each $x \in \ell_\infty(D)$,

(i) g defined on D by $g(\beta) = h(\alpha, \beta) x(\beta)$ belongs to $\ell_1(D)$.

(ii) $\lim_{k \to \infty} (HE_k x)(\alpha)$ exists and is finite.

(iii) We have

$$(Hx)(\alpha) = \sum_{\beta \in D} h(\alpha, \beta) x(\beta) + \lim_{k \to \infty} (HE_k x)(\alpha).$$

Proof

Let $\alpha \in D$ and $x \in \ell_\infty(D)$ be given. By the linearity of H and the definition of Q_k,

$$\sum_{\beta \in c_k} h(\alpha,\beta)u(\beta) = H\Big(\sum_{\beta \in c_k} u(\beta)\delta_\beta\Big)(\alpha) = H(Q_k u)(\alpha) \tag{5.15}$$

for each k and each $u \in \ell_\infty(D)$. In particular, for u given by $u(\beta) = \text{sgn}[h(\alpha,\beta)]\ \forall\beta$,

$$\sum_{\beta \in c_k} |h(\alpha,\beta)| = H(Q_k u)(\alpha)$$

for each k. Thus, by (5.14), $h(\alpha,\cdot)$ belongs to $\ell_1(D)$ and so does g of (i) of the theorem. By (i) the extreme left and right sides of (5.15) with $u = x$ converge as $k \to \infty$, and one has

$$\sum_{\beta \in D} h(\alpha,\beta)x(\beta) = \lim_{k\to\infty} H(Q_k x)(\alpha). \tag{5.16}$$

Since

$$H(E_k x)(\alpha) = (Hx)(\alpha) - H(Q_k x)(\alpha)$$

for each k, it is clear that (ii) holds. Finally,

$$\lim_{k\to\infty} H(E_k x)(\alpha) = (Hx)(\alpha) - \lim_{k\to\infty} H(Q_k x)(\alpha),$$

which, together with (5.16), completes the proof.

We note that for $D = \mathcal{Z}_+^m$ and H causal in the usual sense (see Section G.2), the term $\lim_{k\to\infty}(HE_k x)(\alpha)$ is always zero. In the next section an extension result for shift-invariant maps defined on $\ell_\infty(\mathcal{Z}^m)$ is given from which it follows that there are maps H for which the additional term is not always zero.

G.2: An Extension Proposition

We begin with some additional preliminaries: Let M denote a linear manifold in $\ell_\infty(\mathcal{Z}^m)$ that is closed under translation in the sense that $T_\beta M = M$ for each $\beta \in \mathcal{Z}^m$, where T_β is the usual shift map defined on M for each $\beta \in \mathcal{Z}^m$ by $(T_\beta x)(\alpha) = x(\alpha - \beta)$, $\alpha \in \mathcal{Z}^m$. Assume also that $y \in M$ implies that $z \in M$, where $z(\alpha) = y(\alpha)$ for $\alpha_j \leq 0\ \forall j$, and $z(\alpha) = 0$ otherwise. We do not rule out the possibility that $M = \ell_\infty(\mathcal{Z}^m)$.

Let A be a linear map of M into $\ell_\infty(\mathcal{Z}^m)$. Such an A is *shift-invariant* if

$$(Ax)(\alpha - \beta) = (AT_\beta x)(\alpha), \quad \alpha \in \mathcal{Z}^m$$

for each $\beta \in \mathcal{Z}^m$ and $x \in M$. The map A is *causal* if

$$x(\alpha) = y(\alpha) \text{ whenever } \alpha_j \leq \beta_j\ \forall j \quad \Longrightarrow \quad (Ax)(\beta) = (Ay)(\beta)$$

for each $\beta \in \mathcal{Z}^m$ and every x and y in M. It is *bounded* if $||A||_M := \sup\{||Ax|| : x \in M, ||x|| \leq 1\} < \infty$, in which $||\cdot||$ is the norm in $\ell_\infty(\mathcal{Z}^m)$. Our result is the following.

Proposition: Let A be shift invariant and bounded. Then there exists a bounded linear shift-invariant map B from $\ell_\infty(\mathcal{Z}^m)$ into itself that extends A in the sense that B is causal if A is causal and $Bx = Ax,\ x \in M$.

Proof

By the shift invariance of A, we have $(Ax)(\alpha) = (AT_{-\alpha}x)(0)$ for all α and all $x \in M$. The map $(A\cdot)(0)$ is a bounded linear functional on M, because

$$|(Ay)(0)| = |(AT_{-\alpha}T_\alpha y)(0)| = |(AT_\alpha y)(\alpha)| \leq \sup_\beta |(AT_\alpha y)(\beta)| \leq ||A||_M \cdot ||T_\alpha y|| = ||A||_M \cdot ||y||$$

for $y \in M$. When A is causal, $(A\cdot)(0)$ has the property that $(Ay)(0) = 0$ for any $y \in M$ for which $y(\alpha) = 0$ for $\alpha_j \leq 0\ \forall j$. By the Hahn-Banach theorem [24, p. 178] there is a bounded linear functional $\mathcal{F}$ that extends $(A\cdot)(0)$ to all of $\ell_\infty(\mathcal{Z}^m)$. Set $\mathcal{G} = \mathcal{F}$ if A is not causal, and if A is causal define $\mathcal{G}$ on $\ell_\infty(\mathcal{Z}^m)$ by $\mathcal{G}y = \mathcal{F}Py$ where P is the linear operator given by $(Py)(\alpha) = y(\alpha),\ \alpha_j \leq 0\ \forall j$ and $(Py)(\alpha) = 0$ for $\alpha_j > 0$ for some j. Define B on $\ell_\infty(\mathcal{Z}^m)$ by $(Bx)(\alpha) = \mathcal{G}T_{-\alpha}x$. It is easy to check that B is a linear shift-invariant bounded map into $\ell_\infty(\mathcal{Z}^m)$, that B extends A to $\ell_\infty(\mathcal{Z}^m)$, and that B is causal if A is causal.[19] This completes the proof.

Since the set L of elements x of $\ell_\infty(\mathcal{Z}^m)$ such that $x(\alpha)$ approaches a limit as $\max_j\{\alpha_j\} \to -\infty$ is a linear manifold that is closed under translation, and since

$$(Ax)(\alpha) = \lim_{\max_j\{\beta_j\} \to -\infty} x(\beta)$$

defines a shift-invariant bounded causal linear map of L into $\ell_\infty(\mathcal{Z}^m)$, it follows from our proposition that there exist maps H, even causal time-invariant maps H, of the kind addressed by our theorem for which the term $\lim_{k\to\infty}(HE_kx)(\alpha)$ is not always zero. More explicitly, the associate B via our proposition of the A just described satisfies $\lim_{k\to\infty}(BE_kx)(\alpha) = \lim_{\max_j\{\beta_j\}\to-\infty} x(\beta)$ for $x \in L$. And an example of an H of the type addressed by the theorem for which the additional term is not always zero and H is not shift invariant is obtained by adding to this B any linear bounded map of $\ell_\infty(\mathcal{Z}^m)$ into itself that is not shift invariant and has a representation without an additional term.

---

[19]It is also true that B can be chosen so that it preserves the norm of A, in the sense that $||B|| = ||A||_M$.

A proposition similar to the one above can be given to show that there are bounded linear continuous-time time-invariant input-output maps that do not possess certain convolution representations.[20]

Referring now to Section 5.2, the map B described above provides an example of a linear G that is myopic for no w. To see this, let S be the unit open ball in $\ell_\infty(\mathcal{Z}^m)$ centered at the origin and let B be as described. With w arbitrary, suppose that B is such that given some $\epsilon \in (0, 1/2)$ there is a $\delta > 0$ with the property that $x \in S$ and $\sup_{\alpha \in \mathcal{Z}^m} \|w(\alpha)x(\alpha)\| < \delta$ imply $|(Bx)(0)| < \epsilon$. Let y be any element of $S \cap L$ such that $y(\beta)$ approaches 3/4 as $\max_j\{\beta_j\} \to -\infty$. Let z denote $E_k y$, with k chosen so that $\sup_{\alpha \in \mathcal{Z}^m} \|w(\alpha)z(\alpha)\| < \delta$. Here $z \in S$ and $\sup_{\alpha \in \mathcal{Z}^m} \|w(\alpha)z(\alpha)\| < \delta$ is satisfied, but $|(Bx)(0)| = 3/4 > \epsilon$.[21]

APPENDIX H: Notes on Uniform Approximation of Shift-Varying Systems

It is shown in [27] there that the elements K of a large class of continuous-time causal input-output maps can be uniformly approximated arbitrarily well using a certain structure if and only if K is uniformly continuous. For the case addressed the system inputs and outputs are defined on a finite interval $[0, t_f]$. In this appendix we give corresponding results for the case in which K is not necessarily causal and inputs and outputs are defined on a discrete set $\{0, 1, \ldots, a_1\} \times \ldots \times \{0, 1, \ldots, a_m\}$, in which $a_1, \ldots, a_m$ are positive integers. As in Section 5.2 our approximating structure involves certain functions that can be chosen in different ways. When these functions are taken to be certain polynomial functions, the input-output map of the structure is a generalized shift-varying Volterra series.

---

[20] As suggested above, results along the lines of the theorem in Section G.1.2 hold also in continuous-time settings. More specifically [25], for continuous single-input single-output causal input-output maps H that take bounded Lebesgue measurable inputs defined on $\mathbb{R}$ into bounded outputs on $\mathbb{R}$ such that a certain often-satisfied condition is met, $(Hx)(t) = \int_{-\infty}^{t} h(t, \tau)x(\tau)\, d\tau + \lim_{a \to -\infty}(HP_a x)(t)$ for all t, in which h has the usual continuous-time impulse-response interpretation and $(P_a x)(\tau) = x(\tau)$ for $\tau \leq a$ and $(P_a x)(\tau) = 0$ otherwise. An example is given in [25] of an H for which $\lim_{a \to -\infty}(HP_a x)(t)$ is not always zero. Corresponding results for noncausal input-output maps can be proved too. In this case the extra term involves also the behavior of the input for large values of its argument, as it does in the discrete-time case addressed by our theorem. And corresponding results in the setting of inputs and outputs of a finite number of variables can be proved starting with the approach described in [26].

[21] A similar example can be given of a a linear G in the continuous-time setting of [14] that is myopic for no w.

5.3.1 H.1. Preliminaries and the Approximation Result

With $a_1, \ldots, a_m$ any positive integers, and with $A = \{0, 1, \ldots, a_1\} \times \ldots \times \{0, 1, \ldots, a_m\}$, let $L(A)$ denote the linear space of maps x from A to $\mathbb{R}^n$. For each $\alpha \in A$, let P_α denote the map from $L(A)$ into itself defined by $(P_\alpha x)(\beta) = x(\beta)$, $\beta_j \leq \alpha_j \ \forall j$ and $(P_\alpha x)(\beta) = 0$ otherwise.

We view $L(A)$ as as a normed space with some norm $\|\cdot\|_{L(A)}$ that is regular in the sense that $\|P_\alpha x\|_{L(A)} \leq \|x\|_{L(A)}$ for all α and x.[22] The norm $\|\cdot\|_{L(A)}$ can be chosen in many ways. One choice is given by

$$\|x\|_{L(A)} = \max_{\alpha \in A} \|x(\alpha)\| .$$

Let $M(A)$ stand for the space of real-valued maps y defined on A, with some norm $\|y\|_{M(A)}$. It will become clear that the choice of this norm is unimportant, because all norms on finite dimensional linear spaces are equivalent.

With c any positive number, let E denote $\{x \in L(A) \colon \|x\|_{L(A)} \leq c\}$. Let K map E to $M(A)$. K is the input-output map that we wish to approximate. Of course the set E is the set of inputs corresponding to the map K. By K is *causal* we mean that

$$x(\beta) = y(\beta), \ \beta_j \leq \alpha_j \ \forall j \quad \Longrightarrow \quad (Kx)(\alpha) = (Ky)(\alpha)$$

for each $\alpha \in A$ and every x and y in E.

In the theorem below, which is the main result in this appendix, ℓ denotes the dimension of $M(A)$ (of course $\ell = (1 + a_1) \times \cdots \times (1 + a_m)$, but this fact is not used in what follows), $v_1, \ldots, v_\ell$ is any set of basis elements for $M(A)$, and $\mathcal{H}$ stands for the set of maps H from E to $\mathbb{R}$ such that for each H there are an $n_0 \in \mathbb{N}$, maps $g_1, \ldots, g_{n_0}$ from A to $\mathbb{R}^n$, and an $N \in D_{n_0}$ such that $Hx = N[y_1(x), \ldots, y_{n_0}(x)]$ where each y_j is defined on E by

$$y_j(x) = \sum_{\alpha \in A} \langle g_j(\alpha), x(\alpha) \rangle .$$

Theorem: The following two statements are equivalent.

(i) K is continuous (i.e., continuous with respect to the norms $\|\cdot\|_{L(A)}$ and $\|\cdot\|_{M(A)}$).

(ii) For each $\epsilon > 0$, there are elements $H_1, \ldots, H_\ell$ of $\mathcal{H}$, such that

$$\left| (Kx)(\alpha) - \sum_{j=1}^{\ell} v_j(\alpha)(H_j x) \right| < \epsilon, \quad \alpha \in A \tag{5.17}$$

---

[22] As is well known, norms are typically regular. However, it is not difficult to give an example of a norm that is not regular.

for all $x \in E$.

In addition, $H_j x$ in (ii) can be replaced with $H_j P_\alpha x$ if K is causal.

Proof

We use two lemmas. More specifically, using the hypothesis that E is compact in $L(A)$, the following can be proved using direct modifications of material in Appendix D.

Lemma 3: Let $R : E \to I\!R$ be continuous with respect to $\|\cdot\|_{L(A)}$. Then for each $\epsilon > 0$, there is an $H \in \mathcal{H}$ such that $|Rx - Hx| < \epsilon$, $x \in E$.

In order to describe our second lemma, let $\mathcal{Q}$ be any set of $I\!R$-valued continuous maps defined on E that is dense in the set $U(E)$ of continuous real functionals on E, in the sense that for each $R \in U(E)$ and each $\epsilon > 0$ there is a $Q \in \mathcal{Q}$ such that $\max_{x \in E} |R(x) - Q(x)| < \epsilon$. Let $\mathcal{P}$ denote the set of all maps from E to $M(A)$ of the form $\sum_{j=1}^{\ell} Q_j(\cdot) v_j$ where the Q_j belong to $\mathcal{Q}$. Our second lemma[23], which is proved in Section H.2 is the following.

Lemma 4: Let F be a map from E to $M(A)$. Then F is continuous if and only if for each $\epsilon > 0$ there exists a $P \in \mathcal{P}$ such that $\|Fx - Px\|_{M(A)} < \epsilon$ for $x \in E$.

Continuing with the proof of the theorem, suppose that (i) is satisfied. Using Lemmas 3 and 4, we see that for each $\epsilon > 0$ there are elements $H_1, \ldots, H_\ell$ of $\mathcal{H}$ such that

$$\left| (Ky)(\alpha) - \sum_{j=1}^{\ell} v_j(\alpha)(H_j y) \right| < \epsilon, \quad \alpha \in A$$

for all $y \in E$. This is (ii). And if K is causal, (ii) holds with $H_j x$ replaced with $H_j P_\alpha x$ because $P_\alpha x \in E$ whenever $x \in E$ and $\alpha \in A$, and K causal implies that $(Kx)(\alpha) = (KP_\alpha x)(\alpha)$ for all x and α.

Now assume that either (ii) holds or that (ii) holds with $H_j x$ replaced with $H_j P_\alpha x$. Since each v_j belongs to $M(A)$, and since the maps J_j defined on E by either $J_j x = H_j x$ or $(J_j x)(\alpha) = H_j P_\alpha x$ are continuous from E to $I\!R$ or from E to $M(A)$, respectively,[24], it follows that K is the uniform limit of a sequence of continuous maps. And from this it follows that K is continuous. This completes the proof of the theorem.

---

[23]This lemma is a simple version of a result that plays a central role in [28] (see also [27]). The proof given in the appendix is much simpler than the corresponding proofs in the references cited.

[24]The continuity of the J_j is a consequence of the boundedness of the set E, the Schwarz inequality in $I\!R^n$, and the fact that the elements of D_{n_0} are uniformly continuous on compact sets.

H.1. Special case: finite Volterra-like series approximations

Suppose that K is causal. If the D_{n_0} are chosen to be certain sets of polynomial functions, the approximating maps

$$\sum_{j=1}^{\ell} v_j(\alpha)(H_j P_\alpha \cdot)$$

of our theorem are generalizations of finite discrete shift-varying Volterra series. More specifically, suppose that the maps f of each D_{n_0} are such that $f(w)$ is a real polynomial in the components of w, and consider the $n=1$ case. By writing products of sums as iterated sums, it is not difficult to see that $\sum_{j=1}^{\ell} v_j(\alpha)(H_j P_\alpha x)$ has the form

$$k_0(\alpha) + \sum_{j=1}^{n_1} \sum_{\beta_j \in A_\alpha} \cdots \sum_{\beta_1 \in A_\alpha} k_j(\alpha, \beta_1, \ldots, \beta_j) x(\beta_1) \cdots x(\beta_j) \tag{5.18}$$

in which n_1 is a positive integer, $A_\alpha = \{\beta \in A : \beta_j \leq \alpha_j \ \forall j\}$, and each k_j $(j \geq 0)$ is a real-valued map defined on $A^{(j+1)}$.[25] On the other hand, any map $V : E \to M(A)$ defined by the condition that $(Vx)(\alpha)$ is given by (5.18) is continuous.[26] We have therefore proved the following concerning causal K's (see the last part of the proof of our theorem). For $n=1$ and with $\mathcal{V}$ the set of all maps $V : E \to M(A)$ such that $(Vx)(\alpha)$ is given by (5.18) for some $n_1 \in \mathbb{N}$ and some $k_0, \ldots, k_{n_1}$, (i) of the theorem holds (i.e., K is continuous) if and only if: For any $\epsilon > 0$ there is a $V \in \mathcal{V}$ such that $|(Kx)(\alpha) - (Vx)(\alpha)| < \epsilon$ for $x \in E$ and $\alpha \in A$.

In fact, it is not difficult to check that this proposition holds for all $n \geq 1$ with the understanding that each k_j in (5.18) is row n^j-vector valued, and that $x(\beta_1) \cdots x(\beta_j)$ in (5.18) denotes the corresponding Kronecker product (which is column n^j-vector valued, assuming of course that we view the elements of $\mathbb{R}^n$ as column vectors).

A similar proposition holds for K's that are not necessarily causal. In this case (5.18) is replaced with

$$k_0(\alpha) + \sum_{j=1}^{n_1} \sum_{\beta_j \in A} \cdots \sum_{\beta_1 \in A} k_j(\alpha, \beta_1, \ldots, \beta_j) x(\beta_1) \cdots x(\beta_j), \tag{5.19}$$

and K is continuous if and only if $(Kx)(\alpha)$ can be approximated arbitrarily well in the sense described above by a sum of the form (5.19).

[25] Of course for $m=1$, (5.18) takes the more familiar form $k_0(\alpha) + \sum_{j=1}^{n_1} \sum_{\beta_j=0}^{\alpha} \cdots \sum_{\beta_1=0}^{\alpha} k_j(\alpha, \beta_1, \ldots, \beta_j) x(\beta_1) \cdots x(\beta_j)$.

[26] The continuity of V is a consequence of the boundedness of the set E and the identity $c_1 c_2 \cdots c_j - b_1 b_2 \cdots b_j = c_1 c_2 \cdots c_{(j-1)}(c_j - b_j) + c_1 \cdots c_{(j-2)}(c_{(j-1)} - b_{(j-1)}) b_j + \ldots + (c_1 - b_1) b_2 \cdots b_j$ for real numbers $c_1, \ldots c_j, b_1, \ldots b_j$.

H.2: Proof of Lemma 4

Let continuous F from E to $M(A)$ be given, and let $x \in E$ and any $\epsilon > 0$ also be given. Since the v_j form a basis for $M(A)$, we have $Fx = \sum_{j=1}^{\ell} f_j(Fx)v_j$ for $x \in E$, where the f_j are continuous linear functionals. Using the fact that the $f_j(F\cdot)$ are continuous, choose $Q_1, \ldots, Q_\ell \in \mathcal{Q}$ so that

$$\max_j \max_{x \in E} (|f_j(Fx) - Q_j x| \cdot \|v_j\|_{M(A)}) < \epsilon/\ell.$$

Then set $P = \sum_{j=1}^{\ell} Q_j(\cdot)v_j$. This gives

$$\|Fx - Px\|_{M(A)} \leq \sum_{j=1}^{\ell} |f_j(Fx) - Q_j x| \cdot \|v_j\|_{M(A)} < \epsilon$$

for $x \in E$.

On the other hand, if F can be approximated as indicated in the lemma, then F is the uniform limit of a sequence of continuous maps, and F is therefore continuous. This proves the lemma.

5.4 References

[1] V. Volterra, *Theory of Functionals and of Integral and Integro-Differential Equations*, New York: Dover, 1959.

[2] M. Schetzen, *The Volterra and Wiener Theories of Nonlinear Systems*, John Wiley, New York, 1980.

[3] I. W. Sandberg, "A perspective on system theory", *IEEE Transactions on Circuits and Systems*, vol. 31, no. 1, pp. 88-103, January 1984.

[4] G. Cybenko, "Approximation by superposition of a single function," *Mathematics of Control, Signals and Systems*, vol. 2, pp. 303-314, 1989.

[5] H. N. Mhaskar and C. A. Micchelli, " Approximation by superposition of sigmoidal and radial basis functions," *Advances in Applied mathematics*, vol. 3, pp. 350–373, 1992.

[6] T. Chen and H. Chen, "Universal approximation to nonlinear operators by neural networks with arbitrary activation functions and its applications to dynamic systems," *IEEE Transactions on Neural Networks*, vol. 6, no. 4, pp. 911-917, July 1995.

[7] J. Park and I. W. Sandberg, "Approximation and radial-basis function networks," *Neural Computation*, vol. 5, no. 2, pp. 305–316, March 1993.

[8] I. W. Sandberg, "Structure theorems for nonlinear systems," *Multidimensional Systems and Signal Processing*, vol. 2, no. 3, pp. 267–286, 1991. (See also the Errata in vol. 3, no. 1, p. 101, 1992.) A conference version of the paper appears in *Integral Methods in Science and Engineering-90* (Proceedings of the International Conference on Integral Methods in Science and Engineering, Arlington, Texas, May 15-18, 1990, ed. A. H. Haji-Sheikh), New York: Hemisphere Publishing, pp. 92–110, 1991.

[9] ____________, "Uniform approximation and the circle criterion," *IEEE Transactions on Automatic Control*, vol. 38, no. 10, pp. 1450–1458, October 1993.

[10] ____________, "Approximately-finite memory and input-output maps," *IEEE Transactions on Circuits and Systems-I*, vol. 39, no. 7, pp. 549–556, July 1992.

[11] I. W. Sandberg and L. Xu, "Network approximation of input-output maps and functionals," to appear in the Journal of Circuits, Systems, and Signal Processing, 1996.

[12] T. Chen and H. Chen, "Approximations of continuous functionals by neural networks with application to dynamical systems," *IEEE Transactions on Neural Networks*, vol. 4, no. 6, pp. 910–918, November 1993.

[13] S. Boyd and L. O. Chua, "Fading memory and the problem of approximating nonlinear operators with volterra series," *IEEE Transactions on Circuits and Systems*, vol. CAS-32, no. 11, pp. 1150–1161, November 1985.

[14] I. W. Sandberg and L. Xu, "Uniform approximation of multidimensional myopic maps," submitted. (A conference version is to appear in the Proceedings of the Volterra Centennial Symposium, Department of Mathematics, University of Texas at Arlington, May 1996.)

[15] B. De Vries, J. C. Principe, and P. Guedes de Oliveira, "Adaline with adaptive recursive memory," *Proceedings of IEEE-SP Workshop on Neural Networks for Signal Processing*, pp. 101–110, 1991.

[16] I. W. Sandberg and L. Xu, "Gamma networks and the uniform approximation of input-output maps," Proceedings of Circuits, Systems, and Computers '96, Hellenic Naval Academy, Piraeus, Greece, July 1996 (to appear).

[17] M. H. Stone, "A generalized Weierstrass approximation theorem," In *Studies in Modern Analysis*, ed. R. C. Buck, vol. 1 of *MAA Studies in Mathematics*, pp. 30–87, Prentice-Hall, Englewood Cliffs, NJ, March 1962.

[18] W. Rudin, *Principles of Mathematical Analysis*, McGraw-Hill, New York, 1976.

[19] L. A. Liusternik and V. J. Sobolev, *Elements of Functional Analysis*, Frederick Ungar Publishing Co., New York, 1961.

[20] J. Dieudonné, *Foundations of Modern Analysis*, Academic Press, New York, 1969.

[21] J. G. Proakis and D. G. Manolakis, *Digital Signal Processing*, second edition, New York: Macmillan, 1992.

[22] L. B. Jackson, *Signals, Systems, and Transforms*, New York: Addison-Wesley, 1991.

[23] L. V. Kantorovich and G. P. Akilov, *Functional Analysis*, Oxford: Pergamon, 1982.

[24] G. Bachman and L. Narici, *Functional Analysis*, New York: Academic Press, 1966.

[25] I. W. Sandberg, "A Representation theorem for linear systems," submitted.

[26] D. Ball and I. W. Sandberg, " g- and h- Representations for Nonlinear Maps," J. Mathematical Analysis and Applications, Vol. 149, No. 2, July 1990.

[27] I. W. Sandberg , "Notes on uniform approximation of time-varying systems on finite time intervals," submitted. A conference version appears in Proceedings of the Fourth International Workshop on Nonlinear Dynamics of Electronic Systems, Seville, Spain, pp. 149–154, June 1996.

[28] A. Dingankar and I. W. Sandberg, "Network approximation of dynamical systems," *Proceedings of the 1995 International Symposium on Nonlinear Theory and its Applications*, vol. 2, pp. 357–362, Las Vegas, December 1995.

6

Monotonicity: Theory and Implementation

Joseph Sill
Yaser Abu-Mostafa

ABSTRACT We present a systematic method for incorporating prior knowledge (hints) into the learning-from-examples paradigm. The hints are represented in a canonical form that is compatible with descent techniques for learning. We focus in particular on the monotonicity hint, which states that the function to be learned is monotonic in some or all of the input variables. The application of monotonicity hints is demonstrated on two real-world problems- a credit card application task, and a problem in medical diagnosis. We report experimental results which show that using monotonicity hints leads to a statistically significant improvement in performance on both problems. Monotonicity is also analyzed from a theoretical perspective. We consider the class $\mathbf{M}$ of monotonically increasing binary output functions. Necessary and sufficient conditions for monotonic separability of a dichotomy are proven. The capacity of $\mathbf{M}$ is shown to depend heavily on the input distribution.

6.1 Introduction

It is evident that learning from examples needs all the help it can get. When an unknown function f is represented to us merely by a set of examples, we are faced with a dilemma. We would like to use a model that is sophisticated enough to have a chance of simulating the unknown function, yet we want the model to be simple enough that a limited set of examples will suffice to 'tune' it properly. These two goals are often on a collision course.

One established method of tackling this problem is regularization [1]. It is an attempt to start out with a sophisticated model and then restrict it during the learning process to fit the limited number of examples we have. Thus we have a simple model in disguise, and we make it as simple as warranted by the resources. The hope is that the restriction (simplification) of the model has not rendered f impossible to simulate, and the justification is that this is the best we can do anyway given a limited set of examples.

Another method for tackling the problem is the use of hints [2, 3] as a learning aid. Hints describe the situation where, in addition to the set of examples of f, we have prior knowledge of certain facts about f. We wish to use this side information to our advantage. However, hints come in different shapes, and the main difficulty of using them is the lack of a systematic way of incorporating heterogeneous pieces of information into a manageable learning process. This paper concerns itself with the development of a systematic method that integrates different types of hints in the same learning process. In particular, the monotonicity hint is presented from both practical and theoretical perspectives.

The distinction between regularization and the use of hints is worth noting. Regularization restricts the model in a way that is not based on known facts about f. Therefore, f may be implementable by the original model, but not by the restricted model. In the case of hints, if the unrestricted model was able to implement f, so would the model restricted by the hints, since f cannot be excluded by a litmus test that it is known to satisfy. We can apply any number of hints to restrict the model without risking the exclusion of f. The use of hints does not preclude, nor does it require, the use of any form of regularization.

How to take advantage of a given hint can be an art just like how to choose a model. In the case of invariance hints for instance, preprocessing of the input can achieve the invariance through normalization, or the model itself can be explicitly structured to satisfy the invariance [4]. Whenever such a method of direct implementation is feasible, the full benefit of the hint is automatically realized. This paper does not attempt to offer a superior alternative to direct implementation. *However, when direct implementation is not an option, we prescribe a systematic method for incorporating practically any hint in any descent method for learning.* The goal is to automate the use of hints in learning to a degree where we can effectively use a large number of simple hints that may be available in a practical situation.

We start by introducing the basic nomenclature and notation. The *environment* X is the set on which the unknown function f is defined. The points in the environment are distributed according to some probability distribution P. f takes on values from some set Y

$$f : X \rightarrow Y$$

Often, Y is just $\{0, 1\}$ or the interval $[0, 1]$. The *learning process* takes pieces of information about (the otherwise unknown) f as input and produces a *hypothesis* g

$$g : X \rightarrow Y$$

that attempts to approximate f. The degree to which a hypothesis g is considered an approximation of f is measured by a distance or 'error'

$$E(g, f)$$

The error E is based on the disagreement between g and f as seen through the eyes of the probability distribution P.

Two popular forms of the error measure are

$$E = \mathcal{E}[(g(\mathbf{x}) - f(\mathbf{x}))^2]$$

and (for binary classification)

$$E = \mathcal{E}[-f(\mathbf{x})log(g(\mathbf{x})) - (1 - f(\mathbf{x}))log(1 - g(\mathbf{x}))]$$

where $\mathcal{E}[.]$ denotes the expected value of a random variable. The first measure is the well-known mean-squared error measure, while the second is a cross-entropy measure. The underlying probability distribution is P. E will always be a non-negative quantity, and we will take $E(g, f) = 0$ to mean that g and f are identical for all intents and purposes. We will also assume that when the set of hypotheses is parameterized by real-valued parameters (e.g., the weights in the case of a neural network), E will be well-behaved as a function of the parameters (in order to allow for derivative-based descent methods). We make the same assumptions about the error measures that will be introduced in section 2 for the hints.

In this paper, the 'pieces of information' about f that are input to the learning process will be more general than those in the *learning from examples* paradigm. In that paradigm, a number of points $\mathbf{x_1}, \cdots, \mathbf{x_N}$ are picked from X (usually independently according to the probability distribution P) and the values of f on these points are provided. Thus, the input to the learning process is the set of examples

$$(\mathbf{x_1}, f(\mathbf{x_1})), \cdots, (\mathbf{x_N}, \mathbf{f}(\mathbf{x_N}))$$

and these examples are used to guide the search for a good hypothesis. In this paper, we will consider the set of examples of f as only one of the available hints and denote it by H_0. The other hints $H_1, \cdots, H_M$ will be additional known facts about f, such as monotonicity properties for instance.

The paper is organized as follows. Section 2 develops a canonical method for representing different hints. This is the first step in dealing with any hint that we encounter in a practical situation. Section 3 lays the foundations for learning from hints in general. Section 4 presents specific implementations and experimental results for the case of the monotonicity hint. In section 5, we analyze how powerful a constraint monotonicity is from a theoretical viewpoint.

6.2 Representation of hints

We have so far described what a hint is in very general terms such as 'a known property of f' or 'a fact about f'. Indeed, all that is needed to qualify as a hint for our purposes is to have a litmus test that f passes and that can be applied to the set of hypotheses. In other words, a hint H_m is formally a subset of the hypotheses, namely those satisfying the hint.

This definition of H_m can be extended to a definition of 'approximation of H_m' in several ways. For instance, g can be considered to approximate H_m within ϵ if there is a function h that strictly satisfies H_m for which $E(g, h) \leq \epsilon$. In the context of learning, it is essential to have a notion of approximation since exact learning is seldom achievable. Our definitions for approximating different hints will be part of the scheme for representing those hints.

The first step in representing H_m is to choose a way of generating 'examples' of the hint. For illustration, suppose that H_m asserts that

$$f : [-1, +1] \rightarrow [-1, +1]$$

is an *odd* function. An example of H_m would have the form

$$f(-x) = -f(x)$$

for a particular $x \in [-1, +1]$. To generate N examples of this hint, we generate $x_1, \cdots, x_N$ and assert for each x_n that $f(-x_n) = -f(x_n)$. Suppose that we are in the middle of a learning process, and that the current hypothesis is g when the example $f(-x) = -f(x)$ is presented. We wish to measure how much g disagrees with this example. This leads to the second component of the representation, the error measure e_m. For the oddness hint, e_m can be defined as

$$e_m = (g(x) + g(-x))^2$$

so that $e_m = 0$ reflects total agreement with the example (i.e., $g(-x) = -g(x)$). The form of the examples of H_m as well as the choice of the error measure e_m are not unique.

Once the disagreement between g and an example of H_m has been quantified through e_m, the disagreement between g and H_m as a whole is automatically quantified through E_m, where

$$E_m = \mathcal{E}(e_m)$$

The expected value is taken w.r.t. the probability rule for picking the examples. This rule is also not unique. Therefore, E_m will depend on our choices in all three components of the representation; the form of examples, the probability distribution for picking the examples, and the error measure e_m. Our choices are guided by certain properties that we want E_m to have. Since E_m is supposed to measure the disagreement between g and the hint, E_m should be zero when g is identical to f.

$$E = 0 \Rightarrow E_m = 0$$

This is a necessary condition for E_m to be consistent with the assertion that the hint is satisfied by f (recall that E is the error between g and f w.r.t. the original probability distribution P on the environment X).

Why is this condition necessary? Consider our example of the odd function f, and assume that the set of hypotheses contains even functions only. However, fortunately for us, the probability distribution P is uniform over $x \in [0,1]$ and is zero over $x \in [-1,0)$. This means that f can be perfectly approximated using an even hypothesis. Now, what would happen if we try to invoke the oddness hint? If we generate x according to P and attempt to minimize $E_m = \mathcal{E}[(g(x)+g(-x))^2]$, we will move towards the all-zero g (the only odd hypothesis), even if $E(g,f)$ is large for this hypothesis. This means that the hint, in spite of being valid, has taken us away from the good hypothesis. The problem of course is that, for the good hypothesis, E is zero while E_m is not. In other words, E_m does not satisfy the above consistency condition.

There are other properties that E_m should have. Suppose we pick a representation for the hint that results in E_m being identically zero for all hypotheses. This is clearly a poor representation in spite of the fact that it automatically satisfies the consistency condition! The problem with this representation is that it is extremely weak (every hypothesis 'passes the $E_m = 0$ test' even if it completely disagrees with the hint). In general, E_m should not be zero for hypotheses that disagree (through the eyes of P) with H_m, otherwise the representation would be capturing a weaker version of the hint. On the other hand, we expect E_m to be zero for any g that does satisfy H_m, otherwise the representation would be stronger than the hint itself since we already have $E_m = 0$ when $g = f$.

On the practical side, there are other properties of the representation that are desirable. The probability rule for picking the examples should be as closely related to P as possible. The examples should be picked independently in order to have a good estimate of E_m by averaging the values of e_m over the examples. Finally, the computation effort involved in the descent of e_m should not be excessive. In what follows, we will illustrate these ideas by constructing representations for different types of hints.

Perhaps the most common type of hint is **the invariance hint**. This hint asserts that $f(\mathbf{x}) = f(\mathbf{x}')$ for certain pairs $\mathbf{x}, \mathbf{x}'$. For instance, "f is shift-invariant" is formalized by the pairs $\mathbf{x}, \mathbf{x}'$ that are shifted versions of each other. To represent the invariance hint, an invariant pair $(\mathbf{x}, \mathbf{x}')$ is picked as an example. The error associated with this example is

$$e_m = (g(\mathbf{x}) - g(\mathbf{x}'))^2$$

A plausible probability rule for generating $(\mathbf{x}, \mathbf{x}')$ is to pick $\mathbf{x}$ and $\mathbf{x}'$ according to the original probability distribution P conditioned on $\mathbf{x}, \mathbf{x}'$ being an invariant pair.

There is another way to use the invariance hint that is particular to this type of hint. We can transform the examples of f itself in an invariant way, thus generating new examples of f. How well do these examples capture the hint? The transformed examples represent a restricted version of the invariance (restricted to the subset of the environment defined by the examples of f). On the other hand, the transformation takes advantage of the probability distribution that was used to generate the examples of f, which is usually the target distribution P. Using the invariance in this way and using it as an independent hint (represented to the learning process by its own examples) are not mutually exclusive.

Another related type of hint is **the monotonicity hint** (or inequality hint). The hint asserts for certain pairs $\mathbf{x}, \mathbf{x}'$ that $f(\mathbf{x}) \leq f(\mathbf{x}')$. For instance, "$f$ is monotonically nondecreasing in input variable x_i" is formalized by all pairs $\mathbf{x}, \mathbf{x}'$ such that $x_i \leq x_i'$. To represent a monotonicity hint, an example $(\mathbf{x}, \mathbf{x}')$ is picked, and the error associated with this example is

$$e_m = \begin{cases} (g(\mathbf{x}) - g(\mathbf{x}'))^2 & \text{if } g(\mathbf{x}) > g(\mathbf{x}') \\ 0 & \text{if } g(\mathbf{x}) \leq g(\mathbf{x}') \end{cases}$$

It is worth noting that the set of examples of f can be formally treated as a hint, too. Given $(\mathbf{x_1}, f(\mathbf{x_1})), \cdots, (\mathbf{x_N}, f(\mathbf{x_N}))$, **the examples hint** asserts that these are the correct values of f at these particular points $\mathbf{x_n}$. Now, to generate an 'example' of this hint, we independently pick a number n from 1 to N and use the corresponding $(\mathbf{x_n}, f(\mathbf{x_n}))$. The error associated with this example is e_0 (we fix the convention that $m = 0$ for the examples hint)

$$e_0 = (g(\mathbf{x_n}) - f(\mathbf{x_n}))^2$$

Assuming that the probability rule for picking n is uniform over $\{1, \cdots, N\}$,

$$E_0 = \mathcal{E}(e_0) = \frac{1}{N} \sum_{n=1}^{N} (g(\mathbf{x_n}) - f(\mathbf{x_n}))^2$$

In this case, E_0 is also the best estimator of $E = \mathcal{E}[(g(\mathbf{x}) - f(\mathbf{x}))^2]$ given $\mathbf{x_1}, \cdots, \mathbf{x_N}$ that are independently picked according to the original probability distribution P. This way of looking at the examples of f justifies their treatment exactly as one of the hints, and underlines the distinction between E and E_0.

In a practical situation, we try to infer as many hints about f as the situation will allow. Next, we represent each hint according to the guidelines discussed in this section.

6.3 Monotonicity hints

This section describes the application of a particular hint, the **monotonicity hint**, to two noisy real-world problems: a classification task concerned with credit card applications and a regression problem in medical diagnosis. Both databases were obtained via FTP from the machine learning database repository maintained by UC-Irvine [1].

The credit card task is to predict whether or not an applicant will default. For each of 690 applicant case histories, the database contains 15 features describing the applicant plus the class label indicating whether or not a default ultimately occurred. The meaning of the features is confidential for proprietary reasons. Only the 6 continuous features were used in the experiments reported here. 24 of the case histories had at least one feature missing. These examples were omitted, leaving 666 which were used in the experiments.

Intuition suggests that the classification should be monotonic in the features. Although the specific meanings of the continuous features are not known, we assume here that they represent various quantities such as salary, assets, debt, number of years at current job, etc. Common sense dictates that the higher the salary or the lower the debt, the less likely a default is, all else being equal. Monotonicity in all features was therefore asserted.

The motivation in the liver-diagnosis problem is to determine the extent to which various blood tests are sensitive to disorders related to excessive drinking. Specifically, the task is to predict the number of drinks a particular patient consumes per day given the results of 5 blood tests. 345 patient histories were collected, each consisting of the 5 test results and the daily number of drinks. The "number of drinks" variable was normalized to have variance 1. This normalization makes the results easier to interpret, since a trivial mean-squared-error performance of 1.0 may be obtained by simply predicting for mean number of drinks for each patient, irrespective of the blood tests.

The justification for monotonicity in this case is based on the idea that an abnormal result for each test is indicative of excessive drinking, where abnormal means either abnormally high or abnormally low.

In all experiments, batch-mode backpropagation with a simple adaptive learning rate scheme was used [2]. Several methods were tested. The performance of a linear perceptron was observed for benchmark purposes. For the experiments using nonlinear methods, a single hidden layer neural network with 6 hidden units and direct input-output connections was used on the credit data; 3 hidden units and direct input-output connections were used for the liver task. The most basic method tested was simply to train the network on all the training data and minimize the training error as much as possible. Another technique tried was to use a validation set to avoid

[1] They may be obtained as follows: ftp ics.uci.edu. cd pub/machine-learning-databases. The credit data is in the subdirectory /credit-screening, while the liver data is in the subdirectory /liver-disorders.

[2] If the previous iteration resulted in a decrease in error, the learning rate was increased by 3%. If the error increased, the learning rate was cut in half

overfitting. Training error for the credit problem was defined as

$$E_0 = \frac{1}{N}\sum_{n=1}^{N} -c_n log(g(\mathbf{x_n}) - (1 - c_n)log(1 - g(\mathbf{x_n}))$$

where c_n denotes the class label of example n. Mean squared error,

$$E_0 = \frac{1}{N}\sum_{n=1}^{N}(g(\mathbf{x_n}) - t_n)^2$$

where t_n denotes the target output for example n, was used for the liver problem.

The linear perceptrons were trained for 300 batch-mode iterations each, while the nonlinear networks were trained for 5000 batch-mode iterations on the credit data and 1000 batch-mode iterations on the liver data. These amounts of training more than sufficed for the networks to approach regions near minima where the decrease in training error had become very slow.

A leave-k-out procedure was used in order to get statistically significant comparisons of the difference in performance. For each method, the data was randomly partitioned 200 different ways (The split was 550 training, 116 test for the credit data; 270 training and 75 test for the liver data). The results shown in Table 1 are averages over the 200 different partitions.

In the early stopping experiments, the training set was further subdivided into a set (450 for the credit data, 200 for the liver data) used for direct training and a second validation set (100 for the credit data, 70 for the liver data). The classification error on the validation set was monitored over the entire course of training, and the values of the network weights at the point of lowest validation error were chosen as the final values.

Training the networks with the monotonicity constraints was performed by adding a second term, E_1, to the objective function, so that the total objective function minimized was

$$E_0 + \lambda E_1$$

E_1 was a finite-sample approximation to $\mathcal{E}[e_1]$, the expected monotonicity error of the network. For each input variable, 500 pairs of vectors representing monotonicity in that variable were generated. The pairs asserting monotonicity were generated as follows: A single input vector $\mathbf{x}$ was generated from a multivariate gaussian model of the input distribution. The second vector, $\mathbf{x'}$ was defined such that

$$\forall j \neq i, x'_j = x_j \tag{4}$$

$$x'_i = x_i + sgn(i)\delta x_i \tag{5}$$

where $sgn(i) = 1$ or -1 depending on whether f is monotonically increasing or decreasing in variable i, and δx_i was drawn from a $U[0,1]$ distribution.

For the lth pair,

$$e_1^l = \begin{cases} (g(\mathbf{x}) - g(\mathbf{x}'))^2 & \text{if } g(\mathbf{x}) > g(\mathbf{x}') \\ 0 & \text{if } g(\mathbf{x}) \leq g(\mathbf{x}') \end{cases}$$

This yielded a total of L=3000 hint example pairs for the credit problem and L=2500 pairs for the liver problem. λ was chosen, somewhat arbitrarily,to be 5000. No search for the optimal λ was attempted. E_1 was defined as the sum over all pairs of the monotonicity error on each pair:

$$E_1 = \frac{1}{L} \sum_{l=1}^{L} e_1^l$$

The process of training the networks with the monotonicity hints was divided into two stages. Since the meanings of the features were unaccessible, the directions of monotonicity were not known *a priori.* These directions were determined by training a linear perceptron on the training data for 300 iterations and observing the resulting weights. A positive weight was taken to imply increasing monotonicity, while a negative weight meant decreasing monotonicity.

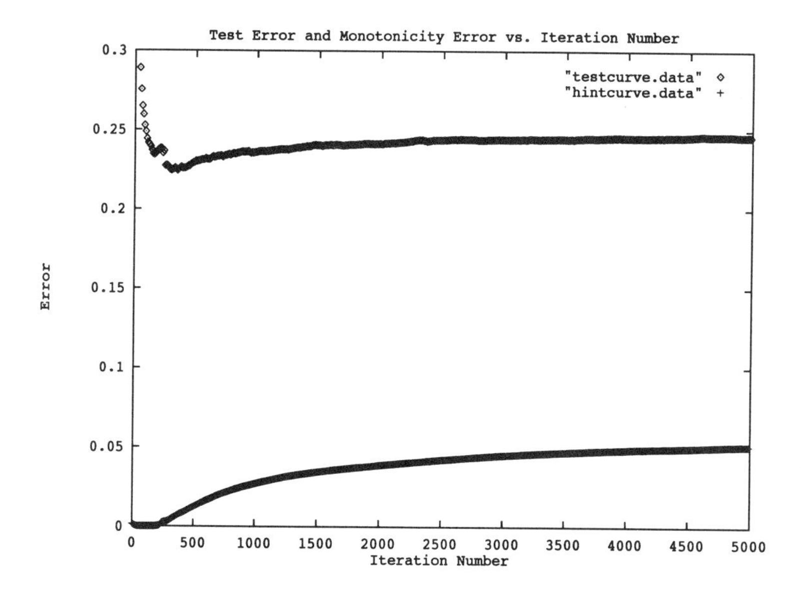

FIGURE 6.1. The violation of monotonicity tracks the overfitting occurring during training

Hint generalization, i.e. monotonicity test error, was measured by using 100 pairs of vectors for each variable which were not trained on but whose monotonicity error was calculated. For contrast, monotonicity test error was also monitored for the two-layer networks trained only on the input-output examples. Figure 1 shows test error and monotonicity error vs. training time for the credit data for the networks trained only on the training data, averaged over the 200 different data splits. The monotonicity error is multiplied by a factor of 10 in the figure to make it more easily visible. The figure indicates a substantial correlation between overfitting and monotonicity error during the course of training.

Method	training error	test error	hint test error
Linear	$22.7\% \pm 0.1\%$	$23.7\% \pm 0.2\%$	-
6-6-1 net	$15.2\% \pm 0.1\%$	$24.6\% \pm 0.3\%$	.005115
6-6-1 net, w/val.	$18.8\% \pm 0.2\%$	$23.4\% \pm 0.3\%$	-
6-6-1 net, w/hint	$18.7\% \pm 0.1\%$	$21.8\% \pm 0.2\%$	.000020

TABLE 6.1. Performance of methods on credit problem

Method	training error	test error	hint test error
Linear	$.802 \pm .005$	$.873 \pm .013$	-
5-3-1 net	$.640 \pm .003$	$.920 \pm .014$	.004967
5-3-1 net, w/val.	$.758 \pm .008$	$.871 \pm .013$	-
5-3-1 net, w/hint	$.758 \pm .003$	$.830 \pm .013$	.000002

TABLE 6.2. Performance of methods on liver problem

The performance of each method is shown in tables 1 and 2. Without early stopping, the two-layer network overfits and performs worse than a linear model. Even with early stopping, the performance of the linear model and the two-layer network are almost the same; the difference is not statistically significant. This similarity in performance is consistent with the thesis of a monotonic target function. A monotonic classifier may be thought of as a mildly nonlinear generalization of a linear classifier. The two-layer network does have the advantage of being able to implement some of this nonlinearity. However, this advantage is cancelled out (and in other cases could be outweighed) by the overfitting resulting from excessive and unnecessary degrees of freedom. When monotonicity hints are introduced, much of this unnecessary freedom is eliminated, although the network is still allowed to implement monotonic nonlinearities. Accordingly, a modest but clearly statistically significant improvement on the credit problem (nearly 2%) results from the introduction of monotonicity hints. Such an improvement could translate into a substantial increase in profit for a bank. Monotonicity hints also significantly improve test error on the liver problem; 4% more of the target variance is explained.

6.4 Theory

Much of learning theory is concerned with determing the number of examples needed for good generalization. Concepts such as capacity [5], VC-dimension [6], and effective number of parameters [7] have been developed in order to get bounds on the generalization (test error) given the training error and the model complexity. The ideas of capacity and VC-dimension also help to explain the benefit of introducing hints [8] into the learning process. By constraining a model to obey a hint, the model's capacity is reduced without an accompanying reduction in its ability to approximate the target function [9].

We present results here concerning the reduction in capacity associated with the use of the monotonicity hint. We showed in the preceding section that the use of monotonicity hints can yield significant improvement in the performance of neural networks on real-world tasks. To quantify how powerful a constraint monotonicity is would therefore be of some interest.

We will consider the class $\mathbf{M}$ of monotonically increasing (non-decreasing) functions from $R^d \rightarrow \{0,1\}$. We will say that $\mathbf{x}' \geq \mathbf{x}$ if $\forall i, 1 \leq i \leq d, x_i' \geq x_i$. $\mathbf{M}$ is the class of all functions such that

$$\mathbf{x}' \geq \mathbf{x} \rightarrow f(\mathbf{x}') \geq f(\mathbf{x})$$

In many applications, prior information may indicate decreasing monotonicity in some variables rather than increasing monotonicity. The analysis which follows will also hold for each of the other $2^d - 1$ function classes where some or all variables have a decreasing monotonicity constraint rather than an increasing one. This equivalence is made clear by observing that decreasing monotonicity may be converted to increasing monotonicity by relabelling an input variable as its negation. There are also many situations where monotonicity only holds for some variables, while the relationship of the other variables to the output is completely unknown a priori. This case is more complex and will not be addressed here. Note that the class $\mathbf{M}$ is not explicitly parametrized by weights, unlike classes such as sigmoidal networks with a fixed number of hidden units. When a finite, parametrized model is further constrained to obey monotonicity in all variables, the resulting class of functions will be some subset of $\mathbf{M}$. Thus, bounds on the capacity of $\mathbf{M}$ will upper-bound the capacity of any parametrized model where monotonicity hints are enforced.

Subsection II derives results about the capacity of $\mathbf{M}$. In particular, the capacity is shown to depend almost completely on the input distribution. Subsection III describes the decision boundaries of monotonic classifiers, comparing and contrasting them to separating hyperplanes.

6.4.1 Capacity results

To make the idea of capacity precise, we must define a few auxiliary concepts. Define a *dichotomy* to be a set of d-dimensional input vectors, each of whom have an associated class label of either 0 or 1. Define a positive example as an input vector labelled 1 and a negative example as an input vector labelled 0. We say that a dichotomy is separable by a function class if there exists at least one function in the class which maps each of the input vectors to its correct class label. Define a random dichotomy as a set of input vectors drawn from the input distribution, with the label 0 or 1 assigned randomly with equal probability. Let $P(n)$ be the probability that a random dichotomy of n examples may be separated by the function class. We call the capacity the integer n^* for which $P(n^*)$ is closest to 0.5. Capacity (unlike VC-dimension) is therefore a quantity which *depends on the input distribution.* The importance of this point will become clear below.

The following theorem holds about monotonic separability:

Theorem 1 *A dichotomy is separable by* $\mathbf{M}$ *if and only if there exists no negative example which is greater than or equal to some positive example*

The necessity of this condition is obvious. Sufficiency may be demonstrated by constructing a function belonging to $\mathbf{M}$ which implements the dichotomy. Consider the function f which classifies as 1 only those input vectors which are greater than or equal to some positive example in the dichotomy. By construction, f obviously separates the dichotomy correctly. It is also clear that f belongs to $\mathbf{M}$. Consider an input vector $\mathbf{x}$ which f classifies positively. $\mathbf{x}$ must be greater than or equal to some vector $\mathbf{x}^*$ in the dichotomy. Any vector $\mathbf{x}'$ which is greater than or equal to $\mathbf{x}$ must also be greater than or equal to $\mathbf{x}^*$. Thus, $f(\mathbf{x}') \geq f(\mathbf{x})$, i.e., f belongs to $\mathbf{M}$.

Capacity results for the perceptron (i.e., the class of linear threshold functions) are well known: the capacity tends to $2d$, where d is the dimensionality of the input space [5]. This result is true for any smooth distribution of continuous input variables, because the number of linearly separable dichotomies of a set of input vectors is independent of how those input vectors are arranged, assuming no degenerate cases. This independence with respect to input distribution is in sharp contrast to the case of $\mathbf{M}$. Here, the number of dichotomies depends *heavily* on how the input vectors are arranged.

Consider figure 2. A little inspection will reveal that this dichotomy is not monotonically separable, even if we are free to take decreasing monotonicity to hold for one or both of the input variables. Figure 3 depicts a drastically different situation. In this case, there exists a monotonically *decreasing* relationship between the two input variables. When n input poin ts are arranged according to such a distribution, any of the 2^n possible dichotomies are separable by $\mathbf{M}$, the class of monotonically *increasing* functions. For a dichotomy not to be separable by $\mathbf{M}$, there must be a negative example which is greater than some positive example. But if the two input

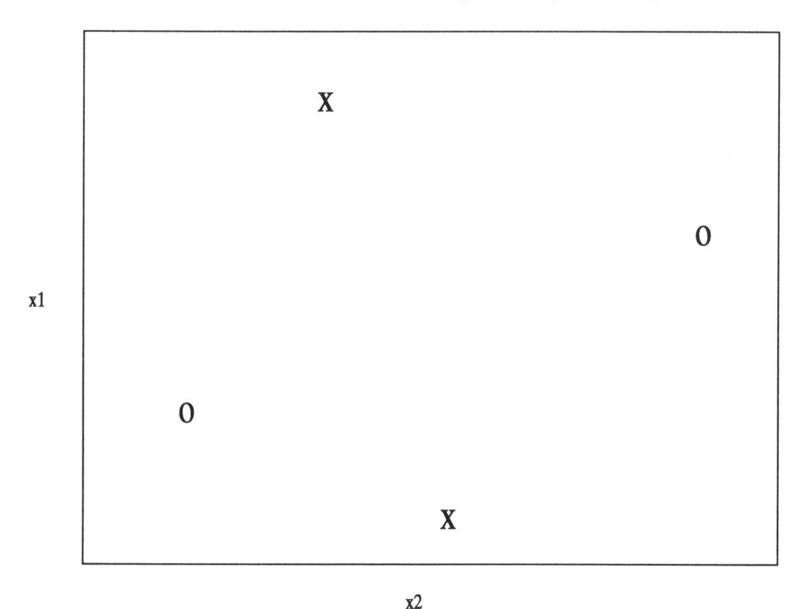

FIGURE 6.2. This dichotomy of four points cannot be implemented by any monotonic function, regardless of whether the monotonicity in each variable is increasing or decreasing. x's indicate positive examples, o's negative examples.

variables are related in a monotonically decreasing way, there are no two examples such that one is greater than the other in the vector sense, i.e., in both input variables. Therefore all dichotomies are separable given this input distribution. Note that even if the input dimensionality is greater than 2, a monotonically decreasing relationship between two of the input variables is sufficient to make all dichotomies separable by **M**. The above argument still holds, and monotonicity in the other $d-2$ variables may be satisfied by ignoring these variables, i.e., by choosing a function which is constant in these variables.

The example depicted in figure 3 establishes a result which explains why the input distribution-dependent notion of capacity is more useful than the concept of VC-dimension for the analysis of monotonicity. Recall that the VC-dimension is the maximum value of n for which the growth function $G(n) = 2^n$. The growth function is defined by the maximum number of separable dichotomies of n points, where this maximum is taken over all possible arrangements of the n points. Figure 3 demonstrates that we may always arrange input points in such a way that all 2^n possible dichotomies may be separated monotonically. It is granted that the figure 3 example is not very realistic- it would be extremely odd to find a problem where the target is believed to increase monotonically with two input variables which appear to be related to each other in a monotonically decreasing way. Nonetheless, such an example is permitted by the definition of the growth

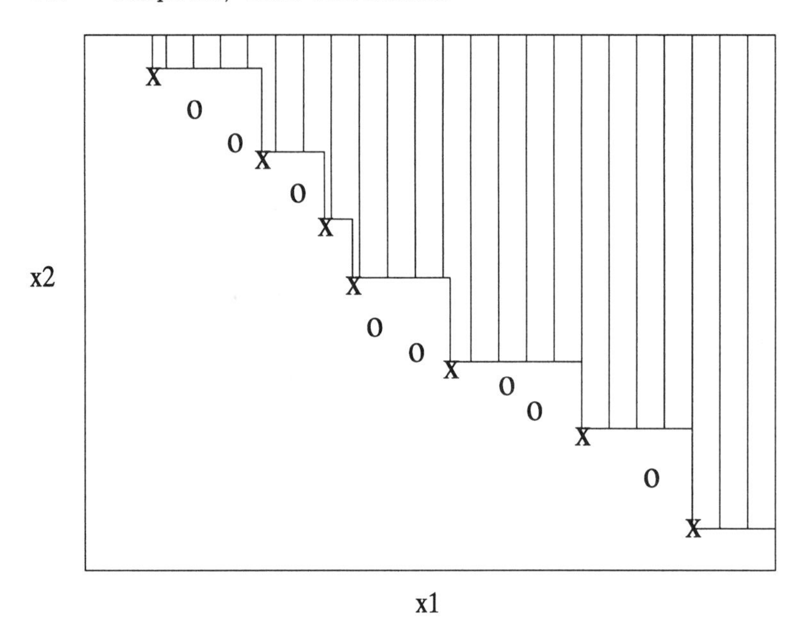

FIGURE 6.3. If two input variables are related in a monotonically decreasing way, then any dichotomy may be separated by M. A particular, arbitrary dichotomy is shown here. The shaded area indicates the region of input space which, by monotonicity, must be classified positively.

function. Thus, the VC-dimension of the class of monotonic functions = ∞ ! This result is misleading, however, since monotonicity is still a very powerful constraint in most cases.

Figure 4 depicts a situation opposite to that of figure 3. In this case, there exists a monotonically *increasing* relationship between the two input variables. **M** has very little separating power in such a situation. The second input variable adds nothing- the only dichotomies separable by **M** are those which were already separable using only one of the input variables. The class **M** is free only to choose a dividing point from among the n input vectors and classify positively those vectors greater than or equal to the dividing vector. Thus, only $n+1$ of the 2^n possible dichotomies is separable by **M**. By a similar argument, the number of dichotomies is separable by **M**. By a similar argument, the number of separable dichotomies remains $n + 1$ if we have d input variables all of which are related to each other in monotonically increasing way, i.e., all of which rise and fall together.

The preceding examples demonstrate that the number of dichotomies separable by **M**, and hence, the capacity of **M** can be arbitrarily large or small depending on the particular input distribution. The second example and especially the third example cannot be dismissed as merely irrelevant, degenerate cases which will never occur in real life. These two examples are the extreme versions of possible real world situations where the input

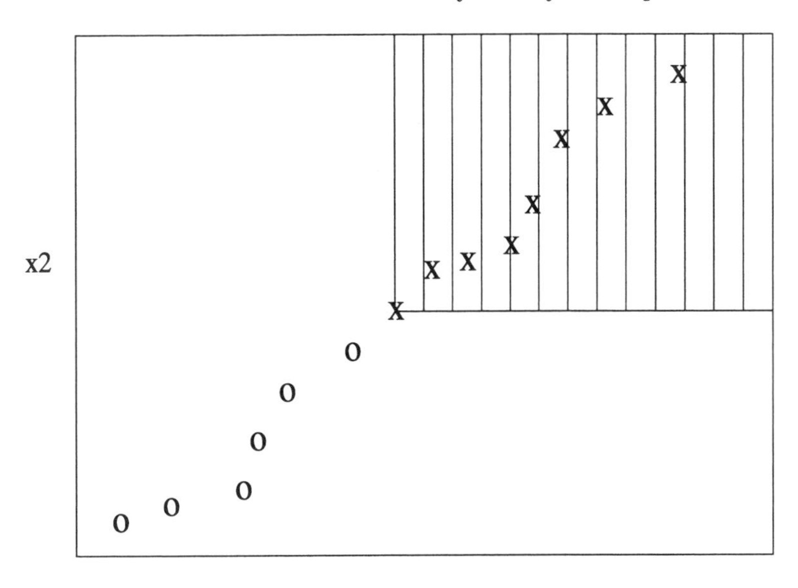

FIGURE 6.4. If two input variables are related in a monotonically increasing way, then only $n + 1$ of the 2^n possible dichotomies are separable by M. A particular dichotomy is pictured here. The shaded area indicates the region of input space, by monotonicity, must be classified positively.

variables do not have strict monotonic relationships but are correlated significantly. It should be clear from figure 4 that if we have two input variables which are highly but not perfectly positively correlated, then the number of separable dichotomies of n points is still likely to be low, although somewhat higher than $n+1$. Likewise, if we have two input variables which have a large, negative correlation, then the number of separable dichotomies is likely to be quite high, although somewhat less than 2^n.

The total dependence of the capacity of **M** on the input distribution makes it somewhat unlikely that much more useful analytic work on this topic can be done. If we have a good model for the input distribution for a given problem, however, the capacity of **M** may be estimated computationally. n input vectors may be drawn from the model of the input distribution and labelled randomly as positive or negative with equal probability. Theorem 1 tells us that we can then check whether or not the dichotomy generated is separable by **M**. This procedure may be repeated many times in order to estimate $P(n)$, the probability that n randomly labelled points are separable by **M**. The capacity of **M** is then that n^* for which $P(n^*) \approx 0.5$.

An estimate of $P(n)$ may also be used to make explicit statements about the generalization of a monotonic model. The VC bounds on generalization, based on the growth function, are well known. Tighter bounds may be obtained by replacing the growth function with the VC entropy, i.e., the expected number of dichotomies [10]. The expected number of dichotomies separable by M is simply $2^n P(n)$. Thus, the estimate of $P(n)$ could be used to get real bounds on generalization error given training error when a monotonic model is used.

6.4.2 *Decision boundaries*

A monotonic classifier may be thought of as a mildly nonlinear generalization of a linear classifier. This relationship is perhaps best demonstrated by considering the decision boundaries corresponding to the two models. It is well known that the decision boundary implemented by a linear perceptron is simply a $d-1$ dimensional hyperplane splitting input space into two regions. In two dimensions, this boundary consists of a straight line dividing the input plane.

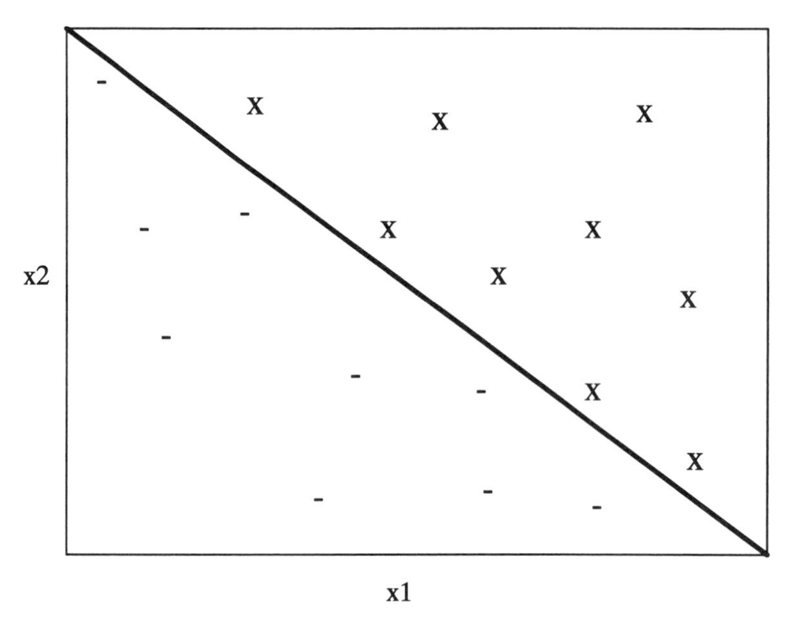

FIGURE 6.5. A linear classifier splits the input plane with a line

Consider a classifier $m(x_1, x_2)$ which is monotonically increasing in both input variables. We will also place one additional, mild constraint on m. Assume that for any x_1, there exists some x_2 such that $m(x_1, x_2) = 0$ and some x_2 such that $m(x_1, x_2) = 1$. Assume also for any x_2, there exists some x_1 such that $m(x_1, x_2) = 0$ and some x_1 such that $m(x_1, x_2) = 1$. If the above conditions hold, then the decision boundary implemented by $m(x_1, x_2)$ must be a one-to-one mapping between x_1 and x_2 [Fig 4]. Let x_1 be fixed, and let $x_2^{min}(x_1)$ be the smallest value x_2 can take such that $m = 1$. By monotonicity, for any $x_2 \geq x_2^{min}$, m also equals 1. Thus, x_2^{min}

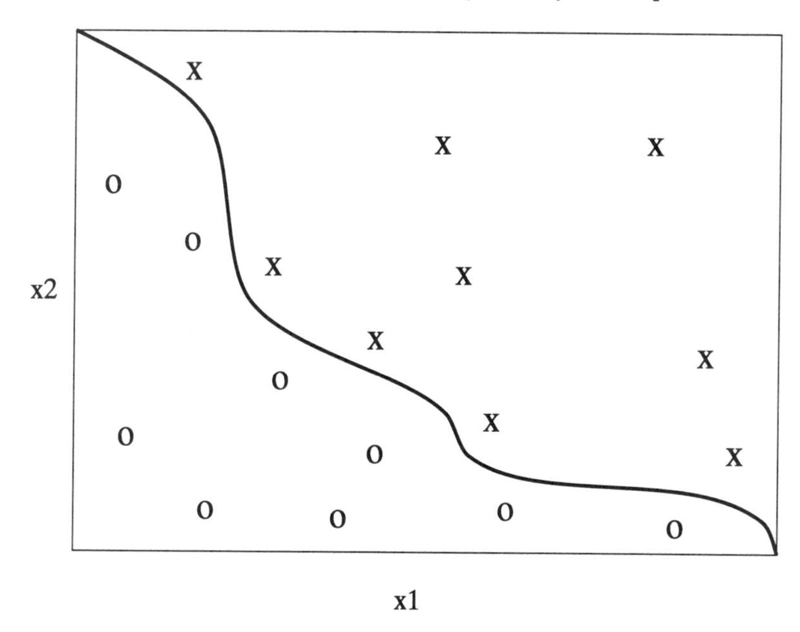

FIGURE 6.6. A monotonic classifier splits the input plane with a decision boundary consisting of a one-to-one mapping between the two input variables. Monotonic classifiers are analogous to but slightly more flexible than linear classifiers, which split the input plane with a straight line.

is the only value of x_2 for x_1 fixed where the classification changes. By the same argument, there exists a single $x_1^{min}(x_2)$ for every x_2. Therefore, the decision boundary must be an invertible function from one input variable to the other. This result makes clear our intuition that a monotonic model is more flexible than a linear one, but is still very severely constrained. The decision boundary of a monotonic classifier in the general case of d dimensions is somewhat harder to summarize, but some statements can be made. If we make analogous assumptions, then the boundary must be one continuous sheet such that specifying the values of any subset of $d-1$ input variables defines a unique value of the dth variable at which the classification changes. Thus, the boundary is a generalization to d dimensions of an invertible function.

6.5 Conclusion

As the use of hints becomes routine, we are encouraged to exploit even the simplest observations that we may have about the function we are trying to learn. Since most hypotheses do not usually satisfy a given hint, the impact of hints is very strong in restricting the learning model in the right direction. The monotonicity hint provides an excellent illustration of the above point. We showed how the monotonicity constraint improved performance on real data. We also analyzed the power of the constraint from a theory persepective, demonstrating how a monotonic decision boundary is much more restricted than an arbitrary decision boundary which might be produced by a neural network with no prior knowledge supplied. Future directions may include the extension of this work to other hints, i.e., both theoretical and practical demonstrations of their benefits.

Acknowledgements

The authors thank Eric Bax, Zehra Cataltepe, Malik Magdon-Ismail, and Xubo Song for many useful comments.

6.6 References

[1] Akaike, H. Fitting autoregressive models for prediction, *Ann. Inst. Stat. Math.* 21:243-247,1969.

[2] Abu-Mostafa, Y. S. Learning from hints in neural networks. *Journal of Complexity* 6:192-198,1990.

[3] Abu-Mostafa, Y. S. A method for learning from hints In S. Hanson et al,editors, *Advances in Neural Information Processing Systems 5*, pp. 73-80 , 1993.

[4] Minsky, M. L. and Papert, S. A. *Perceptrons*, MIT Press , 1969.

[5] Cover, T.M. Geometrical and Statistical Properties of Systems of Linear Inequalities with Applications in Pattern Recognition. *IEEE Transactions on Electronic Computers* 14:326-334,1965.

[6] Vapnik, V.N. and Chervonenkis, A.Y On the Uniform Convergence of Relative Frequencies of Events to Their Probabilities. *Theory of Probability and Its Applications* 16:264-280,1971.

[7] Moody, J.E. The *Effective* Number of Parameters: An Analysis of Generalization and Regularization in Nonlinear Learning Systems. In J.E. Moody, S.J. Hanson, R.P. Lippman, editors, *Advances in Neural Information Processing* 4:847-854. Morgan Kaufmann, San Mateo, CA, 1992.

[8] Abu-Mostafa, Y.S. Hints *Neural Computation* 7:639-671,1995.

[9] Abu-Mostafa, Y.S. Hints and the VC Dimension *Neural Computation* 4:1993

[10] Vapnik, V. *The Nature of Statistical Learning Theory.* Springer-Verlag, New York, NY, 1995.

7

Analysis and Synthesis Tools for Robust SPR Discrete Systems

Carlos Mosquera
J. Ramón Hernández
Fernando Pérez-González

ABSTRACT The robust strict positive real problem arises in identification and adaptive control, where strict positive realness is a sufficient condition for ensuring convergence of several recursive algorithms. The strict positive real (SPR) property of uncertain systems is analyzed in depth here, and some insightful theorems are provided for characterizing the SPRness of different types of uncertain sets. The design of appropriate compensators which enforce the SPRness of a given set is also addressed, through both algebraic and approximate procedures. Those compensators are widely used in the adaptive recursive filters field: they guarantee global convergence in many cases. The robust strengthened strict positive real problem is not excluded from our analysis, given its importance in some recursive algorithms: the problem arises as to how to design a compensator with a given constraint in its norm or its coefficients. Illustrative results are presented at the end of the paper.

7.1 Introduction

Adaptive infinite-impulse response (IIR) filters are desirable in many situations as an alternative for adaptive finite-impulse response (FIR) filters, for their reduced complexity and improved performance. Important applications include adaptive noise cancelling, channel equalization, adaptive differential pulse code modulation, etc. Adaptive techniques for IIR filters have been under investigation during the last years, taking results from the system identification field in most cases, due to the similarities between both areas. Convergence of the algorithms has been the main issue throughout this process; error surfaces are in most cases multimodal, and the analysis of convergence of gradient-based techniques becomes quite hard [1], with convergence to the global minimum not guaranteed in many cases. In addition, most of those procedures need a stability monitoring: otherwise the algorithm may diverge during the adaptation stage. Spurred by the convergence problems, other investigators have borrowed from the control field some tools based on hyperstability, which allow the design of algorithms with proven convergence, provided that a *Strictly Positive Real* (SPR) condition is satisfied. Although various suggestions have been made in order to relax the SPR condition, none of them is completely satisfactory given their suboptimality with output disturbance in some cases [2],[3], or conditions imposed on the input in other cases [4].

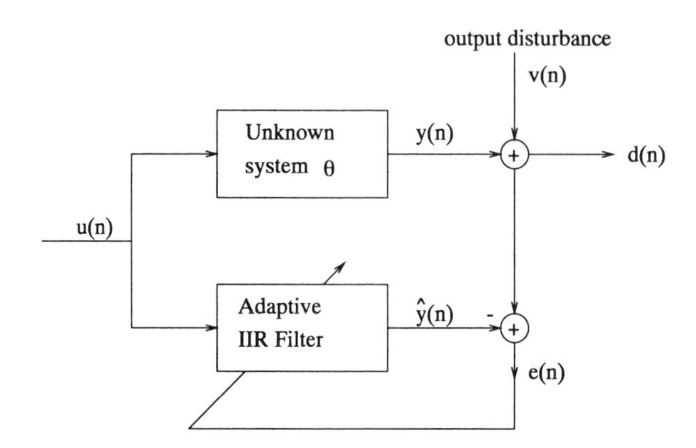

FIGURE 7.1. System Identification Configuration

Many adaptive IIR filtering problems may be addressed in a system identification framework, in which a reference model is hypothesized. The unknown transfer function is assumed rational, and the objective is to construct a rational approximation to the transfer function, based on the input-output measurements, usually noise-corrupted. Figure 7.1 shows the adaptive filter in a system identification configuration, where θ is the unknown parameter vector. The goal is the minimization of a performance criterion of the error $e(n)$. Traditionally there have been two main approaches to the adaptive IIR filtering problem which correspond to different formulations of the error: equation error and output error methods. We will not consider the first family of methods here; equation error methods have well understood properties, given their similarity to adaptive FIR methods. Their main drawback is the bias in the estimate when a disturbance is present in the output.

The output error adaptive IIR filter is characterized by the following recursive equation:

$$\hat{y}(n) = \sum_{k=1}^{N_A} a_k(n)\hat{y}(n-k) + \sum_{k=0}^{N_B} b_k(n)u(n-k) \tag{7.1}$$

with $a_k(n)$ and $b_k(n)$ the adaptive parameters of the filter. The existence of output feedback is the reason why this formulation is more complex than its equation error counterpart.

Two main approaches can be made in the output error problem [5]: minimization viewpoint and stability theory viewpoint. The minimization approach leads to a gradient descent formulation. One of its drawbacks is the sometimes multimodality of the error surface being descended, although in some cases it can be shown to be unimodal [6]. However, the main concern in this approach is the need for on-line stability monitoring of the time-varying AR filter, which filters the regressor vector, and is recomputed at each stage as the estimation of the denominator of the transfer function.

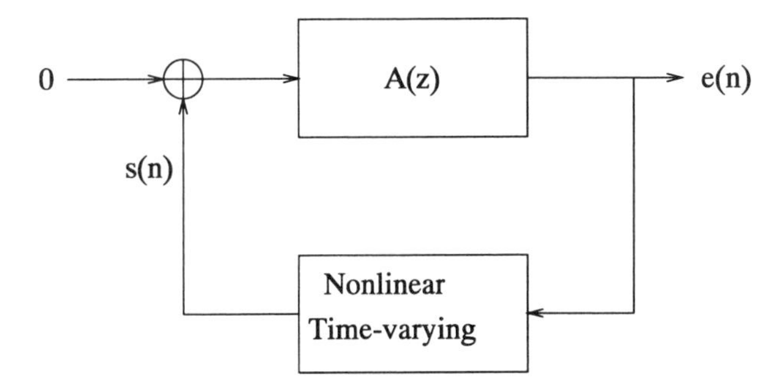

FIGURE 7.2. Nonlinear feedback system

An alternate approach for adapting the parameters of the IIR filter is based on the theory of hyperstability [7], a concept that was developed for the stability analysis of time-varying nonlinear feedback. The adaptive IIR process can be viewed as a linear system having time-varying nonlinear feedback, as shown in Figure 7.2, and is chosen on the basis of assuring that the resulting closed-loop configuration is hyperstable, and hence convergent. The standard result states that if the nonlinearity satisfies

$$\sum_{i=0}^{N} s(n) \cdot e(n) \leq K_m < \infty \tag{7.2}$$

for any N, with K_m a positive constant independent of the upper summation limit, and the linear part holds the following conditions

$$\begin{aligned} A(z) \text{ is analytic for } |z| \geq 1, \\ Re\,\{1/A(z) - K\} > 0, \text{ for all } |z| = 1 \end{aligned} \tag{7.3}$$

then global convergence is guaranteed with no need of stability checking. Inequality (7.2) is often called a Popov inequality. In Eq. (7.3), $A(z) =$

$1 - \sum_{k=1}^{N_A} a_k z^{-1}$ is the denominator of the unknown plant and the real scalar K depends on the specifics of the adaptive algorithm. For the Hyperstable Adaptive Recursive Filter (HARF) [7] and its simplified version (SHARF) [8], two of the main adaptive IIR algorithms based on hyperstability concepts, $K = 0$ suffices. Other more complex algorithms, such as the Pseudolinear Regression algorithm (PLR), require $K = \frac{1}{2}$, the same as in the original algorithm based on hyperstability ideas, developed by Landau [9] for identification purposes, and modified for adaptive IIR filtering by Johnson [7].

The main drawback of this type of algorithms in the sufficient order case is that the satisfaction of condition (7.3) is critical for proper algorithm behavior, although convergence can be achieved in some cases for which such a condition is not satisfied [10]. Yet, there is no general method to eliminate the condition entirely, despite the efforts made in that direction [2],[3],[4].

As we will see throughout the paper, there are significant differences in the treatment of the cases $K = 0$ and $K \neq 0$; therefore, it is customary to stress this difference by means of a definition. We will say that an analytic $A(z)$ satisfies the SPR (Strictly Positive Real) condition if (7.3) holds for $K = 0$.[1] On the other hand, an analytic $A(z)$ will be termed an SSPR (Strengthened Strictly Positive Real) plant if (7.3) is satisfied for $K \neq 0$. Of course, the SSPR property is always accompanied with a corresponding set of values of K for which it holds. Now, a fundamental question comes unexpectedly: if we are trying to identify the denominator of the unknown plant $A(z)$, how can we foreknow that it is SPR or SSPR?. This is undoubtedly a weak point of the hyperstability approach but it should not be overstated because in many cases it is possible to provide a judicious approximation of $A(z)$, based on some knowledge of a local vicinity of the denominator polynomial. For instance, a first strategy would be to start with a different identification scheme, such as least squares, to obtain an approximation of $A(z)$ [11]. In some other cases, an a priori confidence set can be established for either the coefficients or the roots of $A(z)$ based on some knowledge of the system to be identified. In any case, a reasonable starting point would be to consider an uncertain set $\mathcal{A}(z)$ to which the true denominator $A(z)$ belongs and study the properties above in a "robust" context. For example, if all the members of $\mathcal{A}(z)$ are SPR, it is clear that $A(z)$ will be SPR and, therefore, we will have guaranteed global convergence. The problem then becomes how to check any of the properties above for every member in the family, an apparently impossible task in the light of the infiniteness

[1] In this case, condition (7.3) is equivalent to $Re\{A(z)\} > 0$, for all $|z| = 1$, which is the form that we will use throughout the paper.

of $\mathcal{A}(z)$. In section 7.2 we will show how extraordinary computational simplifications can be derived. We should mention here a different use of the a priori information, exposed in [12], where that information is used for selecting a fixed set of poles, thus allowing only the zeros to adapt. Some interesting advantages are obtained in terms of the convergence, although they come at the price of fixing the denominator of the IIR adaptive filter. We will not deal with that approach, and we will hereafter refer only to algorithms which simultaneously adapt both zeros and poles.

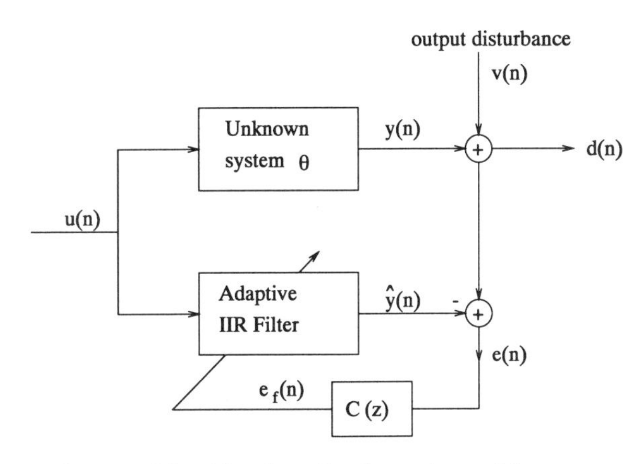

FIGURE 7.3. System Identification Configuration with compensating error smoothing filter

We have to point out that both the SPR and SSPR conditions are rather restrictive and in general are not satisfied.[2] The problem can be partially overcome by using $e_f(n)$ instead of $e(n)$ to drive the adaptive algorithm [7]. $e_f(n)$ is a filtered version of $e(n)$ through a linear and time invariant (LTI) filter called *compensating error smoothing filter*, or *compensator* for short (see Figure 7.3). It can be shown that the two conditions above transform into

$$C(z)/A(z) \text{ is analytic for } |z| \geq 1. \tag{7.4}$$

$$Re\{C(z)/A(z) - K\} > 0, \text{for all } |z| = 1 \tag{7.5}$$

This suggests designing a stable and minimum-phase[3] $C(z)$ to guarantee both conditions and, therefore, global asymptotic convergence. As in the previous paragraphs, we have to deal with an uncertain set $\mathcal{A}(z)$ instead

[2]This obstacle is due not only to the "amount" of uncertainty but also to the fact that these conditions are critical even for a single polynomial.

[3]A rational function is minimum-phase if all its zeros are inside the open unit disk. $C(z)$ must be minimum-phase in order to guarantee stability of its inverse, thus nulling $e(n)$ as $e_f(n)$ goes to zero.

of the actual value of the plant denominator to be identified. This forces us to design $C(z)$ such that condition (7.5) is satisfied for every member $A(z)$ in the set $\mathcal{A}(z)$. The robust SPR design problem ($K = 0$) was raised by Dasgupta and Bhagwat in [13] and then by Anderson *et al.* in [14]. The latter paper provides a necessary and sufficient condition for the existence of the compensator $C(z)$, namely,

$$\max_{\omega\in[0,2\pi)} \left| \max_{A(z)\in\mathcal{A}(z)} \arg(A(e^{j\omega})) - \min_{A(z)\in\mathcal{A}(z)} \arg(A(e^{j\omega}))\right| < \pi, \tag{7.6}$$

where arg(.) denotes unwrapped (continuous) phase. This crucial condition is the basis for further design methods that will be discussed in some detail in section 7.3. Some comments are now pertinent regarding the design of a $C(z)$ for making an uncertain set SSPR. First, it is necessary to understand that the SSPR condition is equivalent to the SPR condition if no further requirements are imposed on $C(z)$. To see this, simply consider that $Re\left\{C(e^{j\omega})/A(e^{j\omega}) - K\right\} \geq \delta > 0$ for all $\omega \in [0, 2\pi)$ if $Re\left\{C(e^{j\omega})/A(e^{j\omega})\right\} \geq \delta + K$ for all $\omega \in [0, 2\pi)$, so the uncertain set can be made SSPR by multiplying $C(z)$ by a sufficiently large constant once that the SPR problem has been solved. This "artificial" solution brings about some troubles, such as disturbance amplification that slows down convergence. It is then reasonable to introduce some constraint in the "size" of the compensator in order to avoid that undesirable behavior. In [15] the constraint is the monicity of C. It is shown there that for the continuous-time counterpart of the SSPR problem, equation (7.6) is still necessary and sufficient for the existence of a monic C that makes the entire family SSPR for $0 < K < 1$. Unfortunately, we show in [18] that this is not true for the discrete-time robust SSPR problem. In any case, we think that it seems more logical to impose some constraint on the norm of $C(z)$, either $\| C \|_\infty$ or $\| C \|_2$, for disturbance rejection purposes. The existence of necessary and sufficient conditions à la Eq. (7.6) with any of the mentioned constraints is presently unknown for the discrete-time case and therefore constitutes an open research problem. In any case, the desire to use a fixed compensator has remained as a severe practical limitation of the hyperstability-based schemes, and its practical design has only been satisfactorily solved in a very restricted set of cases [19]. Section 7.3 is devoted to the synthesis of practical compensators in some useful cases.

To conclude this section, we can summarize the discussion above by excising and formally posing the problems that we have considered so far. We will make the assumption that the family $\mathcal{A}(z)$ has constant degree.

Problem 1a (Robust SPR problem): Given the set of polynomials $\mathcal{A}(z)$, determine whether every member in the family is SPR.
Problem 1b (Robust SPR problem with compensator): Given the set of polynomials $\mathcal{A}(z)$ and a stable and minimum-phase rational function

$C(z)$, determine whether every member in the family $C(z)/\mathcal{A}(z)$ is SPR.
Problem 2 (Robust SPR-izability): Given the set of polynomials $\mathcal{A}(z)$ determine whether there exists a stable and minimum-phase rational function $C(z)$ such that the family $C(z)/\mathcal{A}(z)$ is SPR.
Problem 3a (Robust SSPR problem): Given the set of polynomials $\mathcal{A}(z)$, determine whether every member in the family is SSPR for a certain $K \in (0, 1)$.
Problem 3b (Robust SSPR problem with compensator): Given the set of polynomials $\mathcal{A}(z)$ and a stable and minimum-phase rational function $C(z)$, determine whether every member in the family $C(z)/\mathcal{A}(z)$ is SSPR for a certain $K \in (0, 1)$.
Problem 4 (Robust SSPR-izability): Given the set of polynomials $\mathcal{A}(z)$ determine whether there exists a stable and minimum-phase rational function $C(z)$ such that the family $C(z)/\mathcal{A}(z)$ is SSPR.

For the latter problem, only necessary conditions are known for the existence of $C(z)$ when some constraints are imposed on it. If the limitation is on $\| C \|_\infty$, a lower bound is obtained for $\| C \|_\infty$ as

$$\min_{C} \| C \|_\infty = K \max_{A \in \mathcal{A}} \| A \|_\infty \tag{7.7}$$

If the desired compensator $C(z)$ is to be monic, an upper bound on K is provided in [18], such that no monic solution exists for K above that upper bound.

The regions in the complex plane where $C(e^{j\omega})$ must lie can be derived, and are obviously smaller than their counterparts for Problem 2 ($K = 0$), although no necessary and sufficient conditions for the existence of C are known.

The study of the SPRness of uncertain systems is detailed in the following section. The construction of appropriate compensators is discussed in section 7.3, with both algebraic and approximate schemes described. Finally, some numerical results are presented before the concluding remarks.

7.2 SPR Analysis of Uncertain Systems

The purpose of this section is to present some results concerning the verification of the properties identified in the previous section. An important observation from (7.5) is that the SPR property is studied more conveniently in the frequency domain. However, in many occasions the description of the uncertainty set is given in the time domain. For this reason, the so-called *value-set* techniques have gained great interest in the past decade as they allowed to solve some crucial problems posed in the robust stability arena. The basic idea behind the value-set concept is a transformation from a multidimensional uncertain set into a two-dimensional set on which the desired operations are performed. Assume that we have the following uncertain description of the unknown plant denominator

$$\mathcal{A}(z) = \{A(z) = 1 - \sum_{i=1}^{N_A} a_i z^{-i}, \mathbf{a} = [a_1, \cdots, a_{N_A}] \in \mathbf{A}\} \tag{7.8}$$

where $\mathbf{A}$ is a compact set of $\mathbf{R}^{N_A}$. Then, for a given frequency $\omega \in [0, 2\pi)$ the value-set is simply defined as

$$\mathcal{A}(e^{j\omega}) = \{A(e^{j\omega}), A(z) \in \mathcal{A}(z)\} \tag{7.9}$$

Note that $\mathcal{A}(e^{j\omega}) \subset \mathbf{C}$. In fact, for a given frequency, we have defined a (linear) mapping from $\mathbf{R}^{N_A}$ to $\mathbf{R}^2$. Except in some pathological cases, the set $\mathcal{A}(e^{j\omega})$ moves smoothly as we sweep along the frequency segment $\omega \in [0, 2\pi)$. This allows to solve Problem 1a by testing an equality and the SPR property for a single polynomial, as stated next.

Theorem 1 *Assume that the family of polynomials $\mathcal{A}(z)$ contains at least one SPR polynomial and* $\mathbf{A}$ *is pathwise connected. Then, the entire family of polynomials is SPR if and only if*

$$Re\{A(e^{j\omega})\} \neq 0, \text{ for all } \omega \in [0, 2\pi) \text{ and all } A(z) \in \mathcal{A}(z) \tag{7.10}$$

The proof can be found in [20]. According to Theorem 1, we have to verify that the set $\mathcal{A}(e^{j\omega}) \subset \mathbf{C}$ does not touch the imaginary axis for every $\omega \in [0, 2\pi)$ in order to conclude that the entire family satisfies the SPR property. In doing so, we avoid to check for the phase-minimality of $\mathcal{A}(z)$ that is a necessary condition for robust SPRness. The reason for this saving is that condition (7.10) is more restrictive than phase-minimality. The solution to Problem 1b involves only a minor modification to Theorem 1, that is, the consideration of $C(z)/A(z)$ instead of $A(z)$ so we will devote our efforts to Problem 1a.

With regard to the solution of Problem 2, this is given by condition (7.6) that explicitly deals with the value-set. However, when the set $\mathbf{A}$ is convex, the following result will prove its usefulness.

Lemma 1 *If the set* $\mathbf{A}$ *is convex, then the following two statements are equivalent:*
i) $|\max_{A\in\mathbf{A}}\{arg(A(e^{j\omega}))\} - \min_{A\in\mathbf{A}}\{arg(A(e^{j\omega}))\}| < \pi$
ii) $\mathcal{A}(e^{j\omega}) \neq 0$

Proof: If **A** is a convex set, then $\mathcal{A}(e^{j\omega})$ will also be convex because we are using a linear mapping. The equivalence of the two statements then follows from the separating line principle.

The previous result should not be overlooked; condition ii) is the so-called *zero-exclusion principle* which is exactly what it is required for the entire family $\mathcal{A}(z)$ to be minimum-phase. The following Corollary is then at hand,

Corollary 1 *If the family $\mathcal{A}(z)$ is minimum-phase and the set* **A** *is convex then condition (7.6) holds.*

Whenever possible we will use the zero-exclusion principle for solving Problem 2 in view of the vast amount of results that are available for robust stability determination [20],[21].

The value-set can also be used for finding out the solution to Problem 3 (we prefer here to use instance 3b, i.e., to account for a compensator). In this case, we will use the following theorem:

Theorem 2 *Assume that the family $\mathcal{A}(z)$ has one member $A_*(z)$ such that $C(z)/A_*(z)$ is SSPR. Then, the entire family is made SSPR by $C(z)$ if and only if [15]*

$$\left|\mathcal{A}(e^{j\omega})/C(e^{j\omega}) - 1/(2K)\right| \neq 1/(2K) \tag{7.11}$$

for every $\omega \in [0, 2\pi)$.

Now that we have shown how the value set can help us in solving various analysis problems, we will devote the remainder of this section to discuss how these theorems can be particularized for some significant cases of practical value.

7.2.1 The Polytopic Case

We will assume now that the set **A** is a polytope.[4] Let $\{\mathbf{a}^k = [a_1^k, \cdots, a_{N_A}^k] \in \mathbf{R}^{N_A}\}$, $k = 1, \cdots, M$ be a set of generators of **A**. Then, the set $\mathcal{A}(e^{j\omega})$ is a convex polygon with generators

$$1 - \sum_{i=1}^{N_A} a_i^k e^{-j\omega i}, k = 1, \cdots, M \tag{7.12}$$

[4] A polytope $\mathbf{P} \in \mathbf{R}^n$ is the convex hull of a finite set of points $\{p^1, p^2, \cdots, p^m\}$ called generators. This set of generators does not need to be unique. A remarkable property of a polytope is that every point can be expressed as a convex combination of the generators, that is, there exist real nonnegative numbers $\lambda_1, \lambda_2, \cdots, \lambda_n$ such that for every $p \in \mathbf{P}$, $p = \sum_{i=1}^{n} \lambda_i p^i$. Note that a polygon is a polytope in $\mathbf{R}^2$.

From this result and Theorem 1 it is quite simple to see that in order to check the SPR property for a polytopic family of polynomials it is necessary and sufficient to check only the generators. The importance of this statement is perhaps recognized when we consider that we have reduced the analysis of an infinite member family to the analysis of a finite set. This advantage is more dramatic when it is used together with the results in [24] that show that the SPR property of a single polynomial can be checked with a finite number of operations. In the polytopic context it is generally said that an *extreme-point* or *vertex* result holds. However, in other cases with not so clear geometric connotations the term extreme-point is still used meaning a reduction from the original infinite to a finite test set.

A particularly simple yet illustrative type of polytope is a hyperrectangle.[5] This is quite a reasonable choice for the uncertainty description when the coefficients of the denominator are unknown but bounded from above and below and no information correlating them is available. Here, the set **A** can be described as follows

$$\mathbf{A} = \{[a_1, a_2, \cdots, a_{N_A}], a_i^- \leq a_i \leq a_i^+, i = 1, \cdots, N_A\} \tag{7.13}$$

It can be easily confirmed that a set of generators is described by the vectors **a** obtained by combining any of the two values a_i^- and a_i^+ for $i = 1, \cdots, N_A$. These generators are merely the vertices of the hyperrectangle just described. There are 2^{N_A} of such vertices, thus the initial optimism raised by a finite SPR test has to be softened by this exponential increase (with N_A) of the number of polynomials to check.

It is natural then to ask whether in the process of mapping the hyperrectangle in $\mathbf{R}^{N_A}$ into $\mathbf{C}$ there is a possible reduction of generators. The positive answer to this question comes from the fact that some of the hyperrectangle vertices are mapped into the interior of the resulting polygon, no matter which frequency ω we are considering; on the other hand, some of the vertices become generators of $\mathcal{A}(e^{j\omega})$ only for certain frequencies. It is not difficult to see that, for a fixed frequency ω, the minimal set of generators of $\mathcal{A}(e^{j\omega})$ has $2(N_A + 1)$ members although, once again, this set is frequency dependent. We refer the interested reader to [25] where details are given on how to obtain the polynomials belonging to the minimal set of generators as well as the cardinality of this set. In any case, with this idea, the reduction on the number of polynomials that have to be checked for SPRness is noteworthy. For instance, for $N_A = 10$ the reduction factor is about 50.

---

[5]In the robust stability literature, also known as an interval family of polynomials.

7.2.2 The l^p-Ball Case

In this section we will consider an uncertain set given by the following expression

$$\mathcal{A}(z) = \{A(z) = \sum_{i=0}^{M} p_i(z)q_i, \ \mathbf{q} = [q_0, q_1, \cdots, q_M] \in \mathbf{Q}\} \tag{7.14}$$

where $p_i(z)$, $i = 0, \cdots, M$ are real-coefficients polynomials of degree N_A and the set $\mathbf{Q}$ is an l^p-normed ball of parameters, i.e.,

$$\| \mathbf{q} \|_p = \sum_{i=0}^{M} \left| \frac{q_i - q_i^0}{\alpha_i} \right|^p \leq \gamma^p \tag{7.15}$$

where $1 \leq p < \infty$ (the case $p = \infty$ corresponds to the polytopic case and therefore we will not study it here) and all the weights α_i are positive. For convenience, we have given in (7.14) a new definition of the set $\mathcal{A}(z)$ in terms of $\mathbf{Q}$ instead of $\mathbf{A}$ used so far; nevertheless there is a linear mapping relating both sets.[6] Defining $p_c(z) = \sum_{i=0}^{M} p_i(z)q_i^0$ and $r_i = (q_i - q_i^0)/\alpha_i$, we can write (7.14) and (7.15) as follows

$$\mathcal{A}(z) = p_c(z) + \sum_{i=0}^{M} p_i(z)\alpha_i r_i, \tag{7.16}$$

$$\sum_{i=0}^{M} |r_i|^p \leq \gamma^p \tag{7.17}$$

It can be observed that the family description depends on the real parameter γ that controls the size of the ball and therefore can be regarded to as its radius. Assuming that $p_c(z)$ is SPR, once Problem 1 is solved for this family, it is immediate to compute the maximum value of γ that guarantees that the whole family is SPR. This value, called *SPRness margin*, is useful when $p_c(z)$ is considered to be a "nominal" value of $A(z)$, the remaining term in equation (7.16) is a structured perturbation and we want to measure how much can we "inflate" it such that the SPR property is preserved. This idea is even more important when Problem 1b is brought into consideration, i.e., when we want to compare two different compensators in terms of their robustness margin.

The solution to Problem 1 is simple with the value set concept: we have to compute the minimum of the real part of $\mathcal{A}(e^{j\omega})$ for every $\omega \in [0, 2\pi)$

---

[6] The coefficients of $p_i(z)$ can be tuned in order to include any information modifying the shape of the ball, such as a covariance matrix in the l^2 case.

and check whether this value is positive. Therefore, for a fixed ω we have to minimize

$$Re\{p_c(e^{j\omega})\} + \sum_{i=0}^{M} Re\{p_i(e^{j\omega})\}\alpha_i r_i \tag{7.18}$$

subject to (7.17). To solve the problem, the Lagrange multipliers technique is used in the same way as in [26] to yield the following.

Theorem 3 *The family of polynomials $\mathcal{A}(z)$ described in (7.14) is SPR if and only if*

$$Re\{p_c(e^{j\omega})\} > \gamma \left(\sum_{i=0}^{M} |Re\{p_i(e^{j\omega})\}\alpha_i|^{p/(p-1)} \right)^{(p-1)/p} \tag{7.19}$$

for every $\omega \in [0, 2\pi)$.

In order to solve Problem 2 for the l^p-ball family, using the results of Lemma 1, we have instead to check the robust stability of the set of polynomials. This can be done with the following result [26]:

Theorem 4 *The family of polynomials $\mathcal{A}(z)$ described in (7.14) is SPR-izable if and only if for every $\omega \in [0, 2\pi)$*

$$|p_c(e^{j\omega})| > \gamma \left(\sum_{i=0}^{M} |\alpha_i F_i(\phi) + \eta^* \alpha_i G_i(\phi)|^{p/(p-1)} \right)^{(p-1)/p} \tag{7.20}$$

where $\phi = arg(p_c(e^{j\omega}))$, $F_i(\phi) = Re\{p_i(e^{j\omega})e^{-j\phi}\}$, $G_i(\phi) = Im\{p_i(e^{j\omega})e^{-j\phi}\}$ and $\eta^* \in \mathbf{R}$ is the solution to the equation

$$\sum_{i=0}^{M} \alpha_i^{p/(p-1)} G_i(\phi)|F_i(\phi) + \eta G_i(\phi)|^{1/(p-1)} \mathrm{sgn}[F_i(\phi) + \eta G_i(\phi)] = 0 \tag{7.21}$$

The value-set tool can also be utilized when attempting to solve Problem 3. In this case, assuming that $p_c(z)$ is SSPR, we have to check whether the inverse of the value-set lies within the disk of radius $1/(2K)$ and center $1/(2K)$, according to Theorem 2. The next theorem can be used for this purpose.

Theorem 5 *The family of polynomials $\mathcal{A}(z)$ described in (7.14) is SSPR if and only if for every $\omega \in [0, 2\pi)$ and every $\phi \in [0, 2\pi)$*

$$\left| \left(p_c(e^{j\omega}) + e^{j\phi}\gamma \left(\sum_{i=0}^{M} |\alpha_i F_i(\phi) + \eta^* \alpha_i G_i(\phi)|^q \right)^{1/q} \right)^{-1} - 1/(2K) \right| \neq 1/(2K) \tag{7.22}$$

where $q = p/(p-1)$ and $\eta^* \in \mathbf{R}$ is the solution to equation (7.21).

An important comment is afforded by carefully examining the conditions required by Theorems 3-5. Robust SPR determination can be solved with a unique (ω) parameter search, robust SPR-izability needs in general a two-parameter search (ω and η) and, finally, robust SSPR analysis demands three parameters (ω, η and ϕ) even though the question whether the latter problem could be further simplified remains open. We also remark that in some particular (but most practical) cases, namely, $p = 1, 2$, the solution to (7.21) can be computed analytically thus allowing to drop this parameter and simplify the test. The algorithmic form of the above theorems has been barely investigated, but we refer the interested reader to [26] where some clues are given.

7.2.3 The Roots Space Case

In some practical cases the a priori information that we have concerning the uncertain set $\mathcal{A}(z)$ is expressed in the roots space rather than the coefficients space. More formally, we will deal with descriptions given in the following form

$$\mathcal{A}(z) = \{A(z); A(z_i) = 0, z_i \in \Omega_i,\ 1 = 1, \cdots, N_A\} \tag{7.23}$$

where $\Omega_i \in \mathbf{C}$ is a domain. Note that it is still possible to think of a set $\mathbf{A}$ in the coefficients space but its description is now very involved.[7] As a matter of fact, the computation of the value-set appears to be a very difficult task although many ideas have appeared in the past years that aim at reducing its computational complexity [22],[23]. To make things even worse, the value-set is not necessarily convex[8] although one may work with its convex hull at the cost of a certain degree of conservativeness. This non-convexity of the value-set precludes the use of some of the results we have presented above; however, we must take into account that the value-set consists of mostly "redundant" information that we can discard because we are interested in answering a very particular set of questions.

Consider for instance Problems 1a/1b and 2. It is clear that the SPR property can be expressed exclusively in terms of the maximum and minimum phases of $\mathcal{A}(e^{j\omega})$ provided that the family $\mathcal{A}(z)$ is minimum-phase.[9] If we make the assumption that the roots of $A(z)$ are independent,[10] we

[7]This type of dependence is called ***multilinear*** in the robust stability jargon. This term comes from the fact that if all the parameters but one are kept fixed, then we have a linear dependence.

[8]It can be more of a nightmare, since it is not necessarily simply connected, i.e., the value set may contain holes!, as it has been pointed out in [23].

[9]In this context this simply implies that Ω_i is contained in the open unit disk.

[10]Actually the roots have to be grouped pairwise if the domain Ω_l does not lie on the real axis, since we are considering real coefficients polynomials.

can concentrate on a single domain, say Ω_l, where a single root of $A(z)$ is known to lie, and then resort to the additive property of the phase to arrive at the final result.

Therefore, we restrict our attention to sets of second order polynomials of the form

$$\mathcal{A}_l(z) = \{A_l(z) = (1 - z_l z^{-1})(1 - z_l^* z^{-1}); z_l \in \Omega_l\} \tag{7.24}$$

so that we can write

$$\mathcal{A}(z) = \prod_{l=1}^{N_A/2} \mathcal{A}_l(z) \tag{7.25}$$

where the product of sets is defined in the usual way.

Let $\phi_l^+(\omega)$ and $\phi_l^-(\omega)$ denote, respectively, the maximum and minimum of the phase of $\mathcal{A}_l(e^{j\omega})$ for a fixed ω. Then, one important result [28] is that both $\phi_l^+(\omega)$ and $\phi_l^-(\omega)$ are attained at some $A_l(e^{j\omega})$ with roots on the boundary of Ω_l, so it is only this closed curve what is needed to be searched for the frequency-dependent phase extremes. This way, for a N_A-th degree family, with N_A even,[11] we have to compute $N_A/2$ phase extremes and add them up in order to solve Problems 1 and 2. Moreover, when the sets Ω_l happen to be identical, the problem is solved by finding the maximum and minimum phases and multiplying them by $N_A/2$.

When studying different reasonable domains Ω_l some interesting results were found. These basically imply the existence of extreme-point results that allow to avoid the frequency sweeping otherwise needed for computing the phase extremes. For instance, suppose that Ω_l is a disk of radius r and center $c_0 \in (-1, 1) \subset \mathbf{R}$. Then, the phase extremes are the following two members of the family: $(1 - (c_0 + r)z^{-1}))^2$ and $(1 - (c_0 - r)z^{-1}))^2$. Another example is Ω_l being a rectangle (of course, contained in the unit disk). In this case, the phase extremes are found at either one of the four corners of the rectangle.[12] Refer to [28] for more examples and technical details.

The SSPR (Problem 3) case seems at first sight to be more evasive because it is not possible to work with a single parameter in the same fashion that working with the phase reduces enormously the complexity of the

[11] A similar comment holds for N_A odd. The only difference is that at least one of the root sets has to lie necessarily on the real axis.

[12] Unfortunately, for the rectangular domain there is a combinatorial growth (with the degree of the family) of the number of polynomials that are needed to compute the phase extremes, although some frequency-dependent reduction techniques may be exploited.

robust SPR tests. However, by including a requirement on the maximum modulus of the value set it is possible a computational reduction similar to that given above for Problems 1 and 2. The result is summarized in the following [18].

Lemma 2 *Assume that there exists a finite subset*

$$\mathcal{B}(z) = \{A_l^k(z); A_l^k(z) \in \mathcal{A}_l(z), k = 1, \cdots, M\} \tag{7.26}$$

such that for any fixed $\omega \in [0, 2\pi)$ *the following two conditions hold*
i) $\max_{A_l(z) \in \mathcal{A}_l(z)} |A_l(e^{j\omega})| = |A_l^k(e^{j\omega})|$ *for some* k, $1 \leq i \leq M$.
ii)Either $\phi_l^+(\omega) = \arg(A_l^k(e^{j\omega}))$ *or* $\phi_l^-(\omega) = \arg(A_l^k(e^{j\omega}))$ *for the same* k *as in i).*
Then the set $\mathcal{A}_l(z)$ *is SSPR if and only if the subset* $\mathcal{B}(z)$ *is SSPR.*

See [18] for a proof. Note that with the previous lemma and some specific results it is possible to arrive at exactly the same extremal results as we have just discussed for the SPR case (sufficiency of the boundary of Ω_l and extreme-point results for disk and rectangular root sets).

7.3 Synthesis of LTI Filters for Robust SPR Problems

In this section we address the problem of designing a minimum-phase linear time invariant filter $C(z)$ such that it makes all the members in the family $\mathcal{A}(z)$ SPR. Of course, we assume that condition (7.6) is satisfied, as this can be checked beforehand with the results given in the previous section. Logically, if the answer to Problem 2 is negative for a given family $\mathcal{A}(z)$, the only loophole is to try to reduce the size of the uncertainty, since condition (7.6) is both necessary and sufficient.

As a starting point for the following discussion, we will assume that the bounding functions $\phi^+(\omega) = \max_{A(z) \in \mathcal{A}(z)} \arg(A(e^{j\omega}))$ and $\phi^-(\omega) = \min_{A(z) \in \mathcal{A}(z)} \arg(A(e^{j\omega}))$ are available for every $\omega \in [0, 2\pi)$, although we will see that some existing design methods will make use of a finite number of frequencies.

7.3.1 Algebraic Design for Two Plants

We will assume first that the family $\mathcal{A}(z)$ admits a two-member extreme-point result. This is for example the case of straight line segments in parameter space, disks in root space, and horizontal and vertical line segments in root space, as we discussed in [29]. Then, finding $C(z)$ that makes $\mathcal{A}(z)$ SPR is equivalent to find $C(z)$ that makes $A_1(z), A_2(z) \in \mathcal{A}(z)$ SPR. Even though this problem might look quite simple, we will see that this is not the case. Besides its practical interest, the simultaneous "SPR-ization" of a two-member family will hopefully provide a considerable insight to higher-order problems, in the same way as the simultaneous stabilization problem, now a classic in robust control, paved the way for further related results.

Prior to the development of the formalities required to solve algebraically the two-plants problem, it is reasonable to rule out some trivial but interesting cases. First, if the ratio $A_1(z)/A_2(z)$ is SPR[13] both $C(z) = A_1(z)$ and $C(z) = A_2(z)$ are obvious solutions. Furthermore, any convex combination $C(z) = \alpha A_1(z) + (1-\alpha)A_2(z)$, $0 < \alpha < 1$, is a valid solution.

Second, let $A_i(z)$, $i = 1, 2$, be of the form

$$A_i(z) = \prod_{k=1}^{M} (1 - z_{i,k} z^{-1})^{N_k} (1 - z_{i,k}^* z^{-1})^{N_k} \tag{7.27}$$

with N_k even for every $k = 1, \cdots, M$. Then,

$$C(z) = \prod_{k=1}^{M} (1 - z_{1,k} z^{-1})^{N_k/2} \cdot (1 - z_{1,k}^* z^{-1})^{N_k/2} \cdot (1 - z_{2,k} z^{-1})^{N_k/2} \cdot (1 - z_{2,k}^* z^{-1})^{N_k/2} \tag{7.28}$$

will make $A_1(z)$ and $A_2(z)$ SPR. The justification for this result comes from the fact that

$$\arg(C(e^{j\omega})) = \frac{\arg(A_1(e^{j\omega})) + \arg(A_2(e^{j\omega}))}{2} \tag{7.29}$$

so it is clear that the phase difference between $C(e^{j\omega})$ and any of $A_1(e^{j\omega})$ and $A_2(e^{j\omega})$ is less than $\pi/2$ in absolute value and therefore the choice for $C(z)$ makes $A_1(z)$ and $A_2(z)$ SPR as desired.

Note that while the first particular case we have just given is consequential for it will appear when the size of the phase uncertainty is small (less than $\pi/2$ radians), the second case is but a rare one.

Now, in order to synthesize an appropriate compensator, we need the following definitions:

$$P_1(z) = \frac{C(z)A_1(z^{-1}) + C(z^{-1})A_1(z)}{2} \tag{7.30}$$

[13] This is equivalent to the phase difference between $A_1(e^{j\omega})$ and $A_2(e^{j\omega})$ not exceeding $\pi/2$ in absolute value.

$$P_2(z) = \frac{C(z)A_2(z^{-1}) + C(z^{-1})A_2(z)}{2} \tag{7.31}$$

$$D(z) = \frac{A_1(z)A_2(z^{-1}) - A_1(z^{-1})A_2(z)}{2} \tag{7.32}$$

$$E(z) = \frac{A_1(z)A_2(z^{-1}) + A_1(z^{-1})A_2(z)}{2} \tag{7.33}$$

Condition (7.6) for the existence of $C(z)$ can be restated as follows [29]:

Theorem 6 *A causal and minimum-phase $C(z)$ making $A_1(z)$ and $A_2(z)$ SPR exists if and only if the polynomial $E(z)$ is real and positive at the roots of $D(z)$ on the unit circle.*

The construction of such a $C(z)$ can be done by finding two symmetric positive functions $P_1(z)$ and $P_2(z)$, that as can be noted from (7.30) and (7.31), share their sign with the real part of $C(z)/A_1(z)$ and $C(z)/A_2(z)$ respectively. With the definitions above we can obtain $C(z)$ as

$$C(z) = \frac{P_2(z)A_1(z) - P_1(z)A_2(z)}{D(z)} \tag{7.34}$$

with $P_1(z)$ and $P_2(z)$ symmetric positive functions. They must satisfy the following interpolation conditions in order to cancel the roots of the denominator:

$$P_2(\alpha_i)A_1(\alpha_i) - P_1(\alpha_i)A_2(\alpha_i) = 0 \tag{7.35}$$

with $\{\alpha_i\}$ roots of $D(z)$. The resulting filter $C(z)$ could be non minimum-phase (and non causal), depending on the degree of the interpolating functions $P_1(z)$ and $P_2(z)$. It can be made causal and minimum-phase by dividing it by a symmetric positive function containing the roots outside the unit circle and their reciprocals.

A recursive algorithm, based on the Youla-Saito interpolation algorithm of SPR functions in the Laplace transform domain, can be used to build $P_1(z)$ and $P_2(z)$ [29]. The following corollary indicates those cases at which a minimum-phase FIR $C(z)$ is obtained with this design algorithm:

Corollary 2 *If the number of roots of $D(z)$ on the unit circle is less or equal than four, then an FIR $C(z)$ can be computed.*

For instance, in the case of disks in root space mentioned above, $D(z)$ is a polynomial with only two roots on the unit circle, so an FIR compensator is obtained. It is worth saying that the previous corollary gives a sufficient condition, but not necessary, for the absence of poles of $C(z)$ in $\mathcal{C}\backslash\{0\}$.

7.3.2 Algebraic Design for Three or More Plants

The extension of the previous results to more than two plants involves the use of additional symmetric positive polynomials:

$$P_i(z) = \frac{C(z)A_i(z^{-1}) + C(z^{-1})A_i(z)}{2} \tag{7.36}$$

for $i = 1, \cdots, n$. If $C(z)$ is obtained as in (7.34), then $P_i(z)$ can be expressed as

$$P_i(z) = \frac{P_2(z)D_{1i}(z) - P_1(z)D_{2i}(z)}{D_{12}(z)} \tag{7.37}$$

for $i = 3, \cdots, n$, and with the new polynomials $D_{ij}(z)$ defined as

$$D_{ij}(z) = (A_i(z)A_j(z^{-1}) - A_i(z^{-1})A_j(z))/2 \tag{7.38}$$

Now, in addition to the interpolation conditions on $P_1(z)$ and $P_2(z)$ shown above, some additional constraints are imposed by the need of having $P_i(z)$, $i = 3, \cdots, n$, also SPR:

$$\frac{P_1(z)}{P_2(z)} = \frac{A_1(z)}{A_2(z)} \tag{7.39}$$

when $D_{12}(z) = 0$ on the unit circle and

$$\frac{P_1(z)}{P_2(z)} \neq \frac{D_{1i}(z)}{D_{2i}(z)} \tag{7.40}$$

at the rest of the unit circle, for $i = 3, \cdots, n$. Note that $\frac{D_{1i}(z)}{D_{2i}(z)} = \frac{A_1(z)}{A_2(z)}$ at the roots of $D_{12}(z)$ on the unit circle (except 1 and -1).

Thus, the general form of the filter which makes simultaneously SPR n plants is of the same form as that for two plants. But additional conditions (avoidance conditions) must be imposed when obtaining $P_1(z)$ and $P_2(z)$ in (7.34).

That problem is particularly interesting for $n = 4$ plants. In [13] it is shown that for a hyperrectangle of polynomials defined in the continuous-time case, the robust SPR problem reduces to the simultaneous SPRization of four polynomials (Kharitonov polynomials). And in [14], the continuous-time problem is shown to be equivalent to a discrete-time problem, by means of the bilinear transformation. Therefore, the solution of the interpolation-avoidance problem presented above for $n = 4$ would provide a solution for the continuous-time interval polynomials case.

7.3.3 Approximate Design Methods

Another approach to the design problem is to try to construct a minimum-phase filter $C(z)$ whose phase approximates the function

$$\phi(\omega) = \frac{\phi^+(\omega) + \phi^-(\omega)}{2} \tag{7.41}$$

The rationale behind this idea is as follows. Suppose there exists a rational, finite degree, stable and minimum-phase $C(z)$ such that $\arg(C(e^{j\omega})) = \phi(\omega)$, then this filter will make the whole set $\mathcal{A}(z)$ SPR, provided that the design problem is solvable, i.e., that Problem 2 has a positive answer. Nevertheless, when $\phi^+(\omega)$ and $\phi^-(\omega)$ do not correspond to the phase of any particular rational function, it is not possible to find a rational $C(z)$ with finite degree and phase exactly matching $\phi(\omega)$, so we have to resort to some kind of approximation.

With the assumption that $\phi(\omega)$ is piecewise continuous, Anderson et al. [14] have proposed the following synthesis method. By means of the Hilbert transform it is possible to find a function $v(z)$ analytic together with its inverse outside the open unit disk such that $\arg(v(z)) = \phi(\omega)$ when $|z| = 1$. However $v(z)$ will not be in general analytic on the unit circle. This can be overcome with a simple transformation $w(z) = v(\rho z)$, with $\rho > 1$, so now $w(z)$ is analytic together with its inverse outside the unit disk. Then it is possible to compute the Laurent series expansion of $w(z)$ and truncate it to the first $N + 1$ terms to obtain $w_N(z)$. With enough terms, $w_N(z)$ will inherit from $w(z)$ the property of having all its zeros inside the unit circle and will preserve the property of making the family SPR. See [14] for details on how the truncation error can be a priori chosen and a formal discussion of the algorithm given here.

An alternative set of algorithms was proposed in [27] and try to find the "best" FIR filter $C(z)$ with an a priori fixed degree N. The first algorithm obtains a weighted least squares approximation of $\phi(\omega)$ at $L > N$ given frequencies. If c_i, $i = 1, \cdots, N_C$ denote the coefficients of $C(z)$, we will have at every frequency ω_k, $k = 1, \cdots, L$

$$\tan\phi(\omega_k) + \epsilon_k = \frac{-\sum_{i=1}^{N_C} c_i \sin i\omega_k}{1 + \sum_{i=1}^{N_C} c_i \cos i\omega_k} \tag{7.42}$$

where ϵ_k accounts for the approximation error with weights α_k. Cross-multiplying in (7.42) and defining the weighted square error, $\sum_{i=1}^{L} \alpha_i \epsilon_i^2$, as objective function, we can solve the resulting problem by means of the pseudoinverse. However, with this approach the equation error is minimized but not the error in the phase. In order to do that, an error in the phase-approximation e_k is introduced, which can be related to ϵ_k as follows

$$\epsilon_k = (\cos\phi(\omega_k)X + \sin\phi(\omega_k)Y)\tan e_k \tag{7.43}$$

where X and Y are, respectively, the numerator and denominator of the right term of (7.42). By noting that the maximum value for e_k is bounded by $\Delta_k = \pi - (\phi^+(\omega_k) - \phi^-(\omega_k))$, we can use

$$\alpha_k = |(\cos\phi(\omega_k)X + \sin\phi(\omega_k)Y)\tan\Delta_k|^{-2} \tag{7.44}$$

as weights. Since these weights depend on $C(z)$, it is necessary to employ an iterative procedure using a tentative solution for $C(z)$, then adjust the weights, solve the least-squares problem and so on. The major difficulty found in the previous method is that the solution proposed for $C(z)$ is not necessarily minimum-phase. Then, the following lemma comes to clear the way,

Lemma 3 *$C(z)$ is minimum-phase if and only if* $\arg(C(1)) = \arg(C(-1))$

This is an immediate consequence of the ***principle of the argument*** [20].

With this result it is possible to monitor the phase of $C(e^{j\omega})$ at a chosen set of frequencies ω_k so that phase-minimality of $C(z)$ can be ascertained. The idea is to avoid "invisible" jumps of π radians that would make a non minimum-phase $C(z)$ satisfy the interpolation constraints. This concept was first introduced by Kaminsky and Djaferis [16] and has been recently applied [17] in the context of SPRness tests of single polynomials.

A different synthesis strategy is afforded by recognizing that we have a finite set of frequencies and a phase constraint for $C(z)$ at each, thus giving the basis for a minimax design. In this situation it is better to work with $C'(z) = z^{N_C}C(z)$, so the condition in Lemma 3 is modified to $\arg(C'(1)) = \arg(C'(-1)) + N_C\pi$. In addition, from the monotonously increasing behavior of the phase of a minimum-phase polynomial [20] it is possible to obtain a meaningful frequency selection procedure. If we write $\varphi^-(\omega) = \phi^+(\omega) - \pi/2 + N_C\omega$, $\varphi^+(\omega) = \phi^-(\omega) + \pi/2 + N_C\omega$, $\varphi_C(\omega) = argC(e^{j\omega})$, then the feasibility condition at each frequency is

$$\varphi^-(\omega_k) < \quad \varphi_C(\omega_k) \quad < \varphi^+(\omega_k) \tag{7.45}$$

or equivalently,

$$Im\{C'(e^{j\omega_k})e^{-j\varphi^+(\omega_k)}\} < 0; \; Im\{C'(e^{j\omega_k})e^{-j\varphi^-(\omega_k)}\} > 0 \tag{7.46}$$

that are two linear inequalities on the c_i. The total number of linear inequalities is then $2L$. They can be arranged so as to solve the problem by means of linear programming. Refer to [27] for full details. We simply note that with this method we have obtained the lowest degree compensators $C(z)$.

7.4 Experimental results

Consider the following second-order transfer function, proposed in [8]:

$$H(z) = \frac{1}{1 - 1.7z^{-1} + 0.7225z^{-2}} \tag{7.47}$$

that we desire to identify. The function has a pair or real poles at 0.85, and we assume an uncertainty circular region for the poles, centered at 0.6 at with radius 0.35. The adaptive algorithm used was SHARF (Simplified Adaptive Recursive Filter), with an adaptation constant of 0.001, a white zero mean input sequence of power 0.25 and $SNR = 20$dB. The adaptive filter was initialized to have its poles at 0.8, very close to the actual poles. Two simulations were made, with and without compensator. The compensator used was designed using the algorithm presented in section 7.3, to make the whole uncertainty region SPR. The results are shown in Figure 7.4. Despite the proximity of the initial values to the desired parameter set,

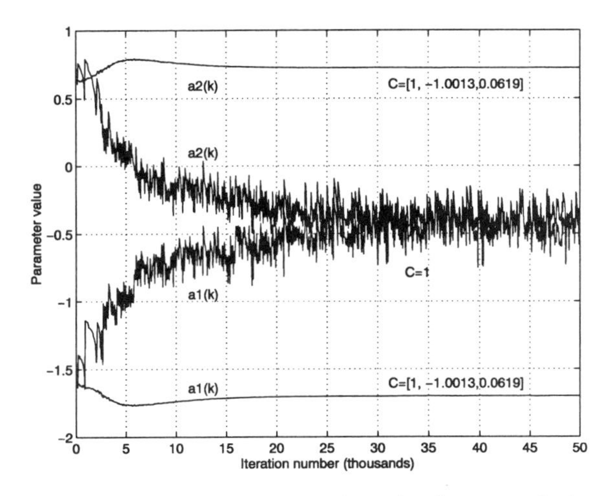

FIGURE 7.4. Parameter trajectories for example 1

the weights converged to a different configuration when SPRness was not satisfied (i.e.,$C(z) = 1$). With the appropriate compensator, the weights converged to their actual values as expected.

As our second example we present the solution of the robust SPR problem in the polytopic case (section 7.2) using the algorithm presented at the end of section 7.3. Consider the following set of interval stable polynomials of degree $N_A = 3$:

$$\mathcal{A}(z) = \{A(z) = 1 - \sum_{i=1}^{N_A} a_i z^{-i}, \mathbf{a} = [a_1, \cdots, a_{N_A}] \in \mathbf{A}\} \tag{7.48}$$

where the set $\mathbf{A}$ can be described as follows

$$\begin{aligned} \mathbf{A} = \{[a_1, a_2, a_3], 1.45 \le a_1 < 1.55, \\ -0.8 \le a_2 < -0.7, 0.1 \le a_3 < 0.13\} \end{aligned} \tag{7.49}$$

If we carry out the algorithm as described, we obtain the following compensator:

$$C(z) = 1 - 0.1213z^{-1} - 0.3339z^{-2} \tag{7.50}$$

which is shown in Figure 7.5 between its upper and lower bounds. Note the degree of the solution, lower than the degree of the plants belonging to the uncertainty set.

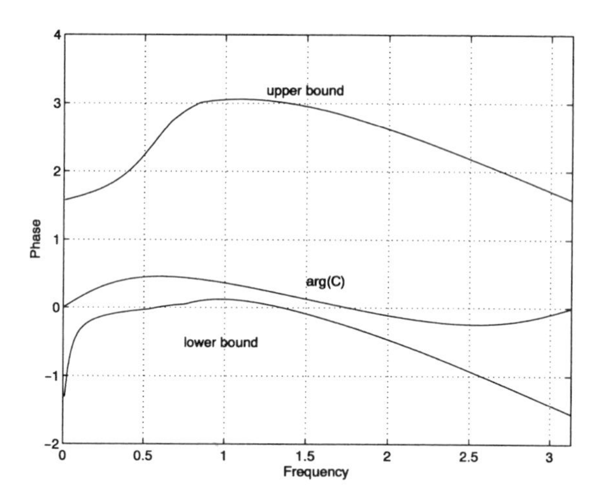

FIGURE 7.5. The feasibility region for C is contained between $\phi^-(\omega) + \pi/2$ and $\phi^+(\omega) - \pi/2$

7.5 Conclusions

The robust SPR property has been studied in depth. The motivation for its study comes from the adaptive IIR filtering field, which for a certain kind of schemes requires satisfaction of the SPR property in order to ensure global convergence. The plant under study is usually unknown, but a local knowledge is sometimes at hand. The study of the SPR property for different types of uncertainty has been made, although the case of uncertainty in the order of the denominator of the plant has not been tackled. In [10] it is shown how the algorithms based on hyperstability may not converge in the undermodelled case, when the order of the tentative solution is lower than the actual order. At present no algorithm is known with proven convergence to the actual set of parameters even in the undermodelled case. New avenues are open for research in this area. One possible alternative is to restrict the set of nonlinearities in the feedback loop (7.2), so that the conditions for convergence are less stringent, and to try to design new algorithms based on those nonlinearities. For instance, the classical Lur'e problem considers the stability of a fixed linear time invariant system perturbed by a family of nonlinear feedback gains: the conditions for convergence are less stringent than the conditions in the hyperstability case [21]. A synthesis study was also made for some interesting cases, showing how to obtain a suitable compensator guaranteeing global convergence. The results can be used for improving the convergence of several recursive algorithms used in identification and adaptive control.

7.6 References

[1] S. D. Stearns. Error surfaces of recursive adaptive filters. *IEEE Transactions on Circuits and Systems*, CAS-28:603–606, June 1981.

[2] I.D.Landau. Elimination of the real positivity condition in the design of parallel MRAS. *IEEE Transactions on Automatic Control*, 23:1015–1020, December 1978.

[3] A. Betser and E. Zeheb. Modified output error identification - elimination of the SPR condition. *IEEE Transactions on Automatic Control*, 40:190–193, January 1995.

[4] M. Tomizuka. Parallel MRAS without compensation block. *IEEE Transactions on Automatic Control*, 27:505–506, April 1982.

[5] C.R. Johnson Jr. Adaptive IIR filtering: Current results and open issues. *IEEE Transactions on Information Theory*, 30:237–250, March 1984.

[6] M. Nayeri. A weaker sufficient condition for the unimodality of error surfaces associated with exactly matching adaptive IIR filters. In *22nd Asilomar Conference on Signals, Systems and Computers*, pages 35–38, November 1988.

[7] C.R.Johnson Jr. A convergence proof for a hyperstable adaptive recursive filter. *IEEE Trans. on Information Theory*, 25:745–759, November 1979.

[8] M.G.Larimore, J.R. Treichler, and Jr. C.R.Johnson. SHARF: An algorithm for adapting IIR digital filters. *IEEE Trans. on Acoustics, Speech and Signal Processing*, 28:428–440, August 1980.

[9] I.D.Landau. Unbiased recursive identification using model reference adaptive techniques. *IEEE Trans. on Automatic Control*, 21:194–202, April 1976.

[10] P. A. Regalia. *Adaptive IIR Filtering in Signal Processing and Control.* Marcel Dekker, 1995.

[11] L. Ljung. On positive real transfer functions and the convergence of some recursive schemes. *IEEE Trans. on Automatic Control*, AC-22:539–551, August 1977.

[12] G.A. Williamson and S. Zimmermann. Globally convergent adaptive IIR filters based on fixed pole locations. *IEEE Trans. on Signal Processing*, 44:1418–1427, June 1996.

[13] S. Dasgupta and A. Bhagwat. Conditions for designing strictly positive real transfer functions for adaptive output error identification. *IEEE Trans. on Circuits and Systems*, 34:731–736, July 1987.

[14] B.D.O. Anderson, S. Dasgupta, P. Khargonekar, F.J. Kraus, and M.Mansour. Robust strict positive realness: Characterization and construction. *IEEE Trans. on Circuits and Systems*, 37:869–876, July 1990.

[15] B.D.O. Anderson and I.D. Landau. Least squares identification and the robust strict positive real property. *IEEE Transactions on Circuits and Systems I*, 41:601–607, September 1994.

[16] R.D. Kaminsky and T.E. Djaferis. The finite inclusions theorem. *IEEE Transactions on Automatic Control*, 40:549–551, March 1995.

[17] B.D.O. Anderson and and M. Mansour and F.J. Kraus. A New Test for Strict Positive Realness. *IEEE Transactions on Circuits and Systems I*, 42:226–229, April 1995.

[18] C. Mosquera and F. Pérez. On the Strengthened Robust SPR Problem for Discrete-Time Systems. Submitted, 1996.

[19] A. Tesi, A. Vicino, and G. Zappa. Design criteria for robust strict positive realness in adaptive schemes. *Automatica*, 30(4):643–654, 1994.

[20] B. R. Barmish. *New tools for robustness of linear systems.* Macmillan, 1994.

[21] S.P. Bhattacharyya, H. Chapellat, and L.H. Keel. *Robust Control: The Parametric Approach.* Prentice Hall, 1995.

[22] M. Mansour, S. Balemi, and W. Truöl Eds. *Robustness of Dynamic Systems with Parameter Uncertainty.* Birkhauser, 1992.

[23] B.R. Barmish, J.E. Ackermann, and H.Z. Hu. The tree structured decomposition: A new approach to robust stability analysis. In *Proc. Conference on Information Sciences and Systems, Johns Hopkins University, Baltimore, Md.*, 1989.

[24] D.D. Siljak. Polytopes of nonnegative polynomials. In *Proc. ACC'89*, pages 193-199, 1989.

[25] F.J.Kraus and M. Mansour. On Robust Stability of Discrete Systems. In *Proc. 29th. Conf. Dec. and Control, Honolulu, Hawaii*, pages 421-422, 1990.

[26] F. Pérez, C. Abdallah, and D. Docampo. Robustness analysis of polynomials with linearly correlated uncertain coefficients in l^p-normed balls. *Circuits, Systems and Signal Processing*, 15:543–554, 1996.

[27] F. Pérez and C. Abdallah. Filter Design to Guarantee Convergence of the Pseudolinear Regression IIR Adaptive Algorithm. In *Proc. Sixth IEEE Digital Signal Processing Workshop, Yosemite Natl. Park, CA*, pages 19–22, 1994.

[28] F. Pérez and C. Abdallah. Phase-convex arcs in root space and its application to robust SPR problems. In *Proc. 33rd IEEE Conf. Dec. and Control, Orlando, FL*, pages 3729–3730, 1994.

[29] F. Pérez and C. Mosquera. Characterization and algebraic solution to the extreme-point robust SPR problem. In *Proc. 13th World Congress IFAC, San Francisco, CA*, pages 391–396, 1996.

8

Boundary Methods for Distribution Analysis

José Luis Sancho
Batu Ulug
William Pierson
Aníbal R. Figueiras-Vidal
Stanley C. Ahalt

ABSTRACT In this chapter we introduce the use of *Boundary Methods* (BM) for distribution analysis. We view these methods as tools which can be used to extract useful information from sample distributions. We believe that Boundary Methods can be used for a number of applications, but here we restrict our attention to three applications. First, we discuss the use of boundary methods for determining the suitability of a particular feature set for pattern classification, i.e. we use the Boundary Methods to perform *feature-set evaluation* (FSE). We present results which establish the correspondence of Boundary Methods and the probability of error (Pe) for normal distributions. Second, we discuss the utility of Boundary Methods as a technique for *sample-pruning* (SP), and show how we can select samples, e.g., for progressive training of neural-networks. Finally, we state a theorem which relates *Fisher's Linear Discriminant* (FLD) and Boundary Methods.

8.1 Introduction

For many investigations of physical processes, scientists and engineers must use samples drawn from the process in order to construct algorithms which model or monitor the underlying process. For example, in the telecommunications industry applications such as equalization (e.g. echo-cancelation), source-coding (e.g. video-coding using vector quantization), and detection (e.g. CDMA decorrelators) require that samples of transmitted signals be analyzed to formulate appropriate signal processing algorithms. For problems such as these we believe that the distribution analysis methods we describe will offer significant advantages in designing and fielding robust and efficient algorithms, particularly those in which classification plays a dominant role in the processing.

Most pattern classification systems have a typical set of components which is shown in Fig.8.1.

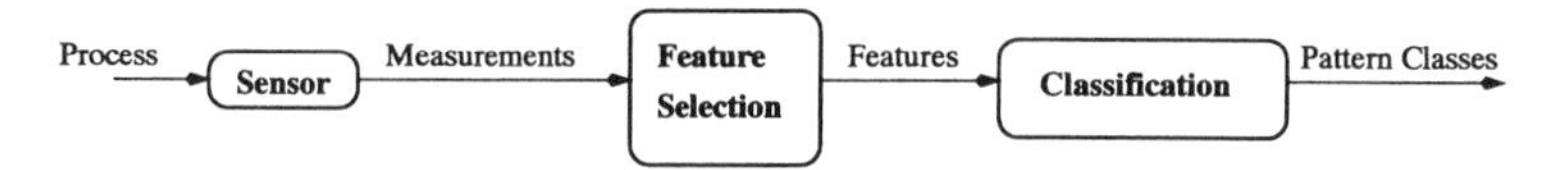

FIGURE 8.1. Classification system components

The *sensor* monitors the process and provides samples suitable for machine processing. These samples can represent many kinds of signals, e.g. seismographic signals, radar returns, speech waveform, images, etc.

The samples may be of high dimensionality and in order to manage the complexity of the classifier it is common for the samples to be passed to a *feature extractor* (also called a property filter or an attribute detector) whose purpose is to reduce the dimensionality of the data while retaining certain "properties" or "features" of the signals that are suitable for obtaining a solution to the classification problem. Thus, each input signal of m-dimensionality, becomes a n-dimensional signal, where $m > n$, and n is the number of features extracted. This new n-dimensional signal can be viewed as an n-dimensional vector called the *feature vector* in an n-dimensional feature space. This vector is then passed to a *classifier* that evaluates the features and makes a final decision by associating a class label with each feature vector. One significant problem that has been investigated for some time is the choice of an optimal set of feature vectors with a) minimum dimensionality and b) which permits an easy solution to the classification problem [1].

After features are selected from the data, the classification problem can be viewed as constructing a partition in the feature space, i.e., dividing the feature space into regions. Each partition is associated with a particular class, and, ideally, each partition contains only feature vectors from one class. Of course, in realistic problems each partition may contain samples that are drawn from a number of classes, and we can view this as the existence of an overlapping partition, or region, in the feature space. In many practical and interesting classification problems the overlapping region exists because of the statistical behavior of the total process. This suggests that we should look for a classification procedure that minimizes the probability of error in a statistical way. Thus, if we know the optimal dimensionality of the feature space, the remaining problem is to find the optimal features which allows one to find the optimal partition to solve the problem. Therefore, it will be important to have available some tool that evaluates the optimality of different features sets, and we will call this task *feature-set evaluation* (FSE).

Note that an ideal classifier would not require the use of a feature extractor. Conversely, an ideal feature extractor would produce features that could be trivially partitioned, say using linear discrimination boundaries. However, in practice, the problem of feature extraction is much more problem dependent than the problem of classification. This means that a feature extractor designed to solve one problem may be of little use in solving a different problem. Subsequently, the role of FSE techniques are quite important.

8.1.1 Building a Classifier System

If an investigator has a reasonably complete understanding of the physical process, a mathematical model of the process can be constructed and the samples can be used to estimate the parameters of the model. If the number of samples are sufficient in number with respect to the dimensionality and the statistics of the problem, then the needed pdf's can be estimated, and an optimal Bayesian classifier can be constructed [2, 3].

However, in many practical situations, there are problems with this approach. First, constructing a model can be time consuming, and verification of the model can be problematic. Second, as the dimensionality of the data increases, exponentially larger numbers of samples are required to accurately estimate the class conditional probabilities. It is often either physically impossible or financially prohibitive to obtain the needed data. Thus, accurate estimates of the probability density functions can not be obtained – which implies that the Bayes error cannot be accurately estimated. Third, determining the estimates for the prior probabilities becomes especially difficult when the number of classes is large – which is common for many practical applications. One approximation is to assume a uniform distribution for the prior class probabilities [2] which simplifies the analysis. However, for optimal performance, the class probabilities need to be estimated accurately in order to apply Bayesian analysis. Consequently, for these reasons, investigators must turn to other alternatives for many practical problems.

For those cases in which the process cannot be readily modeled, either supervised or unsupervised learning can be employed. In either case, the learning process can be viewed as distribution analysis. If only unlabeled data is available, e.g. when the number of classes is unknown, unsupervised learning techniques, usually based on clustering, are used to discover a model which captures the structure of the data in the data-space. Clusters thus formed can be evaluated, e.g. using Indices of Partitional Validity (IPVs) [4, 5], in an attempt to measure how well the clusters capture the structure of the data. Usually these indices use some combination of measures which quantify the compactness and isolation of each of the discovered clusters. While these methods have proven to be useful [6, 5], they require the use of an explicit distance metric. The choice of this distance metric can have a significant impact on the reliability or utility of the analysis.

When labeled data is available, as we assume here, it is standard practice to use mixture decomposition techniques to allocate each pattern to a particular cluster, and then estimate the cluster parameters. These techniques generally require that the number of clusters be known and adopt a density model which assumes that the clusters are multivariate normal. Mixture decomposition techniques then focus on 1) assigning each data sample to the correct cluster, and 2) estimating the mean and covariance matrices of each cluster. A particularly complete discussion of these techniques can be found in [5].

8.2 Motivation

Our research is focused on determining techniques we believe are of significant importance to designers of pattern classifiers. For this paper we restrict ourselves to two techniques:

- *Feature-Set Evaluation* (FSE) techniques which, given alternative ways of deriving feature sets from observations, order those sets by classification - fitness. Of course, this measure of fitness should be related to Pe or other pertinent measures.
- *Sample-Pruning* (SP) techniques which support the development of classifiers which are:
  - quickly constructed,
  - execution-efficient,
  - generalized, and
  - robust.

We observe that an ideal measure of "classifiabilty" would analyze all data contained in the sample population and yield one absolute value to denote how separable the two classes are - regardless of the feature set. We take a more relaxed approach in which an FSE analyzes at least two sample populations and yields two values which can be compared to determine which of the two populations consist of better features.

In contrast, SP techniques operate on one population and yield another population which is always a subset of the original population. We observe that a) in order to quickly construct a classifier, we need to minimize the use of samples that provide little useful classification information, b) classifiers that are execution-efficient are those constructed such that a small number of parameters need to be evaluated in order to reach a classification decision, c) generalized classifiers reliably estimate the classification mapping using noisy training samples, and d) robust classifiers reliably estimate the classification mapping when perturbation processes (e.g., noise, obscuration, etc.) imposed on the training samples differ from the perturbation processes which affects the testing samples.

While we do not claim to have solved any of the above problems, we believe the distribution analysis technique we discuss here, Boundary Methods, does have a significant benefit in meeting the objectives of FSE and SP.

We note that the statistics community has an extensive literature on the use of linear methods, particularly principal component analysis. However, we are not aware of more directly related FSE work in the statistics community. The pattern recognition community has investigated upper bounds on the Bayes error for two distributions using both Chernoff and Bhattacharyya bounds, as well as by using asymptotic nearest neighbor error (see, e.g., Sec. 3.4 of [2]. It is not clear from the literature how well each of these bounds perform for distributions other than those that belong to the family of gamma distributions. Similar work in the Neural Networks and pattern recognition communities typically rely on identification of individual features. An excellent review of some of these techniques for identification of discriminately redundant features and discriminately informative features can be found in [7].

8.3 Boundary Methods as Feature-Set Evaluation

Boundary Methods exploit a collection of distributions in a controlled way in order to extract useful information about how the distributions are composed relative to each other. Suppose we have a hypothetical population consisting of two distributions drawn from two classes, as shown in Fig.8.2(a).

We enclose the classes with a boundary, according to some criteria. In the case shown in Fig.8.2(b) we have drawn the enclosing boundaries as closed interpolated splines with approximately 15 control knots. Generally, some criteria is used in forming the boundaries so that obvious outliers are excluded and relatively compact boundaries are formed.

While the boundaries shown are quite complicated, it is reasonably easy to specify and manipulate complicated boundaries using methods such as interpolated splines. Indeed, the boundaries can be determined in any way that is computationally feasible, and can be of arbitrary shape - with suitable modifications to the basic algorithm. In the examples presented later we use elliptical boundaries since they are very simple to specify and manipulate.

We then collapse the boundaries until the boundaries are just touching. For complicated boundaries, the amount of shrinking necessary to effect tangential, or "just-touching" boundaries can be relatively difficult calculate and there can be multiple possible solutions. However, for many boundaries - such as the one shown - a straight-forward search process over the scale of the boundary, e.g. about the center point, can be used

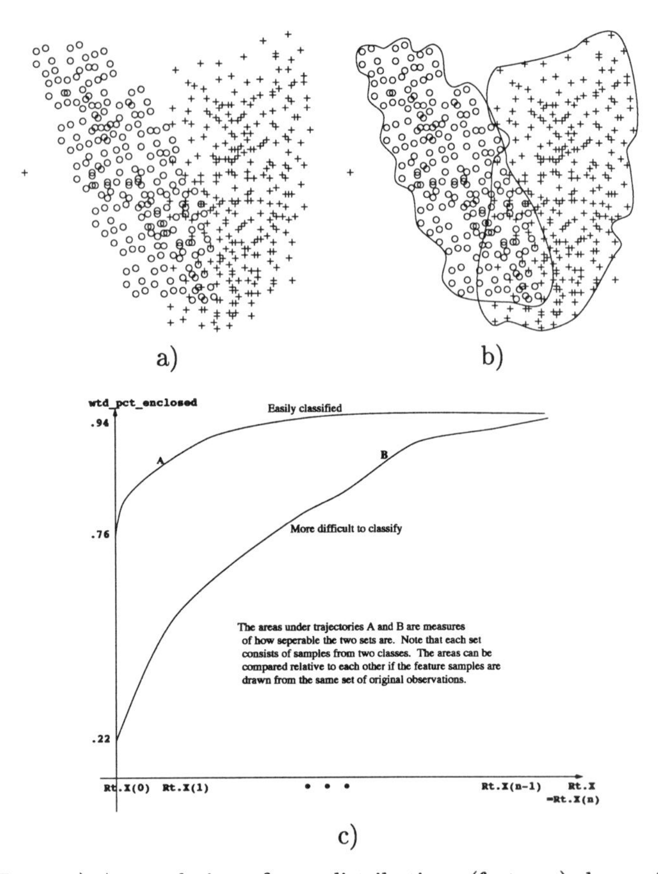

FIGURE 8.2. a) A population of two distributions (features) drawn from two classes; b) Distributions enclosed by arbitrary boundaries; c) Trajectory of samples enclosed by increasing boundaries.

to establish the tangential boundaries. In any case we always establish a canonical starting point at which the boundaries enclose most of the samples from each class, but without overlap of the volumes enclosed within the boundaries.

We choose to use tangential, or zero-volume-overlap, boundaries for the following reasons. First, shrinking the boundaries until they are tangential establishes a specific point to begin our calculations and allows us to normalize our results, as explained later. Second, the tangential boundaries enclose subsets of the class samples that, we believe, most reliably represent the class distributions. Finally, as briefly discussed later, tangential elliptical boundaries have a direct relationship to *Fisher's Lineal Discriminant* analysis (Fisher's LDA) projection axis.

At this point we have established the two values to be used as the endpoints in our final calculations: the original enclosure size, labeled Rt.X(n);

and a minimum, non-overlapping size, labeled Rt.X(0).

We now begin to grow the boundaries. We grow the boundaries gradually for a number of steps, say n, that is sufficient to obtain the desired trajectory, as described below. As the boundaries grow, the number of samples that are enclosed from each class increases, and the number of samples that are enclosed in the region common to both boundaries increases. We keep track, via a count, of how the number of samples in either the overlap region, or within both boundaries, increases, regardless of class. We refer to the area under the trajectory we form as the *Trajectory Area* (TA). We observe, however, that we can weight the samples such that the closer the samples are to the boundary the more they contribute to the count.

Once we have expanded the boundaries out to their original position, we have captured, in the count, a measure of how the samples of the two classes are distributed in the space. We now plot the measured count as a function of the boundary volume to obtain a trajectory. In Fig.8.2(c) we have plotted two hypothetical trajectories for two different populations of distributions (features) of two classes.

We claim that areas under these trajectories is a measure of how the samples are distributed in the following sense. The areas for two different sample populations can be quantitatively compared to one another, and the relative ordering of the areas is invariant with respect to linear transformations applied to the data distributions. We are currently working on analytic verification that a) the areas under the trajectories are proportional to the probability of error (for normal distributions, and b) the calculated areas are proximity indices for the distributions.

For population A the samples that are enclosed in the minimal-boundary, Rt.X(0) are relatively compact (76% are enclosed when the ellipses are tangential) and the number of enclosed samples quickly asymptotes to the number of samples enclosed by the final (i.e. the original) boundary, thus the classes are, to some degree, isolated.

In contrast, population B has a trajectory that indicates that the distributions are not very compact (only 22% are enclosed by the tangential boundaries) and the trajectory indicates that the two classes are fairly interspersed, as the trajectory climbs slowly to the final value - hence the classes are not well isolated.

By comparing the areas (TA) under the two trajectories we have a qualitative measure of how good the two distributions are for classification. Population A, which has the bigger area, will be more easily classified than Population B. We have, effectively, a method for FSE.

8.3.1 Results

In this section we show two simple experimental results using Boundary Methods. The first is shown in Fig.8.3, where we have selected two test populations in which one population is more separable than the other.

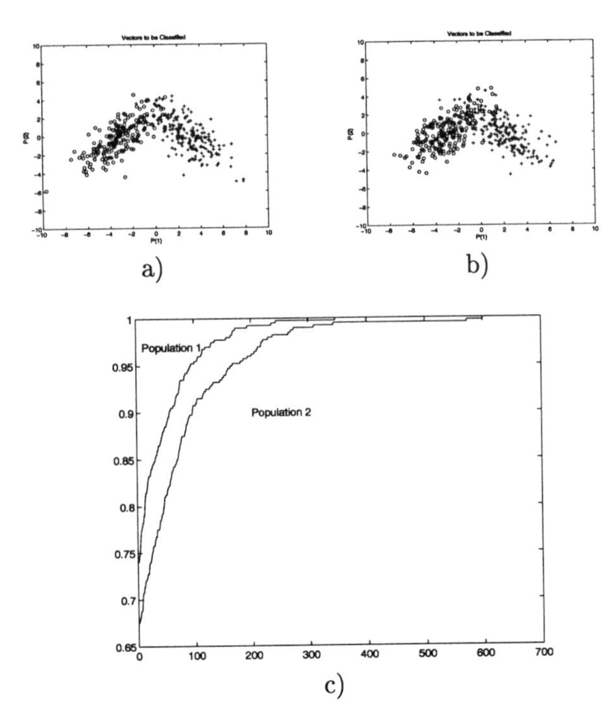

FIGURE 8.3. Figures a) and b) show two test populations (200 samples of each class) with similar separability; Figure c) shows the trajectory areas of populations a) and b). We can see they have very similar trajectories, but that the relative areas are discriminable.

For this population we have used elliptical boundaries, so we call this method an *Elliptic Boundary Method*. Here we calculate the Trajectory area (TA) using the counts of samples within the overlapping boundaries. The initial boundary for each class is formed by an ellipse, where the size of the ellipse is determined using a Chi-square test and fixing the percentages of enclosed samples. Since we know that the data consists of two classes, we have simply estimated the means and covariances directly. If the data was not unimodal we could either use a single boundary (which would result in a different trajectory) or optionally employ unsupervised techniques to determine the number of clusters-per-class to use when estimating these parameters. Additional tests are now underway for multi-modal and multi-class problems.

The use of ellipses is attractive for many reasons. First, the assumption is reasonably satisfied for many realistic cases [7]. Second, the assumption that the data is distributed normally simplifies analysis, giving rise to ellipsoidal boundaries because Gaussian distributions have elliptical constant-density contours. Quadratic forms such as ellipses lend themselves to formal analysis and are closely tied to Bayesian formalisms. Third, ellipses are computationally attractive because they are easily manipulated as they can be specified with a mean, a covariance (matrix), and a volume. Fourth, for ellipses, the amount of shrinking necessary to effect tangential, or "just-touching" boundaries is relatively easy to calculate (although there are multiple possible solutions).

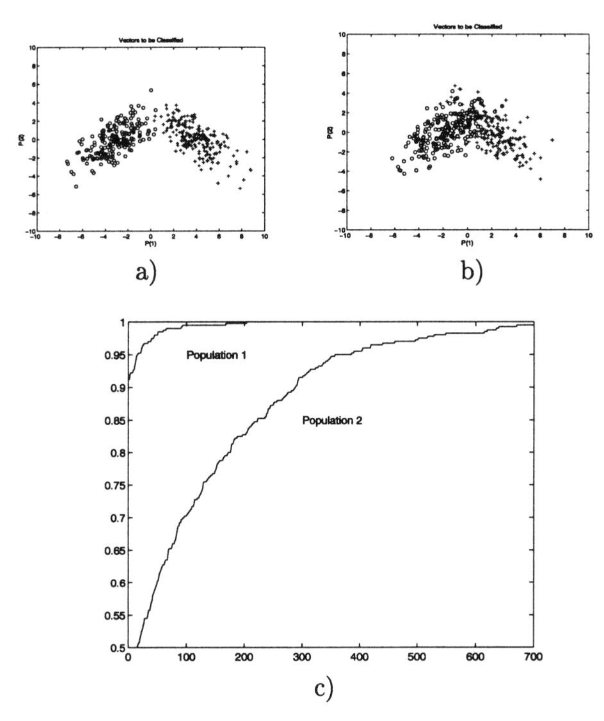

FIGURE 8.4. The Figures a) and b) show two test populations (200 samples of each class) with more widely separated means than the Figure 3; the Figure c) shows the trajectory areas of populations a) and b). Note that the difference in the areas is greater in this second example because of the more widely separated means and differing covariances.

There are a number of ways the ellipses can be collapsed. We typically collapse each boundary so that the constant-density contours of each class, which fixes the volume of each ellipse, are kept equal. A simple search procedure will yield this solution quickly, since only one parameter needs to be varied. However, an alternative is to vary the density contour of each of the class distributions separately in order to determine tangential ellipses with other properties. For example, it is possible to find tangential ellipses which have equal magnitude gradients at the tangent point, and that gradient is equivalent to Fisher's Linear Discriminant projection vector. See the appendix for the proof of this claim.

Our first example, shown in Fig.8.3, demonstrates how the areas under the trajectories (TA) correctly indicate that the first test population is more readily separated that the second.

Another example of the technique is shown in Fig.8.4. For this case the means of the distributions are more widely separated in the data space. Note the trajectories of this case, shown in Fig.8.4(c). Their associated trajectory areas correctly indicate that the first test population is more readily separated that the second and that the differences in the separability between the two test populations are more pronounced than in our first example. This is visually apparent in looking at the data.

As can be seen, our preliminary tests indicate that the Boundary Method works well for these simple distributions. We are now working on more extensive tests, as well as more formal analysis.

The second experiment demonstrates the correlation between the trajectory (TA) area and the probability of error (Pe). Fig.8.5 shows the correlation between TA and the Pe for a number of distributions with different means and covariance matrices. For this experiment, different scenarios were used to vary 1) the distance between the means of the distributions, 2) the relative orientations of the major axes of the two distributions, and to 3) simultaneously vary the means and the covariances of the distributions. We show two equivalent simulations differing only in that Fig.8.5(a) uses the actual mean and covariance matrices, while Fig.8.5(b) uses estimated means and covariance matrices. As can be seen from these figures, the TA and the Pe is highly correlated.

8.3.2 Feature Set Evaluation using Boundary Methods: Summary

We have presented a new method for Feature Set Evaluation (FSE) which provides information useful in determining how separable one feature set population is with respect to others drawn from the same source. Using Boundary Methods we form arbitrary boundaries to investigate various separating surfaces. The relationship among the classes is captured in a number called the Trajectory Area (TA). We have given results of simulations using Gaussian distributions and elliptical boundaries which show that the Boundary Method-TA has a correlation factor of near one with the Pe.

8.4 Boundary Methods as a Sample-Pruning (SP) Mechanism

In this section we show how Boundary Methods may be used to improve training of neural networks designed to operate as nonlinear classifiers. In particular, we present a classification problem which compares the efficiency of the backpropagation (BP) algorithm (working in block mode) and the BP algorithm combined with a sample selection mechanism which is controlled by the elliptic boundary method (EBM Method). We demonstrate the advantage of this second method to determine a more optimal border.

The BP algorithm is a widely used algorithm employed to train neural networks (NNs). The popularity of BP is due to its ease of implementation and because it can be shown to solve a range of problems [8]. It is a gradient-descent algorithm that changes the weights of the network in order to minimizes a global error associated with the training set. The training set consists of samples which are representative of the problem we want to solve and that are presented to the NN to change its weights in the opposite direction to the gradient error. A more comprehensive study of the NNs and the BP algorithm can be found in [9, 10].

In many practical problems the samples to be classified are perturbed by noise due electronic components, means of transmission, etc. Not uncommonly, the noise that is present in the training samples has a significant impact on a) those samples that fall in the intersection (confusable) regions of different classes, as well as b) causing an overall spread in the samples, which emphasizes the impact of outliers during training. In both cases, these effects of the noise requires that the classifier find a set of weights that instantiates a more complex border that might otherwise be required. There exist many different mechanisms of selection samples in order to improve the efficiency of the training according to some criterium of relevance and usefulness of the samples. Some of the alternative techniques which have been presented can be found in [11, 12].

For this example we compare the classification border derived using 1) an MLP trained with the BP algorithm operating in batch mode with 2) an MLP trained using BP, but with samples progressively defined using elliptic boundaries. For simplicity we refer to these two classifiers as BP and EBM-BP respectively. For the first case all available samples form the training set while in the second case each class has an associated ellipse and only the samples inside the corresponding ellipse belong to the training set at time t during the training process.

The ellipses are expanded from the initial (tangential) configuration in which only a subset of high-confidence samples are enclosed, to the final configuration in which the enclosed samples comprise a high percentage (around 95%) of the total samples. This method of sample selection has two important benefits. First, outliers have a significantly diminished impact on training the classification border. Second, we solve the problem in a progressive fashion, starting with a problem that is as separable as possible (just touching situation), and then gradually inducing a more complex border as additional samples are taken into account.

8.4.1 Description of the simulations

The problem consists of two classes, both uniformly distributed inside a two dimension rectangular region. Fig.8.6 shows the training samples of the two classes and the ideal border. Note that we have specifically devised a problem in which a "spur" in the data requires the formation of a relatively complex set of boundaries. The total number of samples in the training set is 600. Another set of 6000 samples is used for testing. Both algorithms use the same training and test sets, initial weights, and both have a learning rate parameter of 0.005. Also, no noise is added during training or testing.

We will use two different procedures for the EBM-BP algorithm. The first procedure is to establish a fixed number of epochs, say 100, for each value of the Rt parameter which defines the size of the ellipses (remember that an ellipse is defined as a quadratic form where the value of the constant term defines the size or volume of the ellipse). The second procedure uses the error of the test set to control the number of epochs that are used during training, when the slope of this error is near zero, the Rt parameter is increased. For both procedures the value of the Rt increment is fixed at 0.1% of the previous value. In the second procedure we check the variation of the error produced using the test set every 25 epochs.

8.4.2 Results

We performed ten simulations for each one of the two Elliptic Boundary Method procedures described, and ten simulations of the BP algorithm.

Simulations with the standard BP algorithm highlight the difficulty of obtaining a good decision boundary.

Fig.8.7 shows a sequence of evolving boundaries obtained during training for one simulation and the corresponding error. Effectively, standard BP becomes trapped in a local minima for this example. This phenomena occurred for all of our 10 simulations, regardless of the initial conditions.

Simulations with the first EBM-BP method demonstrate that, using SP, we can achieve the optimal decision boundary. While the BP algorithm never obtains the desired border, the EBM-BP does converge to the desired boundary in approximately 80% of all of the simulations.

Fig.8.8 shows the sequence of boundaries obtained during training and the corresponding error.

The computational cost for the first method of EBM-BP, which is substantial, can be largely reduced if we control the growing of ellipses as a function of the slope of the error of the test set. This control is included in the second EBM-BP method.

Fig.8.9 shows the sequence of the boundaries obtained during the training and the error.

8.4.3 Sample Pruning using Boundary Methods: Summary

We have demonstrated the utility of the EBM-BP to train neural networks to achieve decision boundaries which are clearly better that those achieved using classical BP. When a BP algorithm is used, it seeks a good local minimum of an complex error surface. When the EBM-BP algorithm is used the complexity surface of the problem is modified at each step and the local minimums corresponding to the current error surface are located. In the first step (tangential boundaries and thus a relatively separable problem) the error surface is simpler than when all training samples are used and training converges quickly. In subsequent steps additional pattern samples are included in the training set, which modifies the error surface. In this fashion the network can more readily converge to a good local minimum.

8.5 Boundary Methods as Fisher's Linear Discriminant (FLD)

An interesting question about Boundary Methods employing elliptical boundaries is the following. Can the tangent line specified by the just-touching ellipses be obtained by reducing the ellipses in some specific way so that an optimum linear discriminant is determined according to some specific criterion of separability?

One of the most widely used criterion is that given by Fisher [1], and here we describe the relationship between Fisher's Linear Discriminant and tangential ellipses. More formal results of these discussions are given in the appendix.

Theorem:
a) For any two ellipses whose covariance matrices are positive definite is always possible to find a tangent point equivalent to the FLD solution (*Equivalent Tangent Point, ETP*). Here "equivalent" means that the tangent line to both ellipses at this point is perpendicular to the vector given by the FLD solution. If the ellipses do not have any common axis there exists a unique ETP; on the contrary, for ellipses which have a common axis there exists an infinite number of ETPs.
b) For any two ellipses whose covariance matrices are positive definite there exists a unique ETP, (x,y), which verify $\nabla_{\mathbf{1}}(x, y) = -\nabla_{\mathbf{2}}(x, y)$.

Therefore, it is possible to analytically calculate a canonical starting point (i.e. tangential boundaries) as mentioned in Section 3. This approach yields a solution without the need for searching for the canonical starting point iteratively. However, in many case it seems intuitively appealing to find the tangential boundaries in such a way that constant-density contours are maintained for each distribution independently, which is easily accomplished using an iterative search, i.e., collapsing each of the class boundaries a small amount and then checking to determining if there are samples in any overlap region which are mis-classified. We use the latter technique in our experiments, but acknowledge that further investigation of these alternatives is needed.

8.6 Conclusions

We have introduced the use of *Boundary Methods* for distribution analysis. Boundary Methods are tools which can be used to extract useful information from sample distributions, and we demonstrated the use of the methods for Features Set Evaluation and Sample Pruning for progressive classifier construction. We also demonstrated the relationship between Boundary Methods and Fisher's Linear Discriminant analysis.

We are currently investigating boundary methods on more complex distributions, including multi-mode and multi-class data. Since Boundary Methods allow us to both quantitatively analyze distributions *and* identify sample subsets that are of particular importance with respect to classification, there are a number of other applications of Boundary Methods that we are currently investigating. These include Design of Experiments, Data Fusion, and Sample Dimension Pruning.

Acknowledgments: Partial support for this work was provided to Stan Ahalt by the AMOS Research consortium. Graduate student support was provided to Bill Pierson through the DoD Palace Knight Program. Stan Ahalt also gratefully acknowledges the support of the Dirección General de Investigación Cientifica y Técnica, Ministerio de Educacion y Ciencia of España while working at the Universidad Politécnica de Madrid.

8.7 Apendix: Proof of the Theorem Relating FLD and Boundary Methods

8.7.1 Assumptions and Definitions

We adopt the following nomenclature: **scalars** are given in normal letters, e. g., a; **vectors** are given in bold lower letters, e.g. $\mathbf{x}$; **matrix** are given in bold capital letters, e. g., $\mathbf{X}$.

We assume a population of two multivariate Gaussian distributions. Each distribution is defined by a covariance matrix, $\mathbf{\Sigma_j}$, and a mean vector, $\mathbf{m_j}$, estimated from the same number of the samples, $N_1 = N_2 = N/2$, where N is the total number of samples.

8.7.2 Fisher's Linear Discriminant (FLD) Analysis

Suppose we have a population of two multivariate Gaussian distributions define by

$$\mathbf{\Sigma_j} = \begin{pmatrix} a'_j & b'_j \\ b'_j & c'_j \end{pmatrix} \quad ; \quad \mathbf{m_j} = \begin{pmatrix} \overline{x}_j \\ \overline{y}_j \end{pmatrix} \qquad j = 1,2 \tag{8.1}$$

then

$$\mathbf{\Sigma_j}^{-1} = \begin{pmatrix} a_j & b_j \\ b_j & c_j \end{pmatrix} = \frac{1}{\Delta'_j} \begin{pmatrix} c'_j & -b'_j \\ -b'_j & a'_j \end{pmatrix} \qquad j = 1,2 \tag{8.2}$$

$$\text{where} \qquad \Delta'_j = \det \mathbf{\Sigma_j} \tag{8.3}$$

and

$$a'_j = \frac{c_j}{\Delta_j} \quad b'_j = -\frac{b_j}{\Delta_j} \quad c'_j = \frac{a_j}{\Delta_j} \tag{8.4}$$

$$\text{where} \qquad \Delta_j = \det \mathbf{\Sigma_j}^{-1} \tag{8.5}$$

The FLD Analysis provides us an optimal straight line, defined by the vector $\mathbf{w}$, such that the projection of the all samples along this straight line reduces the dimensionality of the problem and we obtain an optimal projection for the classification [1]. This solution is given by

$$\mathbf{w} = \mathbf{S_w}^{-1}(\mathbf{m_1} - \mathbf{m_2}) \tag{8.6}$$

The matrix $\mathbf{S_w}$ is called the *within-class scatter matrix* and it is given by

$$\mathbf{S_w} = \sum_{j=1}^{2} \mathbf{S_j} \tag{8.7}$$

The matrices $\mathbf{S_j}$ are *scatter matrices* and they are given by

$$\mathbf{S_j} = \sum_{\mathbf{x} \in C_j} (\mathbf{x} - \mathbf{m_j})(\mathbf{x} - \mathbf{m_j})^T \tag{8.8}$$

where every vector $\mathbf{x} = (x, y)$ is a sample of the population, and C_j indicates the class j with $j = 1, 2$. Then we have

$$\begin{aligned} \mathbf{S_w} &= \sum_{j=1}^{2} \mathbf{S_j} = \sum_{j=1}^{2} \sum_{\mathbf{x} \in C_j} \begin{pmatrix} (x - \overline{x}_j)^2 & (x - \overline{x}_j)(y - \overline{y}_j) \\ (y - \overline{y}_j)(x - \overline{x}_j) & (y - \overline{y}_j)^2 \end{pmatrix} = \\ &= \sum_{j=1}^{2} N_j \mathbf{\Sigma_j} = \frac{N}{2} \sum_{j=1}^{2} \mathbf{\Sigma_j} \end{aligned} \tag{8.9}$$

The FLD solution thus takes the following form

$$\begin{aligned}
\mathbf{w} &= [\frac{N}{2}(\boldsymbol{\Sigma_1}+\boldsymbol{\Sigma_2})]^{-1}(\mathbf{m_1}-\mathbf{m_2}) = \\
&= \frac{2}{N}\begin{pmatrix} a_1'+a_2' & b_1'+b_2' \\ b_1'+b_2' & c_1'+c_2' \end{pmatrix}^{-1}\begin{pmatrix} \overline{x}_1-\overline{x}_2 \\ \overline{y}_1-\overline{y}_2 \end{pmatrix} = \\
&= \frac{2}{N}\frac{1}{\Delta''}\begin{pmatrix} c_1'+c_2' & -b_1'-b_2' \\ -b_1'-b_2' & a_1'+a_2' \end{pmatrix}\begin{pmatrix} \overline{x}_1-\overline{x}_2 \\ \overline{y}_1-\overline{y}_2 \end{pmatrix} \qquad (8.10)
\end{aligned}$$

$$\text{where} \qquad \Delta'' = \det(\boldsymbol{\Sigma_1}+\boldsymbol{\Sigma_2}) \qquad (8.11)$$

Not taking into account the multiplied constant, the components of this vector are

$$\begin{aligned}
w_x &= (c_1'+c_2')(\overline{x}_1-\overline{x}_2)+(-b_1'-b_2')(\overline{y}_1-\overline{y}_2) \qquad (8.12)\\
w_y &= (-b_1'-b_2')(\overline{x}_1-\overline{x}_2)+(a_1'+a_2')(\overline{y}_1-\overline{y}_2) \qquad (8.13)
\end{aligned}$$

using equations (8.4) and (8.5) we can use substitution to obtain

$$\begin{aligned}
w_x &= (a_1a_2c_2-a_1b_2^2+a_2a_1c_1-a_2b_1^2)(\overline{x}_1-\overline{x}_2)+ \\
&\quad +(b_1a_2c_2-b_1b_2^2+b_2a_1c_1-b_2b_1^2)(\overline{y}_1-\overline{y}_2) \qquad (8.14)\\
w_y &= (b_1a_2c_2-b_1b_2^2+a_1b_2c_1-b_2b_1^2)(\overline{x}_1-\overline{x}_2)+ \\
&\quad +(c_1a_2c_2-c_1b_2^2+c_2a_1c_1-c_2b_1^2)(\overline{y}_1-\overline{y}_2) \qquad (8.15)
\end{aligned}$$

8.7.3 *Unicity of the Tangent Point Equivalent to FLD*

The gradient at a generic point $\mathbf{x} = (x, y)$ of the ellipse is given by the following relationship

$$\nabla(x,y) = [(a(x-\overline{x})+b(y-\overline{y})),(c(y-\overline{y})+b(x-\overline{x}))] \qquad (8.16)$$

The locus of all points for which *equivalence ellipses* (i.e., those with the same covariance matrix and different *Ro* parameter) have gradient vectors proportional to the FLD solution vector is given by

$$\frac{a(x-\overline{x})+b(y-\overline{y})}{c(y-\overline{y})+b(x-\overline{x})} = \frac{w_x}{w_y} \qquad (8.17)$$

where we have used (8.6) and (8.16). It can be noticed that this locus is a straight line.

It can be proved that a common point of two equivalence ellipses is a tangent point (with the suitable values of the Rt parameters) iff the gradient vectors of the ellipses at this point are parallel, i.e.

$$\nabla_1(x, y) = \pm\alpha\nabla_2(x, y) \tag{8.18}$$

where α is a real number.

In our case, the ETPs are given by the solution of the following system of linear equations

$$\frac{a_j(x - \overline{x}_j) + b_j(y - \overline{y}_j)}{c_j(y - \overline{y}_j) + b_j(x - \overline{x}_j)} = \frac{w_x}{w_y} \qquad j = 1, 2 \tag{8.19}$$

This system has a unique solution when both ellipses do not have any common axis. In the case that the ellipses share an axis the system has an infinite number of possible solutions, i.e., an infinite number of ETPs. Fig.8.10 ilustrates this point.

8.7.4 Elliptic Tangent Point with the Equal Magnitude and Opposite Sign of the Gradient

We are working with Elliptic Boundary Methods. An ellipse is defined by quadratic form, such as the following

$$(\mathbf{x} - \mathbf{m})^T \mathbf{\Sigma}^{-1} (\mathbf{x} - \mathbf{m}) = Ro \qquad j = 1, 2 \tag{8.20}$$

where the vectors $\mathbf{x} = (x, y)$ are now the points belonging to the ellipse, i.e. lying on the ellipse boundary. The center of the ellipse is defined by the mean vector $\mathbf{m}$, the shape is determined by the inverse of the matrix of covariances $\mathbf{\Sigma}$, and the size by the parameter Ro.

The gradient at a generic point of the ellipse $\mathbf{x} = (x, y)$ is given by (8.16). In our case, we are working with two ellipses and we want to find the tangent point (x^*, y^*) (and calculate the slope of the tangent) that satisfies

$$\nabla_1(x, y) = -\nabla_2(x, y) \tag{8.21}$$

This point is really a tangent point because it verifies (8.18). If we consider (8.16) and (8.21), we obtain the following system of linear equations

$$(a_1 + a_2)x + (b_1 + b_2)y + (-a_1\overline{x}_1 - b_1\overline{y}_1 - a_2\overline{x}_2 - b_2\overline{y}_2) = 0 \tag{8.22}$$

$$(b_1 + b_2)x + (c_1 + c_2)y + (-b_1\overline{x}_1 - c_1\overline{y}_1 - b_2\overline{x}_2 - c_2\overline{y}_2) = 0 \tag{8.23}$$

whose solution is (using Cramer's rule)

$$x^* = \frac{D_1}{D} = \frac{B_1C_2 - C_1B_2}{A_1B_2 - B_1^2}$$

$$y^* = \frac{D_2}{D} = \frac{C_1B_1 - A_1C_2}{A_1B_2 - B_1^2} \tag{8.24}$$

where

$$\begin{aligned} A_1 &= a_1 + a_2 \\ B_1 &= b_1 + b_2 \\ B_2 &= c_1 + c_2 \\ C_1 &= -a_1\overline{x}_1 - b_1\overline{y}_1 - a_2\overline{x}_2 - b_2\overline{y}_2 \\ C_2 &= -b_1\overline{x}_1 - c_1\overline{y}_1 - b_2\overline{x}_2 - c_2\overline{y}_2 \end{aligned} \tag{8.25}$$

This solution is unique if $A_1B_2 - B_1^2 \neq 0$. Really, this condition is equal to

$$\left|\boldsymbol{\Sigma_1}^{-1} + \boldsymbol{\Sigma_2}^{-1}\right| \neq 0 \tag{8.26}$$

It can be easily demonstrated that if the covariances matrices, $\boldsymbol{\Sigma_1}$ y $\boldsymbol{\Sigma_2}$, are positive definite then this condition is always verified. (x^*, y^*) is the point at which the two ellipses are tangent and have equal-magnitude, opposite-sign gradients.

The last question to resolve is to obtain the gradient vectors at the point (x^*, y^*). These vectors are expressed according to (8.16) and (8.21). For the first ellipse we have

$$\nabla_{1x} = a_1(x - \overline{x}_1) + b_1(y - \overline{y}_1) \tag{8.27}$$

$$\nabla_{1y} = c_1(y - \overline{y}_1) + b_1(x - \overline{x}_1) \tag{8.28}$$

Substituting (8.24) in this equation we obtain

$$\nabla_{1x} = a_1(D_1 - D\overline{x}_1) + b_1(D_2 - D\overline{y}_1) \tag{8.29}$$

$$\nabla_{1y} = c_1(D_2 - D\overline{y}_1) + b_1(D_1 - D\overline{x}_1) \tag{8.30}$$

From (8.24) and (8.25), we obtain

$$\begin{aligned} D_1 - D\overline{x}_1 &= (\overline{x}_1 - \overline{x}_2)(b_1b_2 + b_2^2 - a_2c_2 - a_2c_1) + \\ &\quad +(\overline{y}_1 - \overline{y}_2)(b_1c_2 - b_2c_1) \end{aligned} \tag{8.31}$$

$$\begin{aligned} D_2 - D\overline{y}_1 &= (\overline{y}_1 - \overline{y}_2)(b_1b_2 + b_2^2 - a_2c_2 - a_1c_2) + \\ &\quad +(\overline{x}_1 - \overline{x}_2)(a_2b_1 - a_1b_2) \end{aligned} \tag{8.32}$$

Substituting (8.31) and (8.32) in (8.29) and (8.30) yields

$$\nabla_{1x} = -w_x \qquad \text{and} \qquad \nabla_{1y} = -w_y \tag{8.33}$$

A graphic representation is given in Fig.8.11. We have thus demonstrated the Theorem.

8.8 References

[1] R. . Duda and P. E. Hart, *Pattern Classification and Scene Analysis.* Wiley-Interscience, 1973.

[2] K. Fukunaga, *Introduction to Statistical Pattern Recognition.* New York: Academic Press, 2nd ed., 1990.

[3] J. T. Tou and R. C. Gonzalez, *Pattern Recognition Principles.* Reading, Massachusetts: Addison-Wesley Publishing Company, 1974.

[4] R. Dubes and A. K. Jain, "Validity Studies in Clustering Methodologies," *Pattern Recognition*, vol. 11, pp. 235–254, 1979.

[5] A. K. Jain and R. C. Dubes, *Algorithms for Clustering Data.* Prentice Hall Advanced Reference Series, 1988.

[6] D. Hermann, S. Ahalt, and R. Mitchell, "Clustering and Compression of High-Dimensional Sensor Data," in *SPIE International Symposium on Optical Engineering: Signal Processing, Sensor Fusion, and Target Recognition II*, pp. 286–297, April 1993.

[7] C. Lee and D. A. Landgrebe, "Feature Extraction Based on Decision Boundaries," *IEEE Transactions on Pattern Analysis and Machine Intelligence*, vol. 15, pp. 388–400, April 1993.

[8] R. P. Lippmann, "An Introduction to Computing with Neural Nets," *IEEE ASSP Magazine*, vol. 4, pp. 4–22, April 1987.

[9] S. Haykin, *Neural Networks. A Comprehensive Foundation.* Ontario: IEEE Press. Macmillan, 1994.

[10] J. Hertz, A. Krogh, and R. G. Palmer, *Introduction to the Theory of Neural Computation.* Addison-Wesley Publishing Company, 1991.

[11] C. Cachin, "Pedagogical Pattern Selection Strategies," *Neural Networks*, vol. 7, no. 1, pp. 175–181, 1994.

[12] T. H. M. Wann and N. N. Greenbaun, "The Influence of Training Sets on Generalization in Feed-Forward Neural Networks," in *Proceedings of the International Joint Conference on Neural Networks*, vol. 3, (San Diego, California), pp. 137–142, June 1990.

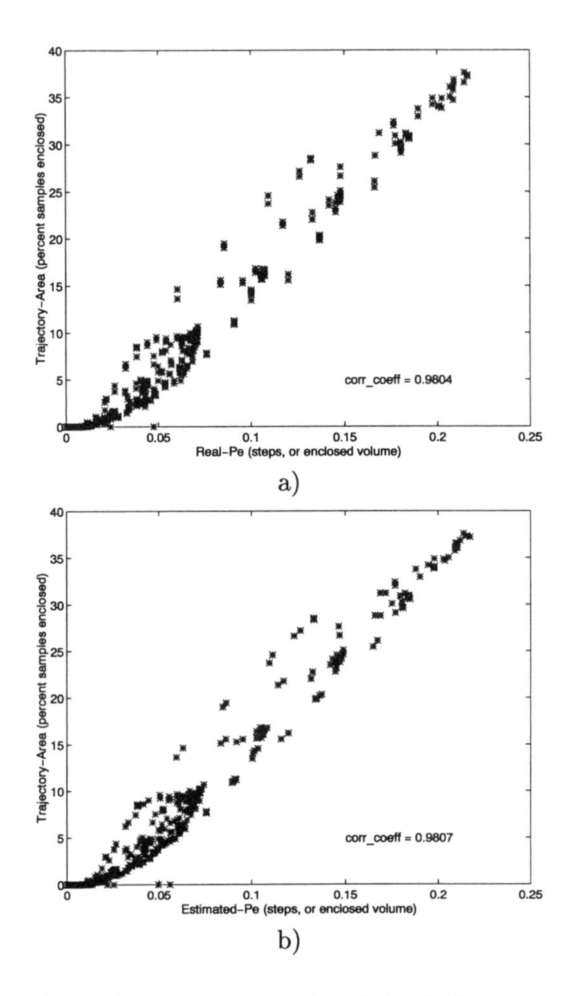

FIGURE 8.5. Relationship between the TA and the Pe for a number of distributions with different means and covariance matrices. a) Simulations with actual means and covariance matrices; b) Simulations with estimated means and covariance matrices.

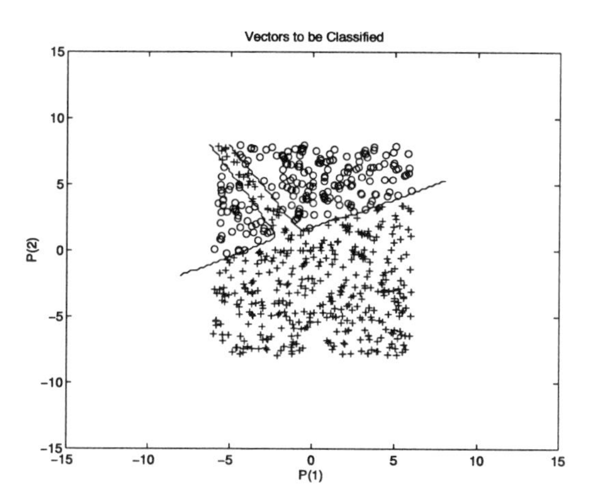

FIGURE 8.6. Problem to be solved

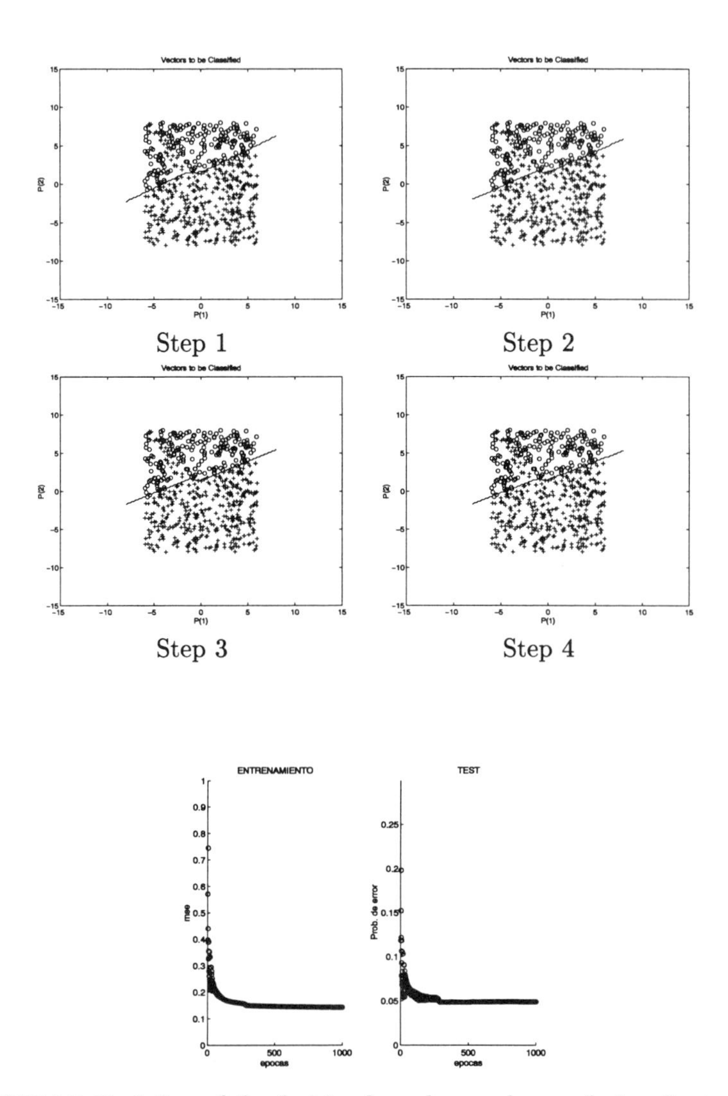

FIGURE 8.7. Evolution of the decision boundary and error during the training process with the BP algorithm.

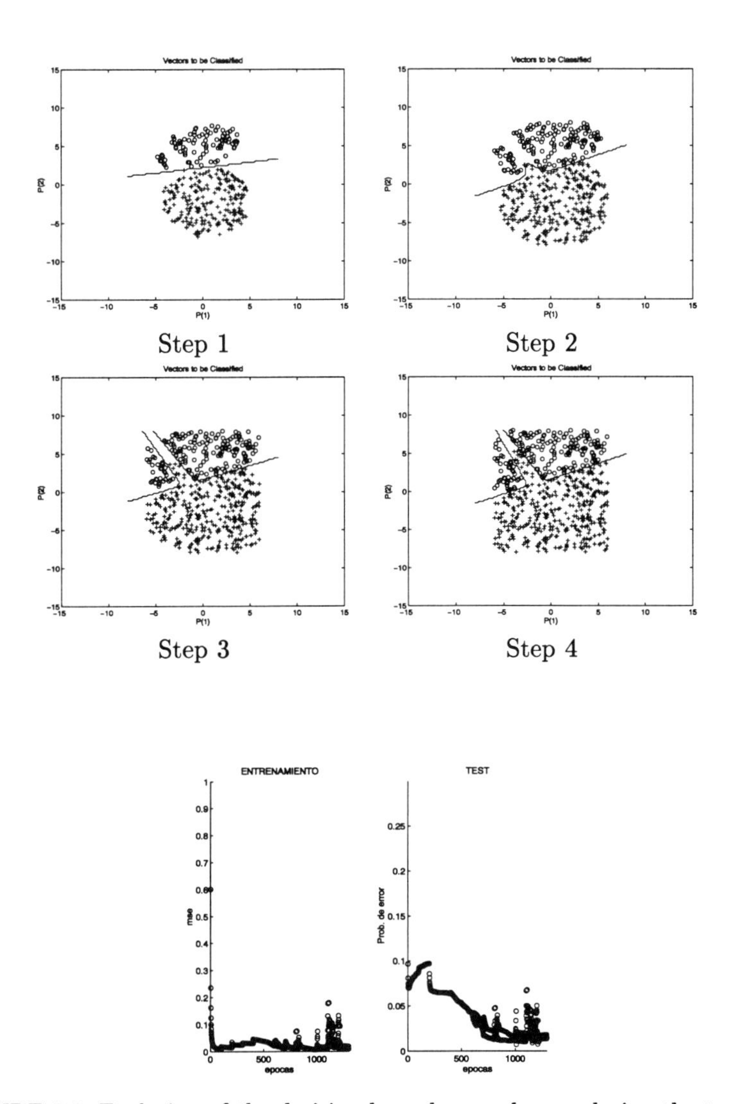

FIGURE 8.8. Evolution of the decision boundary and error during the training process with the first EBM-BP algorithm.

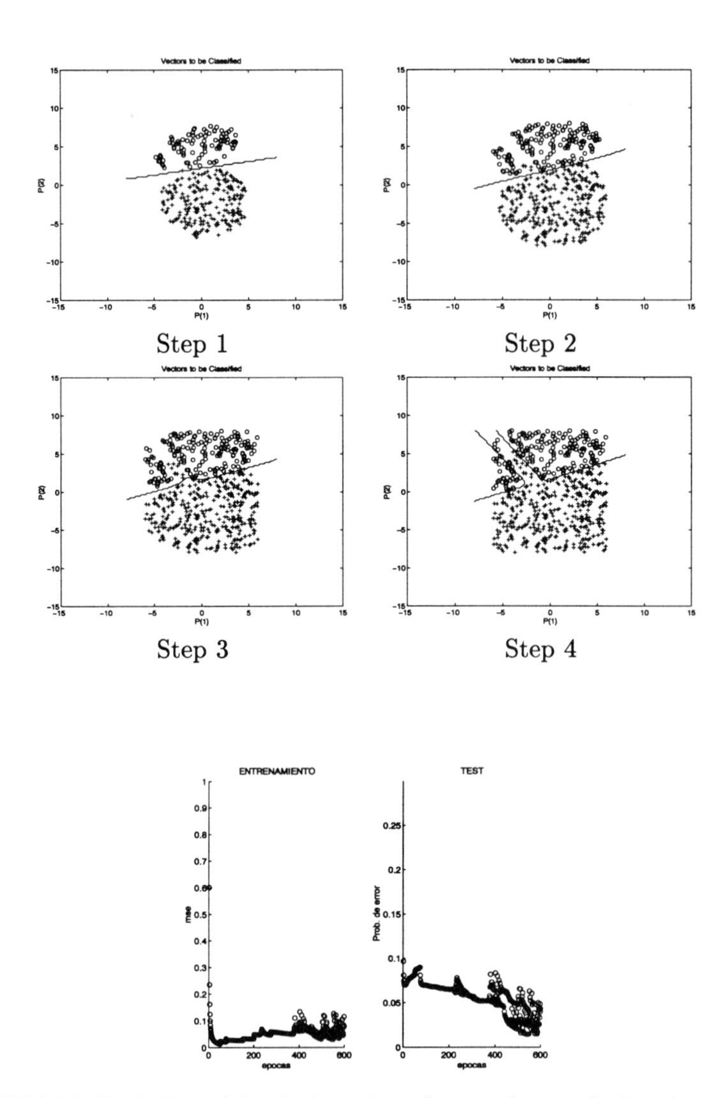

FIGURE 8.9. Evolution of the decision boundary and error during the training process with the second EBM-BP algorithm.

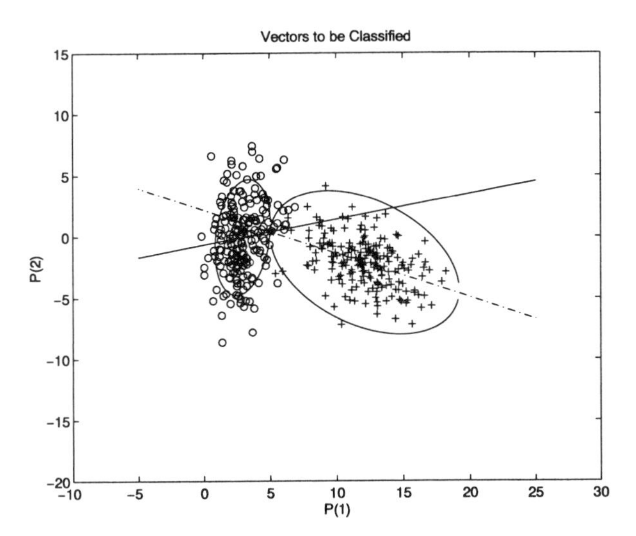

FIGURE 8.10. Shown here are two lines each representing the locus of all points for which equivalence ellipses have tangents with slopes equivalent to the FLD projection vector. The solid line (-) is the locus for leftmost ellipse, and the dash-dot line (-.) is the locus for the rightmost ellipse. As expected, the loci intersect at a unique point.

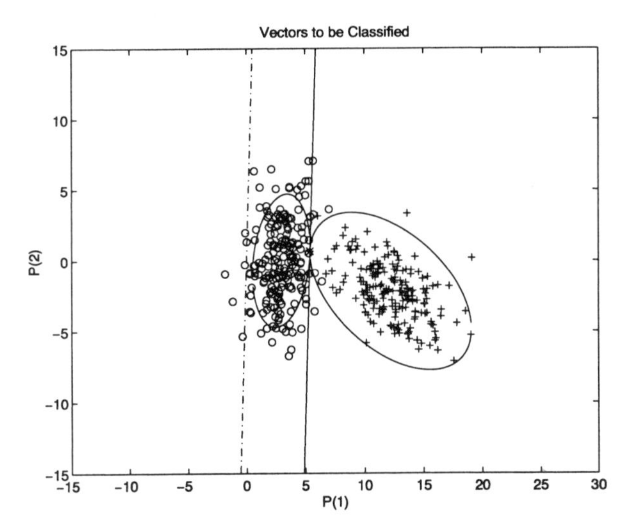

FIGURE 8.11. Graphic demonstration of the Theorem c). The tangent point is defined by (8.21) and tangential ellipses have a common tangent (solid line, -) which is parallel to the FLD projection vector (dash-dot line, -.).

9

Constructive Function Approximation: Theory and Practice

D. Docampo
D.R. Hush
C.T. Abdallah

ABSTRACT In this paper we study the theoretical limits of finite constructive convex approximations of a given function in a Hilbert space using elements taken from a reduced subset. We also investigate the trade-off between the global error and the partial error during the iterations of the solution. These results are then specialized to constructive function approximation using sigmoidal neural networks. The emphasis then shifts to the implementation issues associated with the problem of achieving given approximation errors when using a finite number of nodes and a finite data set for training.

9.1 Introduction

It has been shown that continuous functions on compact subsets of $\mathbb{R}^d$ can be uniformly approximated by linear combinations of sigmoidal functions [11, 20]. What was missing from that result is how the error in the approximation is related to the number of sigmoids used. This can be phrased in a more general way as the problem of approximating a given element (function) f in a Hilbert space H by means of an iterative sequence f_n, and has an impact in establishing convergence results for projection pursuit algorithms [22], neural network training [5] and classification [12]. Moreover, the fact that one will have to achieve the approximation when samples of f are given has been largely forgotten by most papers which quote the results of [11, 20]. The approximation problem can be given a constructive solution where the iterations taking place involve computations in a reduced subset G of H [22, 5]. This leads to algorithms such as projection pursuit. Convergence of the classical projection pursuit regression techniques [13] however, has been shown to be very slow unless the iterate f_{n+1} is chosen to be an optimal combination of the past iterate f_n and a ridge function of elements of the subset G. The bound of the error in this approximation has been refined several times since the initial non-constructive proof given by Maurey, as reported in [23]. Jones [22] provided the first constructive solution to the problem of finding finite convex approximations of a given function in a

Hilbert space using elements taken from a reduced subset. His results have been recently refined by Barron [3], and Dingankar and Sandberg [12]. In this paper we report that the rate of convergence obtained in [22] and [5] is the maximum achievable, and, only under some restricted assumptions, the results in Sandberg and Dingankar can be derived as the optimal convex combination to preserve the desired convergence rate.

In the first part of the paper, we formulate the approximation problem in such a way that we can study the limits of the global error, obtain the best possible trade-off between global and partial errors, and give theoretical bounds for the global error when a prespecified partial error is fixed. We then concentrate on the implementation aspects of the problem, specifically, the problem of achieving a certain approximation error using one approximating function at a time. We then discuss some specific sigmoidal functions and algorithms which have been shown to be efficient in solving a particular step of the approximation problem.

The rest of the paper is organized as follows: We start out by reviewing some theoretical results in section 9.2 where we state the problem and highlight its practical implications. In section 9.3 we review the theoretical solutions to the problem, and provide the framework under which those solutions can be derived. In section 9.4 we analyze the limits of the global error and its relation to the partial errors at each step of the iterative process. In section 9.5 we specialize the constructive functions to sigmoidal functions. Section 9.6 presents the practical issues associated with implementing a constructive algorithm with an eye towards neural network results. Finally, section 9.7 presents our conclusions.

9.2 Overview of Constructive Approximation

In this section, we state and present some theoretical results on the constructive approximation problem. In order to state the results in their full generality, let G be a subset of a real or complex Hilbert space H, with norm $||.||$, such that its elements, g, are bounded in norm by some positive constant b. Let $\bar{co}(G)$ denote the convex closure of G (i.e. the closure of the convex hull of G in H). The first global bound result, attributed to Maurey, concerning the error in approximating an element of $\bar{co}(G)$ using convex combinations of n points in G, is the following:

Lemma 9.2.1 *Let f be an element of $\bar{co}(G)$ and c a constant such that $c > b^2 - ||f||^2 = b_f^2$. Then, for each positive integer n there is a point f_n in the convex hull of some n points of G such that:*

$$||f - f_n||^2 \leq \frac{c}{n}$$

△

The first constructive proof of this lemma was given by Jones [22] and refined by Barron [3]; the proof includes an algorithm to iterate the solution. In the next section, a review of the constructive proof will be presented. We will specifically prove the following in section 9.3.

Theorem 9.2.1 *For each element f in $\bar{co}(G)$, let us define the parameter γ as follows:*

$$\gamma = \inf_{v \in H} \sup_{g \in G} \left\{ \|g - v\|^2 - \|f - v\|^2 \right\}$$

Let now δ be a constant such that $\delta > \gamma$. Then, we can construct an iterative sequence f_n, f_n chosen as a convex combination of the previous iterate f_{n-1} and a $g_n \in G$, $f_n = (1 - \lambda) f_{n-1} + \lambda g_n$, such that:

$$\|f - f_n\|^2 \leq \frac{\delta}{n}$$

Proof: See section 9.3. ■

Note that this new parameter, γ, is related to Maurey's b_f^2, since if we make $v = 0$ in the definition of γ we realize that $\gamma \leq b_f^2$.

The relation between this problem and the universal approximation property of sigmoidal networks was clearly established by [22, 5]; specifically, under certain mild restrictions, continuous functions on compact subsets of $\mathbb{R}^d$ belong to the convex hull of the set of sigmoidal functions that one hidden layer neural networks can generate. Moreover, since the proofs are constructive, an algorithm to achieve the theoretical bounds is provided as well.

Other nonlinear approximation techniques have also benefited from the solution to this problem: approximation by hinged hyperplanes [8], projection pursuit regression [28] and radial basis functions [17]. In all these related approximation problems the solution can always be constrained to fall in the closure of the convex hull of a subset of functions (e.g. hinged hyperplanes, ridge functions or radial basis functions in the examples mentioned above).

9.3 Constructive Solutions

For the sake of clarity and completeness, we include here the proof given in [5] and [12].

Lemma 9.3.1 *Given $f \in \bar{co}(G)$, for each element of $co(G)$, h, and $\lambda \in [0, 1]$:*

$$\inf_{g \in G} \|f - (1 - \lambda)h - \lambda g\|^2 \leq (1 - \lambda)^2 \|f - h\|^2 + \lambda^2 \gamma \tag{9.1}$$

Proof: The proof of the lemma will be carried out for $f \in co(G)$; it extends to elements in $\bar{co}(G)$ because of the continuity of all the terms involved in the inequalities [10].

Since $f \in co(G)$, there exists a convex combination of elements g^* from G, so that $f = \sum_{k=1}^{m} \alpha_k g_k^*$. Let then g^* be a random vector taking values on H with probabilities $P(g^* = g_k^*) = \alpha_k$.

Then: $E(g^*) = f$, $var(g^*) = E(\|g^* - f\|^2) = E(\|g^*\|^2) - \|f\|^2 \leq b_f^2$.

Additionally, for $v \in H$,

$var(g^*) = var(g^* - v) = E(\|g^* - v - (f - v)\|^2) = E(\|g^* - v\|^2) - \|f - v\|^2.$
Thus, $\forall v \in H$,

$$\begin{aligned} var(g^*) &\leq \sup_{g\in G} \|g - v\|^2 - \|f - v\|^2 \Rightarrow \\ var(g^*) &\leq \inf_{v\in H} \sup_{g\in G} \|g - v\|^2 - \|f - v\|^2 = \gamma. \end{aligned}$$

Now, for $\lambda \in [0,1]$ and $d \in H$,

$$E(\|\lambda(g^* - f) + d\|^2) = \lambda^2 E(\|g^* - f\|^2) + \|d\|^2 \leq \lambda^2\gamma + \|d\|^2,$$

and for $\lambda \in [0,1]$

$$\begin{aligned} \inf_{g\in G} \|f - (1-\lambda)h - \lambda g\|^2 &\leq E\left(\|(1-\lambda)h + \lambda g^* - f\|\right)^2 \leq \\ \leq E\left(\|(1-\lambda)(h - f) + \lambda(g^* - f)\|\right)^2 &\leq (1-\lambda)^2\|f - h\|^2 + \lambda^2\gamma \end{aligned}$$

which concludes the proof of Lemma 9.3.1. △

We can now prove Theorem 9.2.1, using an inductive argument.

Proof: At step 1, find g_1 and ϵ_1 so that $\|f - g_1\|^2 \leq \inf_G \|f - g\|^2 + \epsilon_1 \leq \delta$. This is guaranteed by (9.1), for $\lambda = 1$ and $\epsilon_1 = \delta - \gamma$.

Let now f_n be our iterative sequence of elements in $co(G)$, and assume that for $n \geq 2$,

$$\|f - f_{n-1}\|^2 \leq \delta/(n-1)$$

It is then possible to choose among different values of λ and ϵ_n so that:

$$(1-\lambda)^2\|f_{n-1} - f\|^2 + \lambda^2\gamma \leq \frac{\delta}{n} - \epsilon_n \tag{9.2}$$

At step n, select g_n such that:

$$\|f - (1-\lambda)f_{n-1} - \lambda g_n\|^2 \leq \inf_{g\in G} \|f - (1-\lambda)f_{n-1} - \lambda g\|^2 + \epsilon_n \tag{9.3}$$

Hence, using (9.1), (9.3) and (9.2), we get: $\|f - f_n\|^2 \leq \frac{\delta}{n}$, and that completes the proof of Theorem 9.2.1. ■

The values of λ and ϵ_n in [5] and [12] are related to the parameter α, $\alpha = \delta/\gamma - 1$, in the following way:

$$\begin{aligned} [5] \quad &: \quad \lambda = \frac{\|f - f_{n-1}\|^2}{\gamma + \|f - f_{n-1}\|^2}; \qquad \epsilon_n = \frac{\alpha\delta}{n(n+\alpha)} \\ [12] \quad &: \quad \lambda = \frac{1}{n}; \qquad \epsilon_n = \frac{\alpha\gamma}{n^2} \end{aligned}$$

It is easy to check that, in both cases, ϵ_1 is equal to $\delta - \gamma$ as stated in the proof.

Given that the values of the constant λ are different in both cases, we first look for the values of λ which make the problem solvable (i.e. feasible values for the constant λ). Admissible values of λ will have to satisfy inequality (9.2) for positive values of ϵ_n; it is easy to show that those values fall in the following interval, centered at Barron's optimal value for λ:

$$\frac{\|f - f_{n-1}\|^2}{\gamma + \|f - f_{n-1}\|^2} \pm \frac{1}{\gamma + \|f - f_{n-1}\|^2}\sqrt{\|f - f_{n-1}\|^4 - \|f - f_{n-1}\|^2 + \frac{\delta}{n}}$$

To evaluate the possible choices for the bound ϵ_n we need to make use of the induction hypothesis; introducing it in inequality (9.2), values of λ should now satisfy

$$(1-\lambda)^2 \frac{\delta}{n-1} + \lambda^2 \gamma \leq \frac{\delta}{n} - \epsilon_n$$

In this case, admissible values of λ for positive values of ϵ_n fall in the interval (which always contains the value of $\lambda = 1/n$):

$$\frac{1+\alpha}{n+\alpha} \pm \frac{n-1}{n+\alpha}\sqrt{\frac{\alpha(1+\alpha)}{n(n-1)}}$$

In Figure 9.1 we show the bounds of this second interval for λ as a function of n. The bounds are shown in solid lines, the center of the interval using a dotted line, and the value of λ in [12] using a dash dotted line. Note how the dash-dotted line approaches the limits of the interval, which results in a poorer value for ϵ_n, as will be shown later.

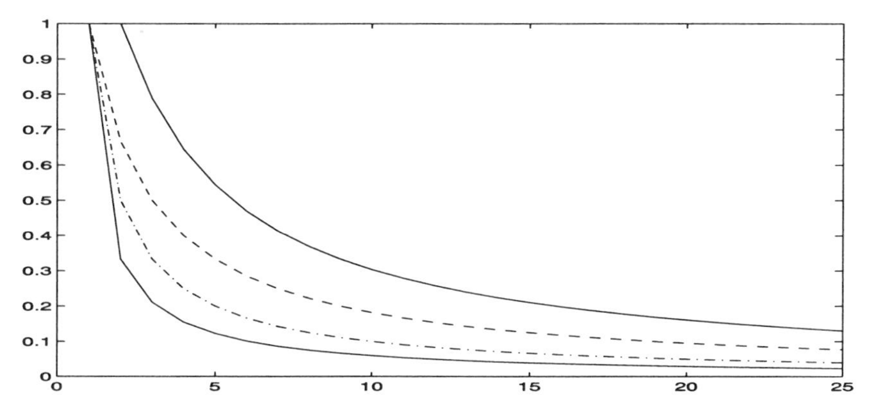

FIGURE 9.1. admissible values of λ for $\alpha = 1$

9.3.1 Discussion

Since the results presented so far achieve a bound of the global error of $O(1/n)$, and, to construct the solution, a partial error ϵ_n of $O(1/n^2)$ is the maximum allowed at each step, it is useful to formulate the following questions:

1. Is there any possibility of achieving a further reduction in the global error using convex combinations of n elements from G? What is the minimum bound for the global error assuming $\epsilon_n = 0$ for all n?
2. What is the optimal choice of λ for a given bound, so that ϵ_n is maximum, making the quasi-optimization problem at each step easier to solve?
3. For that the optimal choice of λ and a prespecified partial error, ϵ_n, what is bound for the global approximation problem?

Based on the assumptions made and in Lemma 9.3.1, let us formulate the problem again in a more general way: Our objective is to look for a constructive approximation so that the overall error using n elements from G satisfies the following inequality:

$$\|f - f_n\|^2 \leq \frac{\delta}{b(n)} \tag{9.4}$$

$b(n)$ being a function of the parameter n which indicates the order of our approximation (i.e. $b(n) = n$ both in [22] and [12]) and δ the parameter related to γ as defined before.

In what follows we will assume that the iterate f_n will be chosen as a convex combination of the previous iterate f_{n-1} and a point in G, g_n; this introduces a loss of generality, since other constructive approaches could be devised in order to re-optimize the coefficients of previous elements from G at each step. The facts that f_n is forced to be a convex combination of n elements from G, and our algorithm has to be constructive, mean that f_n is in the convex hull of $\{g_1, g_2, \ldots, g_n\}$ and f_{n-1} is in the convex hull of $\{g_1, g_2, \ldots, g_{n-1}\}$, but that does not imply that f_n must be a convex combination of f_{n-1} and g_n, as can be easily shown. We leave the more general problem for further investigation and concentrate here on the case where constructiveness of the algorithm is taken as in [22] and [12] to be equivalent to the constraint that, at each step, f_n is in the convex hull of $\{f_{n-1}, g_n\}$.

Before we try to answer the three questions posed at the beginning of this section, let us now set up a framework where the constructive results can be derived.

Let $f_n = (1-\lambda)f_{n-1} + \lambda g_n$, then, in our approximation problem we want to find λ, ϵ_n, and the function $b(n)$ so that:

$$\begin{aligned} \|f - f_n\|^2 &\leq \inf_{0<\lambda<1} \inf_{g \in G} \|f - (1-\lambda)f_{n-1} + \lambda g\|^2 + \epsilon_n \\ &\leq \inf_{0<\lambda<1} (1-\lambda)^2 \|f - f_{n-1}\|^2 + \lambda^2\gamma + \epsilon_n \\ &\leq \inf_{0<\lambda<1} (1-\lambda)^2 \frac{\delta}{b(n-1)} + \lambda^2\gamma + \epsilon_n \leq \frac{\delta}{b(n)} \end{aligned} \tag{9.5}$$

Since $\delta = (1+\alpha)\gamma$, we can rewrite the last inequality in the following way:

$$\inf_{0<\lambda<1} (1-\lambda)^2 \frac{\delta}{b(n-1)} + \lambda^2\delta + \epsilon_n - \lambda^2\alpha\gamma \leq \frac{\delta}{b(n)}$$

This last expression represents the trade-off between the global error we are trying to achieve $\delta/b(n)$ and the error at each of the subproblems, ϵ_n.

We are going to prove the following: if we set $\epsilon_n = \lambda^2\alpha\gamma$, then, for a given λ, the best rate of convergence of the approximation which can be achieved, measured in $b(n)$, is the one given in [12] and [5], and the optimal value of λ which minimizes ϵ_n for that best rate of convergence is precisely the value given in [12]. To see that, let's introduce the value of ϵ_n in (9.5), then:

$$(1-\lambda)^2 \frac{\delta}{b(n-1)} + \lambda^2\delta \leq \frac{\delta}{b(n)}$$

Hence,

$$(1-\lambda)^2 + b(n-1)\lambda^2 \leq \frac{b(n-1)}{b(n)}$$

and then:

$$P(\lambda) = \lambda^2((1+b(n-1)) - 2\lambda + 1 - \frac{b(n-1)}{b(n)} \leq 0 \tag{9.6}$$

$P(\lambda)$ has to have a discriminant greater than or equal than 0 for the inequality (9.6) to hold. So,

$$1 - (1+b(n-1))\left(1 - \frac{b(n-1)}{b(n)}\right) \geq 0$$

and then, finally:

$$\begin{aligned} b(n) &\geq (1+b(n-1))\,(b(n)-b(n-1)) \Leftrightarrow \\ b(n-1) &\geq b(n-1)\,(b(n)-b(n-1)) \Leftrightarrow \\ b(n) &\leq 1+b(n-1) \end{aligned} \tag{9.7}$$

Inequality (9.7) proves that, under the assumption that $\epsilon_n = \lambda^2\alpha\gamma$, there is no better rate of convergence using these kind of convex constructive solutions that the one obtained in references [22] and [12], since the maximum rate is obtained when

$$b(n) = 1 + b(n-1) \Rightarrow b(n) = b(1) + n - 1 = n \tag{9.8}$$

Furthermore, for this rate of convergence there is only one zero of the function $P(\lambda)$, namely, $\lambda = (1/n)$ which is the optimal value and coincides with the one provided in [12].

We will next answer the questions posed at the beginning of this section, concerning the limits and bounds of the approximation.

9.4 Limits and Bounds of the Approximation

If we look back at expression (9.5), we will notice that, after using Lemma 9.3.1, we have at each step a quadratic problem in λ, which consists of minimizing

$$Q(\lambda_n) = (1-\lambda_n)^2 \frac{\delta}{b(n-1)} + \lambda_n^2\gamma$$

provided that the induction hypothesis (9.4) is satisfied for $k < n$. We have introduced the notation λ_n to stress the variation of this parameter along the iterative process.

Taking derivatives, we get

$$\lambda_n\gamma = (1-\lambda_n)\frac{\delta}{b(n-1)} \Rightarrow$$

$$\lambda_n = \frac{(1-\lambda_n)\delta}{\gamma b(n-1)} = \frac{1+\alpha}{1+\alpha+b(n-1)} \tag{9.9}$$

Hence, we get the following expression of the optimal error bound:

$$\begin{aligned} \|f - f_n\|^2 &\leq (1-\lambda_n)^2 \left[\frac{\delta}{b(n-1)} + \frac{(1+\alpha)\delta}{b^2(n-1)} \right] + \epsilon_n \\ &= \delta \left(\frac{b^2(n-1)}{1+\alpha+b(n-1)} \right) \left(\frac{1+\alpha+b(n-1)}{b^2(n-1)} \right) + \epsilon_n \\ &= \frac{\delta}{1+\alpha+b(n-1)} + \epsilon_n = \frac{\delta}{b(n)} \end{aligned} \tag{9.10}$$

From (9.10) we can write the following expression for $b(n)$ and ϵ_n:

$$\frac{1}{b(n)} = \frac{1}{1+\alpha+b(n-1)} + \frac{\epsilon_n}{\delta} \tag{9.11}$$

and then

$$\epsilon_n = \frac{\delta}{b(n)(1+\alpha+b(n-1))} [1+\alpha+b(n-1)-b(n)] \tag{9.12}$$

From this last expression we conclude that there is a fundamental limitation in the rate of convergence that can be achieved under the hypothesis made so far, namely:

$$b(n) - b(n-1) \leq 1 + \alpha = \frac{\delta}{\gamma}$$

9.4.1 Minimum Global Error

Assuming that we can solve the partial approximation problems at each step of the iteration, so $\epsilon_n = 0, n \geq 1$, then

$$b(n) = 1 + \alpha + b(n-1) \quad \Rightarrow \quad b(n) = n(1+\alpha)$$

provided that we make $b(1) = 1+\alpha$, which means that we should find an element g_1 in G so that

$$\|f - f_1\|^2 \leq \frac{\delta}{1+\alpha}$$

which is guaranteed by Lemma 9.3.1. Hence, the best rate of convergence that can be obtained follows the law c/n, since

$$\frac{\delta}{n(1+\alpha)} = \frac{\gamma}{n}$$

We have then reached the minimum value of the constant c, namely $c = \gamma$.

Note that for this minimum to be reached we have

$$\lambda_n = \frac{1+\alpha}{(1+\alpha)n} = \frac{1}{n}$$

so the optimal convex combination would be the average of n elements from G, as in [12].

We have then answered the first of our questions. We now examine the trade-off between the global error and ϵ_n. Specifically, we will find the minimum global error for a specified partial error, ϵ_n, and the maximum bound we can place on the partial error for a prespecified rate of convergence.

9.4.2 Fixing ϵ_n

Given the nonlinear character of the recursion involved in (9.11), there is no analytical procedure to find a closed expression for $b(n)$. However, we can compute the bound of the approximation following the flow diagram of the optimal procedure, and from it derive some asymptotically results.

1. Select a constant δ such that $\delta > \gamma$; let $\delta = (1+\alpha)\gamma$.
2. Find $g_1 \in G$ so that $\|f - g_1\|^2 \leq \delta$. Set $f_1 = g_1$.
3. For $n > 1$, evaluate:
   (a) $\lambda_n = (1+\alpha)/\left(1+\alpha+b(n-1)\right)$ from (9.9)
   (b) Find $g_n \in G$ so that
   $$\|f - (1-\lambda_n)f_{n-1} - \lambda_n g_n\|^2 \leq \inf_G \|f - (1-\lambda_n)f_{n-1} - \lambda_n g\|^2 + \epsilon_n$$
   (c) Make $f_n = (1-\lambda_n)f_{n-1} + \lambda_n g_n$
   (d) Compute $b(n)$ from (9.11)

In order to make the appropriate comparisons with previous results, we will set $\epsilon_n = (\alpha\gamma/n^2)$, as in [12]. Then, again under the induction hypothesis,

$$\frac{1}{b(n)} = \frac{1}{1+\alpha+b(n-1)} + \frac{\alpha}{(1+\alpha)n^2}$$

To predict the asymptotic behavior of $b(n)$, let us assume that, at step $n-1$, $b(n-1) \geq \beta(1+\alpha)(n-1)$, we will prove then that, for some values of the constant β, we can imply that also $b(n) \geq \beta(1+\alpha)n$.

Since $b(n-1) \geq \beta(1+\alpha)(n-1)$, we have:

$$\begin{aligned}
\frac{1}{b(n)} &\leq \frac{1}{(1+\alpha)(1+\beta(n-1))} + \frac{\alpha}{(1+\alpha)n^2} \Rightarrow \\
\frac{1+\alpha}{b(n)} &\leq \frac{1}{1+\beta(n-1)} + \frac{\alpha}{n^2} \Rightarrow \\
\frac{b(n)}{n(1+\alpha)} &\geq \frac{n(1+\beta(n-1))}{n^2+\alpha(1+\beta(n-1))} \Rightarrow \\
\frac{b(n)}{n(1+\alpha)} \geq \beta &\Leftrightarrow n(1-\beta) \geq \beta\alpha(1+\beta(n-1)) \Leftrightarrow \\
n(1-\beta-\beta^2\alpha) &\geq \beta\alpha(1-\beta)
\end{aligned}$$

This last inequality is asymptotically fulfilled for any value of β such that:

$$0 \leq \beta \leq \frac{\sqrt{4\alpha+1}-1}{2\alpha}$$

Then, for the value of ϵ_n selected in [12], the asymptotic value for $b(n)$ is:

$$b(n) = (1+\alpha)n\frac{\sqrt{4\alpha+1}-1}{2\alpha}$$

which is a better rate than the one obtained in [12].

In Figure 9.2 we show $b(n)$ as a solid line, and the straight lines $l(n) = n$ corresponding with the rate in [12], dotted line, and the predicted asymptotic behavior of $b(n)$. The figure clearly supports the asymptotic results, and shows that the constant λ_n found (9.9) always results in a better convergence rate than [12]. The gap between the two lines would be bigger for larger values of the constant α; in other words, the larger the constant δ the worse the convergence rate achieved using $\lambda = 1/n$.

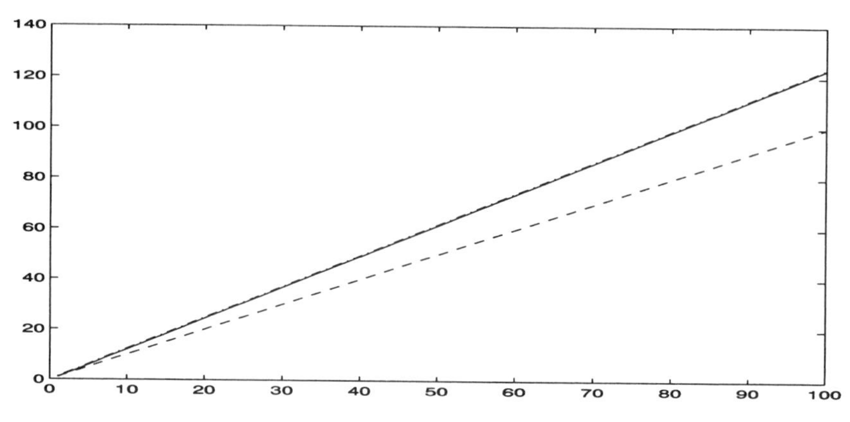

FIGURE 9.2. Optimal convergence rate for $\alpha = 1$

9.4.3 Fixing the rate of convergence

The remaining problem, namely: given the optimal value of λ find the maximum ϵ_n for a fixed convergence rate, thus making the quasi-optimization problem at each step easier to solve, was already explicitly solved in (9.12). Again, to show how our results compare with [5] and [12], we will assume that our desired rate of convergence is given by $b(n) = n$.

The value $\lambda_n = (1+\alpha)/(n+\alpha)$ solves the optimization problem, and:

$$\epsilon_n = \frac{\alpha\delta}{n(n+\alpha)} \tag{9.13}$$

This is the best upper bound we can achieve for the partial error at each step of the iteration process. It is easy to show that it coincides with Barron's bound, and is always greater than the bound found in [12].

Now, in Figure 9.3 we show the bound ϵ_n for $n = 5$ and $\gamma = 1$, as a function of α. The optimal bound is shown using the solid line, while the bound from [12] is shown using a dotted line.

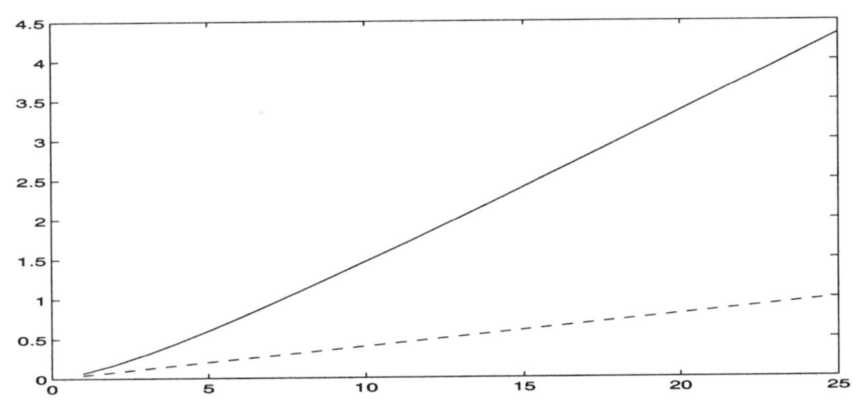

FIGURE 9.3. ϵ_n for $n = 5$

Reference	Activation Function	Approximation In	Proof
[11]	Continuous Sigmoid	$C[K]$	Existential
[11]	Bounded Sigmoid	$L_p[K]$	Existential
[20]	Monotone Sigmoid	$C[K]$	Constructive
[9]	Bounded Sigmoid	$C[\mathbb{R}^n]$	Constructive

TABLE 9.1. Approximation Results

9.5 The Sigmoidal Class of Approximators

When discussing neural networks, we are typically referring to a system built by linearly combining a large collection of simple computing devices (i.e., nodes), each of which performs a nonlinear transformation σ (in general a sigmoid function) on its inputs [18]. A sigmoid is defined here as a bounded function $\sigma(x)$. It is now known that a 1-hidden layer static network, whose nodes are sigmoidal is capable of approximating an arbitrary (continuous) function. Many proofs of this result have appeared of which we recall the ones in [11, 20]. Until recently, these proofs have used the Stone-Weierstrass theorem and required the continuity or even differentiability of the sigmoid (or nonlinearities) in the neural net. Chen et al. [9], building on the research of Sandberg [25, 26, 27] have recently shown however that all is needed is the boundedness of the sigmoid building block. Table 9.5 is taken from [9] and summarizes some available results for the approximation of functions. The set K denotes a compact subset of $\mathbb{R}^n$. Note that even those results labeled "constructive" still ignore the facts associated with the training algorithm and the available data.

The set of popular sigmoids include the *hardlimiting threshold* or *Heaviside* function shown in Figure 9.4(a):

$$\sigma_H(x) = \begin{cases} 1 & x > 0 \\ 0 & x \leq 0 \end{cases} \tag{9.14}$$

In order to derive certain learning techniques, a continuous nonlinear activation

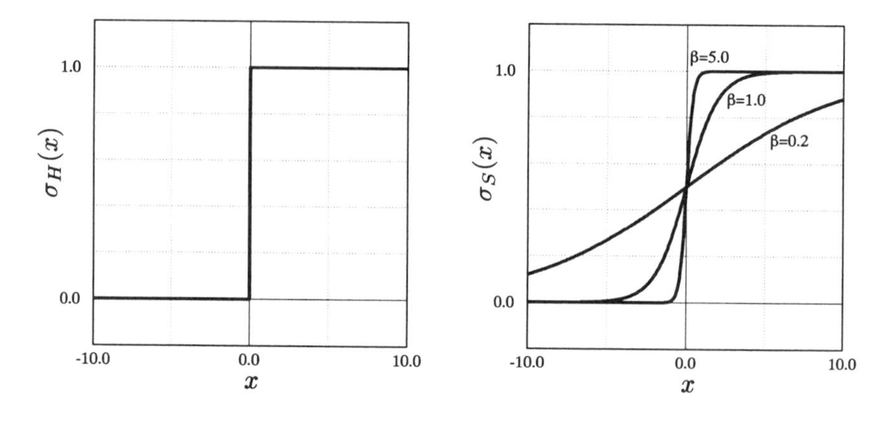

(a) Hardlimiter Nonlinearity (b) Sigmoid Nonlinearities

FIGURE 9.4. Typical Nonlinearities.

function is often required. For example, gradient descent techniques typically require that the sigmoid be differentiable [2]. Thus the threshold function is commonly approximated using the *sigmoid* function shown in Figure 9.4(b):

$$\sigma_S(x) = \frac{1}{1 + e^{-\beta x}} \tag{9.15}$$

The gain of the sigmoid, β, determines the steepness of the transition region. Note that as the gain approaches ∞, the sigmoid approaches a hardlimiting threshold. Often the gain is set equal to one, and β is omitted from the definition in equation (9.15). Later in this paper, we shall use the ramp equation which is another sigmoid defined as

$$\sigma_r(\mathbf{x}) = \begin{cases} 0, & z \le \alpha \\ (z - \alpha)/(\beta - \alpha), & \alpha \le z \le \beta \\ 1, & \beta \le z \end{cases} \tag{9.16}$$

9.6 Practical Considerations

In section 9.2 we recalled that it is possible to approximate an unknown function $f \in \bar{co}(G)$ with a convex combination of functions $g_i \in G$, $i = 1, 2, ..., n$ such that the approximation error is bounded by δ/n. More importantly it was shown that the functions g_i can be found *constructively*, i.e. one at a time. The proof of this result was itself constructive, and thus provides a framework for the development of an algorithm for building these approximations. The purpose of this section is to explore such algorithms. In doing so, our primary concerns are with the *feasibility* and *computational complexity* of these algorithms. Specific statements regarding these two issues are often not possible without considering a specific class of approximating functions. For this purpose we have chosen the class of sigmoidal functions described in the previous section.

Initialization:
$f_0(\mathbf{x}) = 0$

for $n = 1$ **to** n_{max} **do**

1. Compute Residual: $e_n(\mathbf{x}) = f(\mathbf{x}) - f_{n-1}(\mathbf{x})$
2. Fit Residual: $g_n(\mathbf{x}) = \arg\min_{g \in G} ||e_n(\mathbf{x}) - g(\mathbf{x})||$
3. Update Estimate: $f_n(\mathbf{x}) = \alpha f_{n-1}(\mathbf{x}) + \beta g_n(\mathbf{x})$
   where α and β are chosen to minimize $||f(\mathbf{x}) - f_n(\mathbf{x})||$

endloop

FIGURE 9.5. Iterative Approximation Algorithm (IIA).

Perhaps the most straightforward algorithm that can be derived from the proof in section 9.3 is the *iterative approximation algorithm* (IIA) shown in Figure 9.5 [5]. This algorithm is attractive in that the main loop contains only three steps, each of which is conceptually quite simple. However, implementation of this algorithm is not so straightforward. The second step in particular can be quite difficult. This step requires that we find the function g_n that best fits the current residual. Even though this step can be difficult, it is certainly more manageable than finding all n of the g_i simultaneously, which is the more popular (non-constructive) approach used in neural networks.

The final model produced by the IIA algorithm can be expressed in the form

$$f_n(\mathbf{x}) = \sum_{i=1}^{n} a_i g_i(\mathbf{x}) \tag{9.17}$$

where the a_i coefficients are simple deterministic functions of the α and β values computed at each step of the IIA. Thus, the model produced by this algorithm has the familiar "linear combination of basis functions" form, where $\{g_i\}_{i=1}^{n}$ forms the basis set. In this context the g_i are referred to as *tunable* basis functions because they are "tuned" to the data in Step 2 of the algorithm. This is in contrast to the more conventional basis function approach (e.g. polynomial expansions) where the basis are *fixed* a priori, and only their relative weightings (i.e. the a_i's) are tunable.

Let us look more closely at the three steps of the IIA. The first step involves a simple subtraction which is trivial. The third step also turns out to be quite simple when the L_2 norm is used. In this case the solution for α and β has a simple closed form. The second step however, generally does *not* have a closed form solution. This step involves a search over the function space G for the best fitting member g_n. The intrinsic difficulty of this search depends largely on the manner in which the members of G are parameterized, but because they are *nonlinear* and *tunable* there is generally no closed form solution, so the search must be performed algorithmically. Unfortunately, even algorithmic solutions are not guaranteed to produce the optimal function in an efficient manner. All is not lost however, since it is often possible to produce "good approximations"

for $i = 1$ **to** n **do**

1. Compute f_{n-1}: $f_{n-1}(\mathbf{x}) = \sum_{j \neq i}^{n} a_j g_j(\mathbf{x})$
2. Compute Residual: $e_{n-1}(\mathbf{x}) = f(\mathbf{x}) - f_{n-1}(\mathbf{x})$
3. Fit Residual: $g_i(\mathbf{x}) = \arg\min_{g \in G} \|e_{n-1}(\mathbf{x}) - g(\mathbf{x})\|$
4. Update Estimate: $f_n(\mathbf{x}) = \sum_{j=1}^{n} a_j g_j(\mathbf{x})$

   where $\{a_i\}$ are chosen to minimize $\|f(\mathbf{x}) - f_n(\mathbf{x})\|$

endloop

FIGURE 9.6. Refitting Algorithm.

(i.e. near optimal functions) in a computationally efficient manner. This may be adequate, since the function produced in Step 2 does not need to be optimal to achieve the $O(1/n)$ bound in approximation error. In section 9.3 we saw that it was sufficient for Step 2 to produce a function that is within $O(1/n^2)$ of the optimum. A question remains however, as to whether or not this can actually be achieved. Although efficient algorithms exist for Step 2, we know of no *proof* that any such algorithm can produce functions that satisfy the $O(1/n^2)$ tolerance in polynomial time. To our knowledge this is an open question.

In practice it is common to use a *refitting* (or *backfitting*) procedure to "fine tune" the result of the IIA. This procedure can compensate somewhat for the suboptimal result produced at Step 2, and also to some degree for the possible limitations due to the constructive nature of the IIA. A typical refitting procedure is shown in Figure 9.6. The basic idea is to refit each basis function, one at a time, to the residual formed from the approximation using the other $n-1$ basis functions. This algorithm has the same attributes as the IIA: it optimizes individual basis functions by fitting them to a residual, and then reintegrates them into the overall fit. It differs from the IIA in that the residual is computed differently, and that the starting point for each refitting is usually close to its final point. This means that the search in Step 3 is generally very fast compared to its counterpart in Step 2 of the IIA. Because of this, refitting usually runs much faster than the IIA.

9.6.1 Projection Pursuit Methods

This section presents a brief introduction to *projection pursuit* (PP) methods [21]. Projection pursuit can be thought of as a general methodology for data exploitation that can be used in a variety of problem domains including regression, classification, density estimation and data exploration [16, 14, 15, 21]. When used for regression, PP encompasses a popular class of algorithms used to solve Step 2 of the IIA. The motivation for projection pursuit regression (PPR) is the following. Nonlinear regression can be performed accurately and robustly in lower dimensions (e.g. 1 or 2) using a wide variety of techniques (e.g. polynomials, splines, Parzen windows, etc.). However, the natural extension of these

Initialization:
Choose an initial projection, $\mathbf{w}_0$
repeat

1. Project Data: $u = \mathbf{w}_k^T \mathbf{x}$
2. Univariate Fit (find ϕ with $\mathbf{w}$ fixed): $\phi_k = \arg\min_\phi ||y - \phi(u)||$
3. Update Projection (find $\mathbf{w}$ with ϕ fixed): $\mathbf{w}_{k+1} = \arg\min_{\mathbf{w}} ||y - \phi_k(\mathbf{w}^T \mathbf{x})||$

until ($\mathbf{w}_k$ converges) ;

FIGURE 9.7. The Projection Pursuit Regression Algorithm.

techniques to higher dimensional problems is hampered by the *curse of dimensionality.* This curse manifests itself in a variety of ways, one of which is in the rapid growth in the number of free parameters associated with the model. For example, in dimension d the number of free parameters associated with a polynomial model of degree q is $O(d^q)$. This number can become quite large for even modest values of q and d. Projection Pursuit methods attempt to circumvent the curse of dimensionality by projecting the input to a 1–dimensional space before fitting a nonlinear regression function. The projection is usually linear (or affine) and thus defines a "direction of pursuit" in the original d–dimensional space. The key to success with PPR methods is in finding good projection directions. One version of the PPR algorithm is shown in Figure 9.7 [21]. The univariate fit in Step 2 can be performed using any number of nonlinear regression methods (cubic splines are a popular choice). Finding the new projection in Step 3 is the most difficult step in the algorithm. This is a nonlinear optimization problem that can be approached in a variety of ways. We will explore this step in more detail through a specific example in the next section, but for now suffice it to say that it can be computationally expensive. Although PP has become a very popular technique for nonlinear regression, the complexity of Step 3, as well as the potentially large number of iterations required for convergence of the main loop, make it a computationally expensive procedure. In the next section we show how PPR can be used with neural networks. Initially this amounts to little more than using the logistic function for ϕ in PPR. But as we shall see, the computational efficiency can be improved dramatically if we use piecewise continuous functions (called "ramp functions") instead.

9.6.2 Projection Pursuit with Neural Networks

We use the following notation to simplify our development in this section. The symbol $\tilde{\mathbf{x}}$ will be used to represent input vectors $\mathbf{x}$ that have been augmented with a 1 in the first position, i.e.

$$\tilde{\mathbf{x}} = \begin{bmatrix} 1 \\ \mathbf{x} \end{bmatrix} \tag{9.18}$$

Similarly, $\tilde{\mathbf{w}}$ will be used to represent weight vectors $\mathbf{w}$ that have been augmented with a "bias" weight in the first position

$$\tilde{\mathbf{w}} = \begin{bmatrix} w(0) \\ \mathbf{w} \end{bmatrix} \tag{9.19}$$

The dimension of these augmented vectors is $d+1$ and is denoted $\tilde{d}$.

In neural networks the most popular tunable basis function is arguably the logistic function,

$$\sigma_l(\mathbf{x}) = (1 + e^{-(\tilde{\mathbf{w}}^T \tilde{\mathbf{x}})})^{-1} \tag{9.20}$$

This function is smooth, bounded and parameterized by the weight vector $\tilde{\mathbf{w}}$. If we wish to use the logistic function in the PPR algorithm it needs to be scaled and shifted so that it can better fit functions with arbitrary range and position. These scaled and shifted logistic functions form the members of G_s,

$$g(\mathbf{x}) = a_0 + a_1\sigma_l(\mathbf{x}) = a_0 + a_1(1 + e^{-(\tilde{\mathbf{w}}^T \tilde{\mathbf{x}})})^{-1} \tag{9.21}$$

Although one could argue that this scaling and shifting is not needed, since it can be accounted for by the linear weights in (9.17), it is more convenient from the standpoint of algorithmic development to include them separately as we have done here. When viewed from the projection pursuit perspective, $\tilde{\mathbf{w}}$ plays the role of the projection vector and g plays the role of the regression function ϕ. In this case, Step 2 of the PPR algorithm involves updating the coefficients a_0 and a_1.

Using g in (9.21) as the basis, the optimization problem in Step 2 of the IIA (or equivalently Steps 2 and 3 of PPR) takes on the form

$$\theta_l^* = \{a_0, a_1, \tilde{\mathbf{w}}\}^* = \arg\min_{\theta_l} \int (e_n(\mathbf{x}) - a_0 - a_1\sigma_l(\mathbf{x}))^2 \, d\mu(\mathbf{x}) \tag{9.22}$$

where $\mu(\mathbf{x})$ is a suitable probability measure on the input space. In practice we don't have access to $e_n(\mathbf{x})$ (recall that $e_1(\mathbf{x}) = f(\mathbf{x})$), only samples of this function at a finite number points in the input space. This forces us to consider a somewhat different optimization problem, where we seek the θ_l that minimizes the error over the sample data set $\{\mathbf{x}_i, e_n(\mathbf{x}_i)\}, i = 1, 2, ...N$. This new optimization problem takes on the form

$$\theta_{l,N}^* = \{a_0, a_1, \tilde{\mathbf{w}}\}_N^* = \arg\min_{\theta_l} \sum_{i=1}^{N} (e_n(\mathbf{x}_i) - a_0 - a_1\sigma_l(\mathbf{x}_i))^2 \tag{9.23}$$

The solution to this optimization problem is generally different from the solution to (9.22). This in turn introduces error into our estimate of g_n. If we let $g_{n,N}$ represent the function parameterized by $\theta_{l,N}^*$ in (9.23), then there is an error of the form

$$e_{n,N} = g_{n,N} - g_n \tag{9.24}$$

due to the fact that g is estimated using a finite number of samples. This error is referred to as the *estimation error* [4]. The extent to which this error becomes significant depends largely on the quantity and richness of the sample data. For a single sigmoidal function Barron has shown that the estimation error is bounded by the following expression [4]

$$\|e_{n,N}\| = O(d \log N/N) \tag{9.25}$$

where all N samples are assumed to be independent and identically distributed (IID). This result assumes "perfect learning", i.e. that $\theta^*_{l,N}$ is the global optimum of (9.23). Thus, if we hope to estimate g_n closely enough so that $e_{n,N}$ is within $O(1/n^2)$ of e_n, Barron's result tells us that the number of samples N should satisfy $N/\log N = \Omega(dn^2)$. With this in mind we turn to the task of actually solving for $\theta^*_{l,N}$ in (9.23).

The solution to (9.23) has no closed form. This is easy to see because σ_l is a nonlinear function of the weight vector $\tilde{\mathbf{w}}$. But (9.23) is an unconstrained nonlinear optimization problem with a differentiable criterion, and thus lends itself to a wide variety of *local descent* search methods. The most popular method (the backpropagation algorithm [24]) uses a stochastic gradient approach. A number of more sophisticated algorithms such as Levenberg–Marquart, conjugate gradient and modified–Newton have also been proposed (see [6] for an overview of these methods). Unfortunately, the characteristics of the criterion in (9.23) can make it very difficult to optimize. Local minima can be a real problem. Their exact number and location depend largely on the data set, but it is possible to have up to an exponential number of them [1]. In addition, convergence is generally quite slow for local descent methods, and a robust stopping criteria for these searches is especially difficult to come by (see [19] for a perspective on why this is true). Although it is difficult to avoid all of these problems entirely, their degree of severity can be reduced tremendously by making a slight change in the basis function.

By replacing the logistic function with the *ramp function* (described in (9.16) and repeated below), Breiman and Friedman were able to develop a search algorithm that runs orders of magnitude faster than algorithms like backpropagation [7]. A ramp function is defined as follows. Let $z = \mathbf{w}^T\mathbf{x}$, then

$$\sigma_r(\mathbf{x}) = \begin{cases} 0, & z \leq \alpha \\ (z-\alpha)/(\beta-\alpha), & \alpha \leq z \leq \beta \\ 1, & \beta \leq z \end{cases} \tag{9.26}$$

This function can be viewed as a piecewise continuous approximation to the logistic function defined in (9.20). It is a member of G_s defined in section 9.5 and thus satisfies all the properties necessary for the approximation (and estimation) error bounds presented previously.

Following the previous development for the logistic function, the members of G_s derived from the ramp function are of the form

$$g(\mathbf{x}) = a_0 + a_1\sigma_r(\mathbf{x}) \tag{9.27}$$

The optimization problem (corresponding to Step 2 of the IIA) then takes on the form

$$\theta^*_{r,N} = \{a_0, a_1, \alpha, \beta, \mathbf{w}\}^*_N = \arg\min_{\theta_r} \sum_{i=1}^{N} [e_n(\mathbf{x}_i) - a_0 - a_1\sigma_r(\mathbf{x}_i)]^2 \tag{9.28}$$

{ *Let* $y_i = e_n(\mathbf{x}_i)$ *below.* }
Initialization:
Make initial guesses for $\mathbf{w}$, α and β
repeat

1. Compute $z_i = \mathbf{w}^T \mathbf{x}_i$, $i = 1, 2, ..., N$.
2. Partition the input data into S_-, S_l and S_+:
$$S_- = \{(\mathbf{x}_i, y_i) : z_i < \alpha\}$$
$$S_l = \{(\mathbf{x}_i, y_i) : \alpha \leq z_i \leq \beta\}$$
$$S_+ = \{(\mathbf{x}_i, y_i) : \beta < z_i\}$$
3. Perform Least-Squares Fit to S_-, S_l and S_+:
$$\mathbf{R} = (\sum_{S_l} \tilde{\mathbf{x}}_i \tilde{\mathbf{x}}_i^T)/N_l$$
$$\mathbf{r} = (\sum_{S_l} \tilde{\mathbf{x}}_i y_i)/N_l$$
$$\beta_- = (\sum_{S_-} y_i)/N_-$$
$$\beta_+ = (\sum_{S_+} y_i)/N_+$$
$$\tilde{\mathbf{w}} = \mathbf{R}^{-1}\mathbf{r}$$
4. Update α and β:
$$\alpha = (\beta_- - \tilde{\mathbf{w}}(0))/||\mathbf{w}||$$
$$\beta = (\beta_+ - \tilde{\mathbf{w}}(0))/||\mathbf{w}||$$
5. Normalize $\mathbf{w}$:
$$\mathbf{w} = \mathbf{w}/||\mathbf{w}||$$

until ($\mathbf{w}$,α,β converge) ;

Compute the bias and scale parameters:
$a_0 = \beta_-$
$a_1 = \beta_+ - \beta_-$
End

FIGURE 9.8. Breiman/Friedman Ramp Function Algorithm.

subject to the constraint

$$||\mathbf{w}|| = 1 \tag{9.29}$$

An algorithm for approximating $\theta^*_{r,N}$ is shown in Figure 9.8 [7]. Conceptually this algorithm partitions the input samples into three sets, S_- , S_l and S_+, and performs a least-squares fit to the samples in each set separately. The fit to S_- and S_+ is a constant (0th order fit), while the fit to S_l is a hyperplane (1st order fit). It then uses the result of these fits to re–partition the data into new sets S_- , S_l and S_+. This process is continued until the partitions converge. The bulk of the work performed in the main loop of this algorithm involves least-squares fitting, a process for which very efficient and numerically robust algorithms exist. Experience with this algorithm shows that it converges very quickly and scales well to higher dimensions. Its efficiency and robustness is often far superior to that of local descent algorithms (like backpropagation) used with the logistic basis.

9.7 Conclusions

In this paper, we have reviewed some theoretical results on constructive function approximation. We have specifically set up a framework where constructive algorithms based on convex combinations of elements taken from a subset of a Hilbert space can be analized. We have obtained the optimal values for the coefficients in the convex expansions to guarantee a desired convergence rate. We have also studied the trade-off between global and partial errors for those optimal values.

It was recalled that one can achieve an approximation error bound of $O(1/n)$, with n sigmoidal units, obtained one at a time. From a practical standpoint we have revealed several potential barriers to achieving this bound:

1. The function g_n in Step 2 of the IIA must be estimated from a finite number of examples.
2. The solution to Step 2 of the IIA has no closed form, and must be sought algorithmically.
3. No provably efficient algorithm is guaranteed to produce the optimal function at Step 2, or for that matter a function that is within the $O(1/n^2)$ tolerance.

In spite of these barriers, it is reasonable to assume that the $O(1/n)$ bound on approximation error can actually be achieved in practice, as long as the training examples are sufficiently rich. For higher dimensional problems it becomes essential to use algorithms like the one in Figure 9.8 if we hope to achieve good approximations in a reasonable time. In addition, we must often employ the refitting procedure in Figure 9.6 to compensate for the suboptimal functions produced in Step 2 of the IIA. Future work will concentrate on studying Step 2 of the IIA, and on studying approximating results for dynamical neural networks from a constructive point of view.

9.8 Acknowledgments

The research of D. Docampo was supported by CICYT (TIC96-0500-C10-10), ISTEC, and the US Air Force. The research of C.T. Abdallah and D.R. Hush was supported by ISTEC, Xunta de Galicia (Visiting Professors Program) and Boeing Computer Services Contract number W-30045.

9.9 References

[1] P. Auer, M. Herbster, and M.K. Warmuth. Exponentially many local minima for single neurons. In D. Touretzky, M.C. Mozer, and M.E. Hasselmo, editors, *Advances in Neural Information Processing Systems 8*, pages 316–322. Morgan Kaufmann, 1996.

[2] P. Baldi. Gradient descent learning algorithm overview: A general dynamical systems perspective. *IEEE Trans. Neural Nets*, 6(1):182–195, 1995.

[3] A.R. Barron. Statistical properties of artificial neural networks. In *Proceedings of the 28th IEEE Conf. on Decision and Control*, pages 280–285, 1989.

[4] A.R. Barron. Approximation and estimation bounds for artificial neural networks. In L.G. Valiant and M.K. Warmuth, editors, *Proceedings of the 4th Annual Workshop on Computational Learning Theory*, pages 243–249, 1991.

[5] A.R. Barron. Universal approximation bounds for superpositions of a sigmoidal function. *IEEE Transactions on Information Theory*, 39(3):930–945, 1993.

[6] R. Battiti. First- and second–order methods for learning: between steepest descent and newton's method. *Neural Computation*, 4(2):141–166, 1992.

[7] L. Breiman and J.H. Friedman. Function approximation using ramps. In *Snowbird Workshop on Machines that Learn*, 1994.

[8] L. Breiman. Hinging hyperplanes for regression, classification and function approximation. *IEEE Trans. on Inf. Theory*, 39(3), 1993.

[9] T. Chen, H. Chen, and R-W. Liu. Approximation capability in C($\bar{R}^n$) by multilayer feedforward networks and related problems. *IEEE Trans. Neural Nets*, 6(1):25–30, 1995.

[10] E.W. Cheney. Topics in approximation theory, 1992.

[11] G. Cybenko. Approximation by superpositions of a sigmoidal function. *Mathematics of Control, Signals, and Systems*, 2(4):303–314, 1989.

[12] A. T. Dingankar and I. W. Sandberg. A note on error bounds for approximation in inner product spaces. *Circuits, Systems and Signal Processing*, 15(4):519–522, 1996.

[13] J.H. Friedman. Multivariate adaptive regression splines. *The Annals of Statistics*, 19, 1991.

[14] J.H. Friedman and W. Stuetzle. Projection pursuit regression. *J. Amer. Stat. Assoc.*, 76:817–823, 1981.

[15] J.H. Friedman, W. Stuetzle, and A. Schroeder. Projection pursuit density estimation. *J. Amer. Stat. Assoc.*, 79:599–608, 1984.

[16] J.H. Friedman and J.W. Tukey. A projection pursuit algorithm for exploratory data analysis. *IEEE Transactions on Computers*, C–23(9):881–890, 1974.

[17] F. Girosi and G. Anzellotti. Convergence rates of approximation by translates. Technical Report 1288, MIT Art. Intell. Lab., 1992.

[18] Simon Haykin. *Neural Networks: A Comprehensive Foundation*. Macmillan, New York, 1992.

[19] D. Hush, B. Horne, and J.M. Salas. Error surfaces for multi–layer perceptrons. *IEEE Transactions on Systems, Man and Cybernetics*, 22(5):1152–1160, 1992.

[20] K. Hornik, M. Stinchcombe, and H. White. Multilayer feedforward networks are universal approximators. *Neural Networks*, 2(5):359–366, 1989.

[21] P.J. Huber. Porjection pursuit. *The Annuals of Statistics*, 13(2):435–475, 1985.

[22] L.K. Jones. A simple lemma on greedy approximation in hilbert space and convergence rates for projection pursuit regression and neural network training. *The Annals of Statistics*, 20:608–613, 1992.

[23] G. Pisier. Remarques sur un resultat non publié de b. maurey, 1980–1981.

[24] D.E. Rumelhart, G.E. Hinton, and R.J. Williams. Learning internal representations by error propagation. In D.E. Rumelhart and J.L. McClelland, editors, *Parallel Distributed Processing: Explorations in the Microstructure of Cognition*, pages 318–362. MIT Press, Cambridge, MA, 1986.

[25] I.W. Sandberg. Structure theorems for nonlinear systems. *Multidim. Syst. and Sign. Proc.*, 2:267–286, 1991.

[26] I.W. Sandberg. Uniform approximation and the circle criterion. *IEEE Trans. Automat. Control*, 38(10):1450–1458, 1993.

[27] I.W. Sandberg. General structures for classification. *IEEE Trans. Circ. and Syst.-1*, 41(5):372–376, 1994.

[28] Y. Zhao. *On projection pursuit learning.* PhD thesis, Dept. Math. Art. Intell. Lab., MIT, Boston, MA, 1992.

10

Decision Trees Based on Neural Networks

J. Cid-Sueiro
J. Ghattas
A.R. Figueiras-Vidal

ABSTRACT The classification of a data collection using tree structures has been studied by statisticians and psychologists for many years, and it has shown to be an effective way of dividing a complex classification problem in a sequence of simpler decision tasks. The modularity of both learning and classification showed by these structures has attracted the attention of neural network researchers, looking for alternatives to the learning and computational problems of backpropagation networks. This paper overviews the current research on tree classification based on neural networks. Structure and learning algorithms are described; the implications of probabilistic interpretations of the network behaviour are discussed and some new learning rules that can speed up learning and reduce the final misclassification probability are proposed.

10.1 Introduction

Human beings are accustomed to solve complex problems by splitting them into simpler tasks: grouping objects in families and sub-families is often the first step to gain a further insight in their underlying structure. Before making decisions about a possible action, we usually first extract information about their consequences, or about the conditions in which they have to be made; thereafter, the final decision can be reduced to a simple test. Many decision processes can be represented by a decision tree that is a sequence of partial decision steps: branches and subbranches are selected until we arrive to the terminal nodes, where the decision is finally made (Fig. 10.1).

There are also evidences that modularity is a common way of information processing in our brains [8]. Decision trees have also attracted the interests of statisticians for many years [20]; many questions concerning the determination of optimal decision trees has been the research topic in the statistical literature. In Signal Processing, tree classifiers have been proposed for a wide number of applications: speech processing [6][34], equalization of digital communication systems [32, 11, 4], medical diagnosis [17], remote sensing [22, 19], character recognition [25], control systems [28],etc.

More recently, the construction of a decision tree from a set of examples became

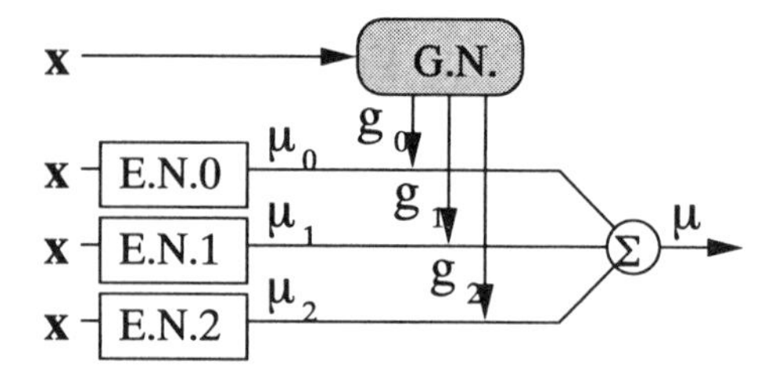

FIGURE 10.1. A decision tree structure

a problem of interest in machine learning and neural networks; the modular approach to learning seems to be a promising alternative to solve many of the problems inherent to other non-linear structures, as the multilayer perceptron (MLP). Several arguments have been given in favor of the decomposition of the network in several modules [8, 21, 33]:

- Speed of learning: in general, learning becomes a harder problem as the network size increases. A simple module is much easier to train.
- Data representation: in a modular structure, the identification of the task carried out by every network module is easier to identify.
- Structure optimization: as a consequence of the previous item, there is a natural way of adapting a tree structure to the environment: for instance, trees can be constructed in such a way that terminal nodes with a high degree of misclassification are grown and those that do not participate actively in the decision process can be pruned.

There are, however, several problems concerning the design of tree classifiers [33]. For instance, although a modular structure allows the decomposition of a complex problem into simpler tasks, modular decision does not imply modular learning; a simple module is easier to train but, to do so, we should be able to modularize learning. On the other hand, many of the difficulties found when growing and pruning is applied to multilayer perceptrons [26] may also appear in tree structures. In general, the design of tree classifiers states three main problems [27, 33]:

- Feature selection; that is, the selection of the attributes that should be used to make a decision at each node of the tree.
- Decision rule learning: in order to adapt the network to a particular spatial distribution of the data points in the space, a learning algorithm should be used to adjust the network weights and thresholds.
- Structure optimization: the selection of pruning and growing algorithms that determines a network structure satisfying some trade off between the global number of nodes and weights and the misclassification rate.

The purpose of this paper is to discuss and overview the problem of training tree classifiers from examples, and its relationship with neural networks. Although we will briefly overview the feature selection and structure optimization problems, we will focus our attention on the selection of the learning algorithm for the different modules. Detailed discussions about structure optimization and feature selection, can be found in other reviews of topic [27] [33].

The paper has been structured as follows: Sect. 2 describes the general tree structure that we will discuss along this paper; Sect. 3 overviews some non-neural approaches used to train decision trees, which will serve to establish some comparisons with the neural trees discussed in Sects. 4 and 5; Sect. 5 focuses the problem of training soft decision trees; in Sect. 6 we propose a novel training algorithm for these structures and, finally, the conclussions and further lines suggested in Sect.7 conclude the paper.

10.2 Adaptive modular classifiers

10.2.1 The classification problem

Let us define a set of data patterns $\chi = \{x_1, x_2, \cdots, x_n\}$, each one of them belonging to a class in a finite set $\boldsymbol{\Omega} = \{a_1, a_2, \cdots, a_c\}$. The patterns can be previously contaminated by noise or by some non-linear operator, and they can even be an infinite set. They are usually represented as arrays with a fixed number, N, of components or *attributes*; in the sequel, these attributes are assumed real numbers; thus, patterns are points in an N-dimensional space.

The classification problem consists of establishing a correspondence between χ and $\boldsymbol{\Omega}$ which maps each pattern x_i into its class a_j; that is, to partition the pattern space into c regions, one per class.

The decision made by the system for an input patterns can be represented in two ways:

- using a one-output network, where the decision is usually an index representing the class assigned to the input; or,
- using a multi-output network, where there is one output for each possible class; only the class output assigned by the classifier to the input pattern is nonzero.

If the regions occupied by the patterns of the different classes are overlapped, or if the classifier is not able to map intricate regions, some patterns may be incorrectly classified, and the system should be designed in order to minimize the number of errors.

10.2.2 Splitting the input space

When the patterns are distributed over the data space in complex regions, the design of a system that separates the different classes becomes difficult. The idea underlying modular classification consists of dividing the input space into subregions, so as to assign a different classifier to every subregion, hoping that the classification subproblems would become simpler than the original one.

Fig. 10.2 represents such a classifier structure. Each expert module should be specialized in a different subregion of the input space; the gating net select which expert is the most adequate for the current pattern. It has so many outputs as expert modules.

Note that the gating net is also a classifier, which has to decide what kind of pattern is at the input, so as to select the expert to manage the data. Thus, it

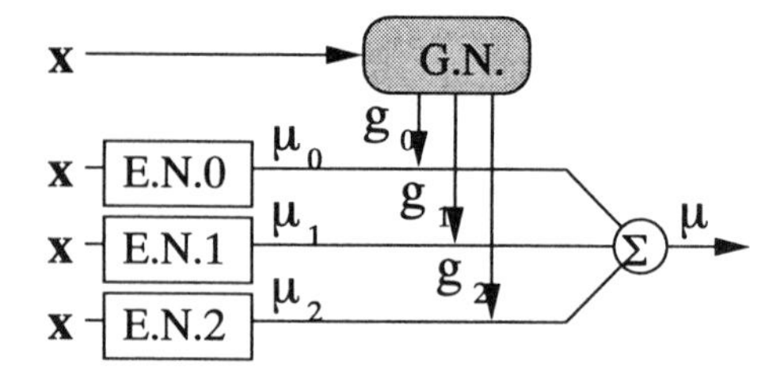

FIGURE 10.2. A modular classifier with 3 expert networks (E.N.); G.N. is the gating network

can also be implemented by means of another modular classifier. In such a case, the modular structure is expanded into a tree with two or more levels of gating nets, as shown in Fig. 10.3. A gating net in the i-th level select one of a sub-group of the gating nets in the $(i+1)$-th level.

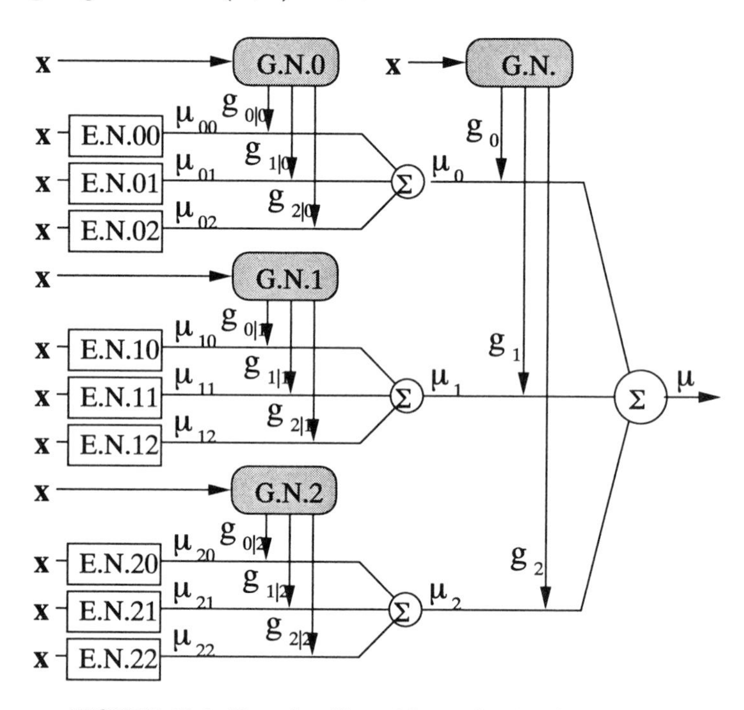

FIGURE 10.3. Tree classifier with two levels of gating nets

Without loss of generality, we assume that all the modules receive the same input. If different inputs are desired, we can consider that every module only uses some of the information at the input, rejecting the rest (for instance, multiplying by zero the non-used components of the input).

Note that we have not imposed any constraint to the modules, so we can add computational capabilities to the network using universal classifiers such as MLP's or any other neural network at each module. However, modularity has been adopted in order to employ simple gating and expert nets, such as threshold detectors or linear modules with a non-linear activation function at their output. The global structure is nonlinear and the boundary decisions of the classifier can be made arbitrarily complex by increasing the number of nodes in the tree. However, if some previous information about the data distribution is available, other kind of modules may be recommended.

Finally, note that, as a linear classifier can just construct hyperplane boundaries, a tree classifier based on linear filters is equivalent to a piece-wise linear classifier.

10.2.3 Supervised and non-supervised learning

Usually, the only available information about the data distribution is the class of the training patterns. The pattern class is mainly useful for adapting the parameters of the experts, becaus e they have to carry out a decision about the class. However, the gating nets do not decide about the class of the pattern, but about the expert that should manage the current data. This suggests that unsupervised learning could be used to train the gating nets [7, 15]; there are, however, some problems associated with this approach, and many algorithms apply supervised techniques even for the gating nets. Moreover, most tree structures proposed in the literature does not make differences between gating and expert networks, and all modules try to classify the data, even if there are not placed in the terminal nodes.

The main difference between the algorithms that we will describe in the following sections are given by the solutions they propose to the problem of dividing the input space, that is, the problem of training the gating nets.

10.3 A survey on tree classification

10.3.1 Hypercubic cells

Perhaps, the simplest way of splitting the input space is to use a fixed and data-independent space partition. This eliminates the need to train the gating nets; only the expert modules should be adapted to the respective subregions.

This idea has been explored by Callender [4] in a digital equalization problem: a digital transmission system sends a sequence $\{s_i\}$ of binary symbols (alphabet $A = \{a_0, a_1\}$), through a dispersive digital channel. Received sample x_k at time k is related to the transmitted sequence according to

$$x_k = h(\mathbf{s_k}) + n_k \tag{10.1}$$

where h is the channel distortion function, and n_k is typically a Gaussian noise process.

A digital feedforward equalizer has to restore the transmitted sequence $\{s_k\}$ computing a function of last received samples $\{x_k, x_{k-1}, \ldots, x_{k-n}\}$, which can be expressed as

$$\mu_{k-d} = g(x_k, x_{k-1}, \ldots, x_{k-n}) \tag{10.2}$$

where d is a time delay imposed to take advantage of using future samples in the decision process, and g is the function implemented by the equalizer. Although the linear equalizer is widely used, the equalization problem is inherently non-linear (even if the channel is linear) and, thus, non-linear networks can reduce the error probability.

In [4], the input space is divided into hypercubic regions using a linear filter for each region. During training, only the filter of the hypercube containing the input vector is adapted.

The main advantage of this scheme is that it can be easily done for a digital implementation. Since the observation vector is usually stored in a binary form, groups of the most significant bits can be used to provide a uniform division of the input space. This is illustrated in Fig. 10.4, which shows the way to select the index of the hypercube (i.e. a square in this two-dimensional example) containing the input vector.

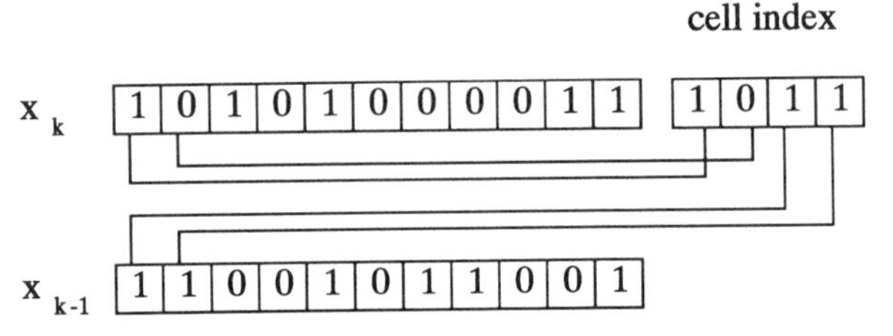

FIGURE 10.4. Selection of a hypercubic cell using the most significant bits of the received samples

The extreme simplicity of the hypercubic partition states several problems:

- If a complex mapping is necessary to separate classes, the required network can be too large.
- Many filters in the partition may be never used, because most hypercubes do not contain samples of more than one class, and their filters are not really necessary. Obviously, a pruning method can reduce the network size, although in applications where the data distribution varies with time (as equalizing time variant channels), the boundary decision may move from some regions to others, making the network pruning and growing difficult.
- Finally, as the hypercubic cells become smaller, the filters of the cells without training samples have no determined parameters. Thus, the patterns falling in these region will be classified randomly, and the network generalization ability is reduced. A large training set is required in order to get a fine parameter adjustment; in some applications, such as the equalization of some digital channels,where a simple linear discriminant may obtain an approximate class separation with a few examples, the increasing training time may not balance the reduction in error probability.

From the previous considerations, one may conclude that the classifier efficiency can be increased by using adaptive gating nets. However, the use of adaptive gating nets increases the complexity of the decision tree and introduces the problem of finding the best partition for a set of expert modules. Thus, while designing tree structures, one should look for a trade off between the increment of the gating net classification capabilities and the extra complexity of learning and classification.

10.3.2 Thresholding attributes

A simple configuration of tree classifiers with adaptive gating nets that does not require a high computational cost consists of using a unique attribute of the input pattern at each node. Each gating net implements a simple comparison of the attribute with some threshold value; the result determines if the input pattern is fed to the left or right node in the next level (non-binary partitions are also possible using several thresholds).

The main problems are the determination of the tree structure and the selection of the feature to be tested at each node. It has been shown that the problem of finding the smallest possible decision tree that correctly classifies the training data is NP-complete [10]; therefore, all the algorithms proposed in the literature are based on heuristics.

The tree structures are usually constructed starting from a root node, which is grown by successive splitting into child nodes. The threshold and the attribute are selected in order to get the classes as much separated as possible. The usual way of determining the efficiency of a splitting is to define some 'impurity' function, which is related to the distribution of the patterns after the split: for instance, the *entropy* function [30, 24]

$$H(p) = -\sum_j p_j \log p_j \tag{10.3}$$

where p_j is the probability of a pattern reaching the node is of class j. If the classes are well separated, most patterns reaching the node will be of the same class, and the entropy is small.

Both attributes and thresholds are selected in such a way that the impurity reduction is maximized. If the root node does not reach a zero misclassification error, another feature is selected and a new partition is carried out. The structure is thus grown up to the moment where all training patterns are correctly classified. The determination of the threshold which produces the best class separation for a given attribute is not difficult (it can be carried out, for instance, by an exhaustive search over the training set).Finally, in order to avoid an excessive network growing, pruning algorithms are commonly applied after growing.

More detailed discussions of feature selection and structure optimization methods can be found in [3, 27, 33].

10.3.3 Linear Combinations of the Attributes

The space partition generated by the application of a threshold to one of the attributes in a tree node is given by an hyperplane that is orthogonal to the coordinate axis corresponding to this attribute in the input space. Breinman [3] was one of the first authors to suggest the use of linear combinations of features in the gating nets, in order to divide the input space with general hypersurfaces.

Gelfand [31, 32] proposed a structure in which the task of selecting a linear filter to make a decision is also carried out by linear filters. It is adapted iteratively as follows: let $\mathbf{x}(k)$ be the input vector to the filter; the nodes of the tree are numbered consecutively from up to down and from left to right, as shown in Fig. 10.1. The linear filter at node i is determined by weights $\mathbf{w}_i(k)$ and offsets $d_i(k)$, in such a way that output $\mu_i(k)$ is related to the input vector according to

$$\mu_i(k) = \mathbf{w}_i(k)^T \mathbf{x}(k) + d_i(k) \tag{10.4}$$

Except those of the terminal nodes, every filter output is compared with an adaptive threshold $\theta_i(k)$; the result determine s whether the input vector is propagated through the left or right child node.

During training, the frequency of an input vector passing through node i is iteratively estimated using

$$p_i(k+1) = p_i(k) + \eta(I_i(k) - p_i(k)) \tag{10.5}$$

where $I_i(k)$ is an indicator that is equal to one if $\mathbf{x}(k)$ reaches node i and zero otherwise; η is the adaptation step.

The filter coefficients and the threshold are updated by means of a stochastic gradient learning rule that minimizes quadratic cost functions,

$$\epsilon_i = E_i \left\{ (y(k) - \mu_i(k))^2 \right\} \tag{10.6}$$

$$\beta_i = E_i \left\{ (y(k) - \theta_i(k))^2 \right\} \tag{10.7}$$

The following rules result:

$$\eta_i(k) = \frac{\eta}{p_i(k+1)} \tag{10.8}$$

$$\mathbf{w}_i(k+1) = \mathbf{w}_i(k) + \eta_i(k) I_i(k)(y(k) - \mu_i(k))\mathbf{x}(k) \tag{10.9}$$

$$d_i(k+1) = d_i(k) + \eta_i(k) I_i(k)(y(k) - \mu_i(k)) \tag{10.10}$$

$$\theta_i(k+1) = \theta_i(k) + \eta_i(k) I_i(k)(y(k) - \theta_i(k)) \tag{10.11}$$

A tree pruning method is also proposed. In [32] a recursive tree is designed for an equalization application; the simulations carried out show that the tree equalizer outperformed linear, polynomial and decision feedback (DFE) equalizers (training with standard LMS) both in learning time and final error probabilities. It is also shown that, as the length of the channel response grows, the complexity requirements of the polynomial equalizers are much greater than those of the tree structure.

10.4 Neural Decision Trees

The first attempts of joining ideas from neural networks and decision trees were based on using neural schemes in the modules. Note that, actually, the modules of the Gelfand's scheme are perceptrons, and the learning rule is similar to the perceptron rule. Sankar et al.[34] use modules with a sigmoidal activation function, although the decision tree is still a hard structure, as only the child node connected to the sigmoid that produced the highest output receives the training pattern. The sigmoid function serves for learning purposes: the L_1 norm between the sigmoid output and the desired neuron output is minimized, resulting the learning rule for weight vector $\mathbf{w}$ given by:

$$\mathbf{w}_{k+1} = \mathbf{w}_k - \eta\mu_j(1-\mu_j)\text{sign}(y_j - \mu_j)\mathbf{x}_k \tag{10.12}$$

where k is the time. The L_2 criterion was also tested, with a poorer classification performance. The weight adaptation is stopped when the average error over some time window does not decrease beyond a small threshold. Thereafter, the hyperplane formed by the neuron is used to split the feature space into two subregions. Each subregion is assigned to a new perceptron, and the algorithm is repeated.

As in Breiman [3], the tree is pruned after growing, in order to keep a moderate network size. This is done by defining a cost function which penalizes the larger networks, in such a way that only if the reduction in misclassification probability is significant, the network is not pruned. Thus, the number of tree nodes can be reduced, at the expense of a higher error rate.

This approach has been applied to the speaker independent vowel recognition, and compared with other adaptive schemes: MLP, radial basis functions (RBF), Gaussian nodes, and the CART algorithm. The test set stated a difficult problem, where the best classifier just achieved a classification performance of 55 percent. The proposed structure, that Sankar et al. call a Neural Tree Network (NTN) achieved a much better performance than the CART algorithm, and a similar performance to the best MLP. The MLP, however, resulted to be very sensitive to the number of nodes, and performance decreased significantly when the number of layers was not appropriate. The NTN does not suffer of this problem as the network size is adapted by the growing and pruning algorithm; moreover, it showed a faster learning than backpropagation networks.

10.5 Hierarchical mixtures of experts

10.5.1 Soft decision classifiers

A natural generalization of the decision trees in the previous section consists of using more general non-linear classifiers at each module of the network. This idea has been explored by several authors [18, 5]; MLP's are used as experts or gating nets, generating non-linear partitions of the input space at each level of the network.

However, in these schemes, a global gradient search technique cannot be used, because the input pattern is passed through the successive levels of the network using logical gates. Partial decisions are made by the network at all levels, so the global function implemented by the tree is not differentiable. Alternatively, the Hierarchical Mixture of Experts (HME) proposed Jacobs and Jordan [15] employs soft decisions at the end of all the network modules. Just a final slicer is used to make the decision about the class; thus, the non-linear function computed by the network before the slicer is differentiable, and a gradient search over the whole net can be done.

There is an additional reason to search for global approximation methods: in the decision trees proposed above, there is no functional difference between experts and gating nets, in the sense that even the later behave as classifiers, trying to separate classes. The learning algorithms proposed for the gating nets pursue a correct classification at every gating net; if it is not achieved, the network is usually grown hanging a subtree down this module. However, the main goal of the gating nets is finding the appropriate partition of the input space, in order to make easier the classification task of the expert modules. They do not need to be good classifiers to achieve a good partition. Consider, for instance, the 'donut' problem discussed in [34], or the 'pac-man' problem represented in Fig. 10.5.

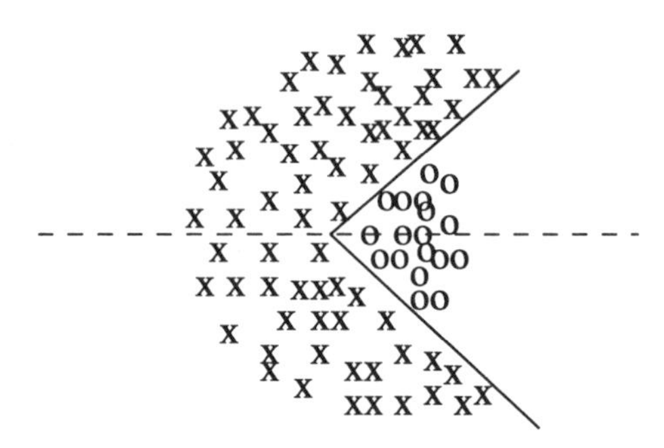

FIGURE 10.5. The 'pac-man' problem

In this example, the boundary minimizing the probability of misclassification at the root node of the tree leaves all points in one of the decision regions, the other being empty. Adding more levels to the tree cannot reduce the error further, because one of the child nodes must solve the same problem as the root node. Note, however, that a decision tree with a unique gating net can separate all samples. Some heuristic must be used to solve this problem. The price paid for using a soft structure, as stated in the introduction of this paper, is that all the tree nodes are used to make the decision, thus increasing the computational load and the decision time in a sequential implementation of the algorithm. In a paralle l architechture, however, the computing time is proportional to the number of levels in the network.

The HME network with general neural modules has been also proposed [14]; however, we will consider here the case where nodes compute linear combinations of the features. Our analysis follows [15] although, for simplicity, we consider binary gating and expert nets.

10.5.2 Training HME classifiers

As the function computed by a HME network is differentiable, stochastic gradient search methods can be applied. In [12], several cost functions for stochastic gradient learning were applied to the equalization of binary symbols in a digital communication context, concluding that the logarithmic cost given by

$$C(y, d) = -d \log y - (1 - d) \log(1 - y) \tag{10.13}$$

is the most adequate. The resulting stochastic learning rules are simple, but they present a slow convergence speed. On the other hand, Jordan *et al.* [15] applied the Expectation-Maximization (EM) algorithm [1] to these structures. They have also proposed an on-line algorithm, based on the application of Recursive Least Squares (RLS) algorithm to every module of the HME, that can be applied to regression problems, but is not suitable for classification. This algorithm works in 2 steps: first, an expectation (E) step, where the expected value of a log likelihood function is taken with respect to some unknown parameters, as we explain below; and a maximization step (M step) in which the expected log likelihood function is maximized. The complexity problems of the maximization step in the EM algorithms can be reduced (at the cost of increasing the training time) by replacing the global maximization by just one iteration of a stochastic learning rule. This is in fact the recommended learning mode in some applications, such as data equalization in digital communications. In [15], a stochastic EM algorithm is derived after some simplifications of the M-step.

In this section, we show that the stochastic EM algorithms can be viewed as a particular case of stochastic learning rules characterized by different cost functions. Moreover, we propose a modification of the EM algorithm that gives an initially faster convergence.

As in [15], let us consider, for simplicity, a HME network with just two levels (the analysis can be easily generalized to an arbitrary number of levels); if there is just a single output, μ, we can write

$$\mu = \sum_{i=0}^{1} \sum_{j=0}^{1} g_i g_{j|i} \mu_{ij} \tag{10.14}$$

where μ_{ij}, $g_{j|i}$ and g_i are the outputs of the experts, the first level gating nets and the top level gating nets, respectively. Let us assume that each gating net has a unique filter and that both expert and gating nets are linear filters with a sigmoidal activation unit; thus

$$\mu_{ij} = \text{sigm}\left(\mathbf{w}_{ij}^T \mathbf{x}\right) \tag{10.15}$$

$$g_{1|i} = 1 - g_{0|i} = \text{sigm}\left(\mathbf{v}_i^T \mathbf{x}\right) \tag{10.16}$$

$$g_1 = 1 - g_0 = \text{sigm}\left(\mathbf{v}^T \mathbf{x}\right) \tag{10.17}$$

where $\mathbf{w}_{ij}$, $\mathbf{v}_i$ and $\mathbf{v}$ are the weights of the different modules.

10.5.3 Applying the EM algorithm

Let us assume that the class of the input has been generated by a stochastic tree with a structure similar to that in Fig. 10.3; consider, however, that the output of expert module (i, j) is not μ_{ij}, as in the figure, but a binary value y_{ij}, which is generated at random, according to a Bernoulli distribution with parameters μ_{ij} $(i = 0, 1,\ j = 0, 1)$ which may depend on $\mathbf{x}$; thus, we can write

$$P_{ij}\left(y_{ij} \mid \mathbf{x}\right) = \mu_{ij}^{y_{ij}} \left(1 - \mu_{ij}\right)^{1 - y_{ij}} \tag{10.18}$$

In a similar way, for every input $\mathbf{x}$, the gating nets at the top and second level generate binary outputs z_i and $z_{j|i}$, respectively, according to Bernoulli distributions with parameters g_i and $g_{j|i}$, respectively. The network output is the class of the input

$$y = \sum_i \sum_j z_{ij} y_{ij} \tag{10.19}$$

where

$$z_{ij} = z_i z_{j|i} \tag{10.20}$$

Binary class y can also be represented by a Bernoulli model; it is easy to show that, if the Bernoulli variables defined above are statistically independent, the distribution of the class is given by:

$$P(y \mid \mathbf{x}) = \mu^y (1-\mu)^{1-y} \tag{10.21}$$

where

$$\mu = \sum_i \sum_j g_{ij} \mu_{ij} \tag{10.22}$$

$$g_{ij} = g_i g_{j|i} \tag{10.23}$$

Therefore, the HME network in Fig. 10.3 computes the probability of the stochastic net generating a class $y = 1$. Thus, if the weight of the network are such that the modules compute the parameters of the Bernoulli distributions described above, the HME network becomes a probability estimator, and it works as the optimal Bayesian detector.

To compute the network weights $\mathbf{w}_{ij}$, $\mathbf{v}_i$ and $\mathbf{v}$ leading to such estimates, note that

$$P(y, z_{ij} \mid \mathbf{x}) = \prod_i \prod_j \left(P(z_{ij} \mid x) P(y \mid \mathbf{w}_{ij}, \mathbf{x})\right)^{z_{ij}} \tag{10.24}$$

and, taking logarithms,

$$L(y, \mathbf{x}) = \sum_i \sum_j z_{ij} \left(\ln g_i + \ln g_{j|i} + \ln P(y \mid \mathbf{w}_{ij}, \mathbf{x})\right) \tag{10.25}$$

If variables z_{ij} were known, the HME weights could be estimated maximizing the log-likelihood function in Eq.(10.25). Unfortunately, this is not usually the case. The EM algorithm [1] solves this problem by taking the expectation of $L(d, \mathbf{x})$ (step E) with respect to these *hidden variables*, before its maximization (step M). It can be shown [15] that

$$E\{L \mid \mathbf{x}, \mathbf{w}_{ij}, \mathbf{v}_i, \mathbf{v}, y\} = \sum_i \sum_j h_{ij} \ln g_i g_{j|i} P(y \mid \mathbf{w}_{ij}, \mathbf{x}) \tag{10.26}$$

where

$$h_{ij} = \frac{g_{ij}(1 - y - \mu_{ij})}{1 - y - \mu} \tag{10.27}$$

Furthermore, it can be shown that, is the HME net computes the Bernoulli parameters, variables h_{ij} are the posteriori probability of expert (i, j) generating class y,

$$h_{ij} = P\{z_{ij} = 1 \mid \mathbf{x}, y\} = h_{ij} = h_i h_{j|i} \tag{10.28}$$

where

$$\begin{aligned} h_i &= P\{z_i = 1 \mid \mathbf{x}, y\} = \frac{g_i \sum_j g_{j|i} P_{ij}(y)}{\sum_i g_i \sum_j g_{j|i} P_{ij}(y)} \\ h_{j|i} &= P\left\{z_{j|i} = 1 \mid \mathbf{x}, y\right\} = \frac{g_{j|i} P_{ij}(y)}{\sum_j g_{j|i} P_{ij}(y)} \end{aligned}$$

The expected log-likelihood is a non-linear function of the weights. In order to find its maximum, Jordan *et al.* propose the application of the Iteratively Re-weighted Least Squares (IRLS) algorithm [23].

In the paper, it is also proposed the application of HME architectures to regression problems. In that case, the expert modules are just linear filters without sigmoidal activations and a Gaussian probability model replace the Bernoulli model describe here. This is the case tested in the simulations. In order to avoid the computational load of the IRLS algorithm, an on-line version is proposed, which is based on the application of the Recursive Least Squares (RLS) algorithm [9]. However, it can not be applied to classification problems.

In order to avoid the computational load required by the global maximization (M step), it is possible to replace it by a local maximization, computing a single iteration of a stochastic gradient search; following Dempster *et al.* [1], this is done assuming that h_{ij} in Eq.(10.26) is *independent* on the network parameters; however, it can be shown that the resulting algorithm is equivalent to a stochastic gradient minimization of the logarithmic cost function (10.13). In spite of this, it is known that this cost is minimized when the networks becomes the optimal Bayesian detector [2] (if it has the required size).

In [13] we tested the stochastic minimization of Eq.(10.26) taking into account the dependence of h_{ij} with respect to the network parameters. The resulting algorithm is derived in Appendix A, and shown below.

$$\begin{aligned} \Delta \mathbf{w}_{ij} &= \eta f_{ij}(y - \mu_{ij})\mathbf{x} \\ \Delta \mathbf{v}_i &= \eta(f_{i1} - f_i g_{1|i})\mathbf{x} \\ \Delta \mathbf{v} &= \eta(f_1 - g_1)\mathbf{x} \end{aligned} \tag{10.29}$$

where η is the adaptation step and

$$\begin{aligned} f_{ij} &= h_{ij}\,(1 + \varepsilon_{ij}) \\ f_i &= \sum_j f_{ij} \end{aligned} \tag{10.30}$$

Variables ε_{ij} are defined as:

$$\varepsilon_{ij} = \ln h_{ij} - \sum_i \sum_j h_{ij} \ln h_{ij} \tag{10.31}$$

These terms make the difference between the previous algorithm and that resulting from the minimization of a logarithmic cost, which can be obtained by assuming $\varepsilon_{ij} = 0$.

Note that in the HME network structure, each gating net output is the result of applying a soft max function to the output of the corresponding gating net

filter. In our case, we considered HME structures where there are only 2 outputs in each gating net. so as to reduce the softmax function into a sigmoid function. The softmax function is non linear non bijective function, that has no unique inverse function. However, we can see that when applying the softmax function to ε_{ij}, the following relations result:

$$\text{softmax}(\varepsilon_{ij}) = \frac{e^{\varepsilon_{ij}}}{\sum_k \sum_l e^{\varepsilon_{kl}}} = h_{ij} \tag{10.32}$$

and

$$\sum_m \sum_n h_{mn} \varepsilon_{mn} = 0 \tag{10.33}$$

Thus, we can see ε_{ij} can as the *inverse* of the softmax function for the set of values $\{\varepsilon_{ij}\}$ that verify Eq.(10.33).

Fig. 10.6 compares the convergence vis time of both algorithms when training a 4 level HME architecture to equalize the linear non-minimum-phase channel with transfer function $H(z) = 0.5 + z^{-1}$, for several SNR values. The inputs to the network are the last 2 received samples. The adaptation step was optimized independently for each algorithm. Initially, the resulting algorithm has a significantly faster convergence but, as can be observed in Fig.10.7, the final bit error rate of the proposed scheme is usually higher than the resulting from using a logarithmic cost function. This suggests the possibility of using this algorithm for a fast start-up while switching to the stochastic minimization of the log cost thereafter. As we have seen, this can be done by simply making $\varepsilon_{ij} = 0$ in Eq.(10.30).

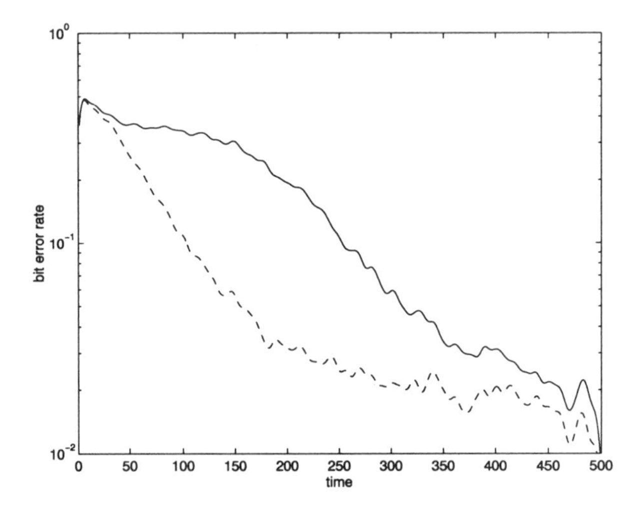

FIGURE 10.6. Evolution of the BER vs time, using a logarithmic cost (continuous) and with the proposed method (dashed)

10.6 Lighting the hidden variables

In [13], an alternative approach to the EM algorithm attempts to reduce the computational complexity of the training algorithm for modular networks by establishing a known training reference for each module of the network. In this section we describe the proposed structure.

Each level of the HME network makes consecutive soft partitions of the input space. Note that, in the EM approach, the hidden variables z_{ij} determine which expert network should make the symbol decision. The EM algorithm solves the lack of knowledge about these hidden variables by computing their expected values. However, if they were known, they could be used as output references for the network modules, and each module could be trained separately in a more efficient way.

In an equalization application, a possible way to do so is to use the past training symbols as references, as we describe below. Let us consider, for example, a 2-level HME network. We train the gating net in the top level (level 2) using symbol y_{k-2} as their output reference; level-1 modules with reference y_{k-1}, and level-0 modules (i.e. the expert networks) using y_k. In each level only one expert is trained at a time depending on the value of the past training symbols, which leads to lower numerical cost. Using the same 3 levels HME network example, in level 2, the unique expert is always trained, while in level 1 the first expert is trained only if $y_{k-2} = 0$ and the other is trained only if $y_{k-2} = 1$; in level 0, expert 1 is trained only if $y_{k-2}y_{k-1} = 00$, expert 2 is trained only if $y_{k-2}y_{k-1} = 01$, expert 3 only if $y_{k-2}y_{k-1} = 10$ and expert 4 only if $y_{k-2}y_{k-1} = 11$.

Working in this way, each module (a sigmoidal perceptron) can be trained separately; moreover, as only one filter at each level is updated at every training step, the computational load is considerably reduced; it increases linearly with the number of levels (i.e. with the logarithm of the number of modules). Also, making a hard partition of the input space (replacing the soft decision devices by hard slicers), each module becomes a linear equalizer which can be trained using standard LMS or RLS algorithms.

Fig. 10.7 shows that the use of the previous symbols of the training sequence to update the gating nets offers significant faster convergence and lower bit error rate than those based on hidden variables. Linear channel $H(z) = 0.5 + z^{-1}$ with 15dB of signal to noise ratio.

10.7 Conclusions

The modular approach to classification has shown to be a promising alternative to the learning and computational problems of backpropagations structures. The application of neural networks to pattern classification with decision trees has been addressed in two directions

- Introducing neural tree nodes, in order to map non-linear boundaries
- Using soft decisions for learning and classification

In general, large neural nodes reduce the advantages of the modular approach, which is based on constructing large networks with simple elements. On the other

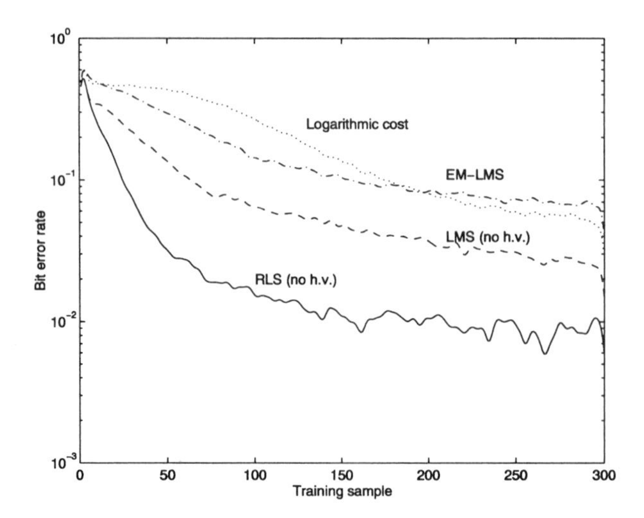

FIGURE 10.7. Evolution of the bit error probability during training for the RLS and LMS with different references for each level and no hidden variables (h.v.) vs the EM-LMS and the standard stochastic gradient search with a logarithmic cost. The curves are the average of 400 simulations, after smoothing with a low pass filter

hand, the use of soft classifiers allows the application of stochastic gradient learning rules to the whole net,at the expense of an increment of the computational cost in a serial implementation, because all nodes are active for every input.

This work is mainly focused on the soft decision trees; the maximum likelihood framework is a usefull tool which opens the way for a probabilistic interpretation of the network behaviour. The stochastic version of the EM algorithm described in this paper yields the same learning rules as the LMS applied to the logarithmic cost function, which is known to converge to the bayesian classifier.

The paper presents two novel algorithms for training neural trees; the first is a modification of the stochastic version of the EM algorithm, which shows a faster startup in the simulations, in an equalization application. The reasons for this behaviour are being explored by the authors. The second approach suggest the use of lateral information as a reference for training the gating nets. Although this is, theoretically, a suboptimal approach, the use of independent references for each module allows the application of standard linear estimation methods. On the other hand, simulations showed that the final bit error rate is reduced with respect to that of the single-reference methods.

Note that, as described above, the gating nets of the hard tree classifiers are also used to separate the classes, which is an inefficient method, as in the 'pacman' problem. The main goal of the gating nets is to make easy the task of the experts. In order to do so, we are exploring the application of non-supervised learning rules for the experts.

Finally, note that the logarithmic cost is not the only cost function which is minimized when the classifier becomes Bayesian. This also happens usingL_2

[29] norm and, also, the family of cost functions described in [16]; the authors are investigating the consequences of selecting different cost functions inside this family.

Appendix A: EM based algorithm

During the M step of the EM algorithm, function $\mathbf{Q} = E\{L \mid \mathbf{x}, \mathbf{w}_{ij}, \mathbf{v}_i, \mathbf{v}, y\}$ given in Eq. (10.26) has to be maximized with respect to the network weights $\mathbf{w}_{ij}$, $\mathbf{v}_i$ and $\mathbf{v}$. Therefore, differentiating Q, we get the following gradients:

$$
\begin{aligned}
\nabla_{\mathbf{w}_{ij}} \mathbf{Q} &= \frac{\partial \mathbf{Q}}{\partial P_{ij}} \cdot \nabla_{\mathbf{w}_{ij}} P_{ij} = \frac{h_{ij}}{P_{ij}} (1 + \varepsilon_{ij}) \nabla_{\mathbf{w}_{ij}} P_{ij} \\
\nabla_{\mathbf{v}_i} \mathbf{Q} &= \sum_k \frac{\partial \mathbf{Q}}{\partial g_{j|i}} \nabla_{\mathbf{v}_i} g_{k|i} = \\
&= \sum_k \frac{h_{ik}}{g_{k|i}} (1 + \varepsilon_{ik}) \nabla_{\mathbf{v}_i} g_{k|i} \\
\nabla_{\mathbf{v}} \mathbf{Q} &= \sum_k \frac{\partial \mathbf{Q}}{\partial g_k} \nabla_{\mathbf{v}} g_k = \\
&= \sum_k \frac{1}{g_k} \sum_n h_{kn} (1 + \varepsilon_{kn}) \nabla_{\mathbf{v}} g_k
\end{aligned}
$$

where ε_{ij} has been defined in Eq. (10.31).

Considering the Bernoulli model for the output of each expert,

$$
P_{ij} = \mu_{ij}^{y} (1 - \mu_{ij})^{1-y} \tag{10.34}
$$

and, thus,

$$
\nabla_{\mathbf{w}_{ij}} P_{ij} = (2y - 1)\mu_{ij}(1 - \mu_{ij})\mathbf{x} \tag{10.35}
$$

If each gating network has a unique filter (with coefficients $\mathbf{v}_i$ for first level gating nets and $\mathbf{v}$ for second level gating nets) and that $g_{0|i} = 1 - g_{1|i}$ and $g_0 = 1 - g_1$, we can write

$$
\begin{aligned}
\nabla_{\mathbf{v}_i} g_{1|i} &= g_{1|i} \left(1 - g_{1|i}\right) \mathbf{x} \\
\nabla_{\mathbf{v}_i} g_{0|i} &= \nabla_{\mathbf{v}_i} \left(1 - g_{1|i}\right) = -g_{1|i} \left(1 - g_{1|i}\right) \mathbf{x}
\end{aligned}
$$

$$
\begin{aligned}
\nabla_{\mathbf{v}} g_1 &= g_1 (1 - g_1) \mathbf{x} \\
\nabla_{\mathbf{v}} g_0 &= \nabla_{\mathbf{v}} (1 - g_1) = -g_1 (1 - g_1) \mathbf{x}
\end{aligned}
$$

Therefore, the learning rules for a 2-level HME network can be written as

$$
\begin{aligned}
\Delta \mathbf{w}_{ij} &= \eta \nabla_{\mathbf{w}_{ij}} \mathbf{Q} = \eta h_{ij} (1 + \varepsilon_{ij}) (y - \mu_{ij}) \mathbf{x} \\
\Delta \mathbf{v}_i &= \eta \nabla_{\mathbf{v}_i} \mathbf{Q} = \\
&= \eta \left(h_{m1} (1 + \varepsilon_{i1}) (1 - g_{1|i}) - h_{m0} (1 + \varepsilon_{i0}) g_{1|i}\right) \mathbf{x}
\end{aligned}
$$

$$\begin{aligned}\Delta \mathbf{v} &= \eta \nabla_{\mathbf{v}} \mathbf{Q} = \\ &= \eta \left(\sum_n h_{1n} (1 + \varepsilon_{1n}) (1 - g_1) - \sum_n h_{0n} (1 + \varepsilon_{0n}) g_1 \right) \mathbf{x}\end{aligned}$$

10.8 References

[1] D.B. Rubin. J.R. A.P. Dempster, N.M. Laird. Maximum likelihood from incomplete data via EM algorithm. *Statistic. Soc.B*, No.19, pp.1–38, 1977.

[2] S. Amari. Backpropagation and stochastic gradient descent method. *Neurocomputing*, (5):185–196, 1993.

[3] L. Breiman, J. H. Friedman, R. A. Olshen, and C. J. Stone. *Classification and Regression Trees*, Wadsworth and Brooks/Cole, Monterey, CA, 1984.

[4] C.F.N. Cowan C.P. Callender. Two novel non-linear appraoches to channel equalization for digital communications. In *Proc. of Second Cost #229 on adaptive Algorithms in Communications*, pp 247–254, Bordeaux, France, Oct.1992.

[5] H. Guo and S. B. Gelfand. Classification trees with neural network feature extraction. *IEEE Transactions on Neural Networks*, 3:923–933, 1992.

[6] V.V.S. Sarma G.R.Dattatreya. Bayesian and desicion tree approaches for pattern recognition including feature measurement costs. *IEEE Trans. Patt. Anal. Mach. Intell.*, PAMI-3:293–298, 1981.

[7] S. Haykin. *Neural Networks. A Comprehensive Foundation.* Macmillan College Publishing, New York, 1994.

[8] Simon Haykin. *Neural Networks: a Comprehensive Foundation.* Maxwell Macmillan International, 1994.

[9] S. Haykin. *Adaptive Filter Theory.* Prentice-Hall International, third edition, 1996.

[10] R. Hyafil and R. L. Rivest. Constructing optimal binary trees is NP-complete. *Information Processing Letters*, 5:15–17, 1976.

[11] A.R. Figueiras-Vidal J. Cid-Sueiro. Digital equalization using modular neural networks: an overview. In *Proc. of the 7th. Int. Thyrrhenian Workshop on Dig. Comm.*, pp.337–345, Viareggio, Italy, September 1995.

[12] A.R.Figueiras-Vidal J. Cid-Sueiro. The role of objective functions in modular classification (with an equalization application). In *Invited paper at Int. Conf. on Neural, Parallel and Scientific Computations*, Atlanta, GE, May 1995.

[13] J. Cid-Sueiro, Johny Ghattas. An EM approach to channel equalization with modular networks. In *Proc. of the 4th Bayona Workshop on Intelligent Methods on Signal Processing and Communications*, Bayona (Vigo), Spain, 1996.

[14] Michael I. Jordan and Robert A. Jacobs. Hierarchies of adaptive experts. In John E. Moody, Steve J. Hanson, and Richard P. Lippmann, editors, *Advances in Neural Information Processing Systems*, volume 4, pp 985–992. Morgan Kaufmann Publishers, Inc., 1992.

[15] M. I. Jordan and R. A. Jacobs. Hierarchical mixtures of experts and the EM algorithm. *Neural Computation*, 6:181–214, 1994.

[16] P. Smyth J.W. Miller, R. Goodman. Objective Functions for Probability Estimation, editor. *Proc. of the Int. Joint Conference on Neural Networks*, pp.881–886, 1991.

[17] K.S.Fu J. Mui. Automated classification of nucleated blood cells using a binary tree classifier. *IEEE Transactions on Pattern Analysis and Machine Intelligence*, 2:429–443, 1980.

[18] A. Isaksson J.E. Stromberg, J. Zrida. Neural trees-using neural nets in a tree classifier structure. *Proc. of the IEEE Int. Conf. on Acoustics, Speech and Signal Processing.*

[19] C. Wu L.A.Bartolucci, P.H. Swain. Selective radiant temperature mapping using a layered classifier. *IEEE Trans. Geosci. Electron.*, GE-14:101–106, 1976.

[20] R.A. Olshen C.J. Stone L. Breinman, J.H. Freidman. *Classification and Regression Trees.* Wadsworth International Group, Belmont, CA, 1984.

[21] R.A. Jacobs M.I. Jordan. Hierarchical mixture of experts and the em algorithm. *Neural Computation*, (6):181–214, 1991.

[22] P.Beaudet P.Argentiero, R.Chin. An automated approach to the design of decision tree classifiers. *IEEE Trans. Patt. Analy. Mach. Intelligence*, PAMI(4):51–57, 1982.

[23] J.A. Nelder P. McCullagh. *Generalized Linear Models.* Chapman and Hall, London, 1983.

[24] J. R. Quinlan. Decision trees and decision making. *IEEE Transactions on Systems, Man and Cybernetics*, 20:339–346, 1990.

[25] G.Nagy R.C.Casey. Decision tree design using a probabilistic model. *IEEE Trans. Inform. Theory*, IT-30:93–99, 1984.

[26] R. Reed. Pruning algorithms: a survey. *IEEE Trans on Neural Networks*, 4(5):740–747, 1993.

[27] B. D. Ripley. *Pattern Recognition and Neural Networks.* Cambridge University Press, Cambridge, 1996.

[28] M.I. Jordan, R.A. Jacobs. Learning piecewise control strategies in a modular neural network architecture. *IEEE Transactions on Systems, Man and Cybernetics*, 23(2):337–345, March/Apr.1993.

[29] D.W. Ruck, S.K. Rogers, M. Kabrisky, M.E. Oxley, B.W. Suter.. The multilayer perceptron as an approximation to a Bayes optimal discriminant function.. *IEEE Transactions on Neural Networks*, 1(4):296–298, Dec.1990.

[30] I.K. Sethi. Entropy nets: from decision trees to neural networks. *Proceedings of the IEEE*, 78:1605–1613, 1990.

[31] E.J. Delp, S.B. Gelfand, C.S. Ravishankar. An iterative growing and pruning algorithm for classification tree design. *IEEE Transactions on Pattern Analysis and Machine Intelligence*, 13:163–174, Feb.1991.

[32] E.J. Delp S.B. Gelfland, C.S. Ravishankar. Tree-structured piecewise linear adaptive equalization. *IEEE Transactions on Communications*, 41(1):70–82, Jan.1993.

[33] S. R Safavian and D. Landgrebe. A survey of decision tree classifier methodology. *IEEE Transactions on Systems, Man and Cybernetics*, 21:660–674, 1991.

[34] A. Sankar and R. J. Mammone. Growing and pruning neural tree networks. *IEEE Transactions on Computers*, 42:291–299, 1993.

[35] P.E. Utgoff Incremental induction of decision trees. *Machine Learning*, 4:161–186, 1993.

11

Applications of Chaos in Communications

Michael Peter Kennedy

ABSTRACT This work reviews the state of the art in techniques for communicating with chaos and highlights some of the engineering challenges in this field.

11.1 Introduction

Chaos provides a means of generating "noiselike" signals which are predictable in the short-term and have clearly defined statistical properties. The recent discovery that chaotic systems can be synchronized [1, 27] has generated an interest in exploiting chaos in signal processing and communication systems [26]. Indeed, the underlying properties of chaos are already widely used in spread spectrum communications, compression, and cryptographic schemes. The goal of this review paper is to introduce the basic concepts of nonlinear dynamics and chaos and to give an overview of research in the area of communicating with chaos.

11.2 Deterministic dynamical systems and chaos

A *deterministic* dynamical system is one whose state evolves with time according to a deterministic evolution rule called a *dynamic* [19]. The time evolution of the state (called a *trajectory*) is completely determined by the initial state of the system, the input, and the dynamic. For example, the state of a digital filter is determined by the initial state of the filter, the input, and a difference equation (the dynamic) which describes the evolution of the state from one time instant to the next.

By contrast with a stochastic dynamical system, which may follow any number of different trajectories from a given state according to some probabilistic rule, trajectories of a deterministic dynamical system are unique. ¿From any given state, there is only one "next" state. Therefore, the same system started twice from the same initial state with the same input will follow precisely the same trajectory through the state space.

Deterministic dynamical systems can produce a variety of steady-state behaviors, the most familiar of which are stationary, periodic, and quasiperiodic solutions. These solutions are "predictable" in the sense that a small piece of a

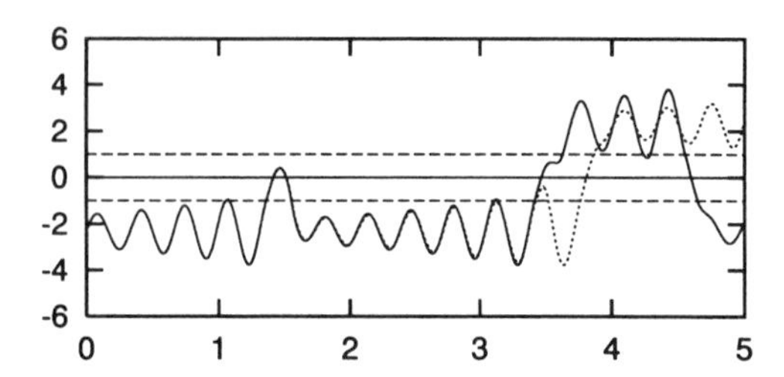

FIGURE 11.1. Sensitive dependence on initial conditions. Two simulated time waveforms $V_1(t)$ from Chua's oscillator (see Fig. 11.2(a)) starting from initial conditions which differ by less that 0.01% in the V_2 coordinate. Note that the trajectories diverge within 5 ms. Horizontal axis: t (ms); Vertical axis: V_1 (V).

trajectory enables one to predict the future behavior along that trajectory. Chaos refers to solutions of deterministic dynamical systems which, while predictable in the short-term, exhibit *long-term unpredictability.*

Since the initial state, input, and dynamic uniquely determine the behavior of a deterministic dynamical system, it is not obvious that any "unpredictability" is possible. Long-term unpredictability arises because the dynamic of a chaotic system persistently amplifies errors in specifying the state. Thus, two trajectories starting from nearby initial conditions quickly become uncorrelated, as shown in Fig. 11.1; this is called *sensitive dependence on initial conditions.* In a chaotic system, the precision with which the initial conditions must be specified in order to predict the behavior over some specified time interval grows *exponentially* with the length of the prediction interval. As a result, long-term prediction becomes impossible.

A chaotic system appears to exhibit "randomness" in the time domain because its initial conditions cannot be specified with sufficient precision to make accurate long-term predictions of its behavior. This long-term unpredictability manifests itself in the frequency domain as a continuous power spectrum. Thus, a chaotic system provides a convenient way of producing a "noiselike" signal which is predictable in the short-term.

In the following section, we illustrate the characteristic properties of chaotic solutions in a physical system—Chua's oscillator [17].

11.3 Chua's oscillator: a paradigm for chaos

Chua's oscillator, shown in Fig. 11.2(a), is a simple example of a continuous-time deterministic dynamical system. It provides a useful paradigm for studying the qualitative behavior of dynamical systems [18]. The circuit contains a linear inductor, two linear resistors, two linear capacitors, and a nonlinear resistor N_R called a *Chua diode* [16]. It is described by three ordinary differential equations:

$$\frac{dV_1}{dt} = \frac{G}{C_1}(V_2 - V_1) - \frac{1}{C_1}f(V_1),$$

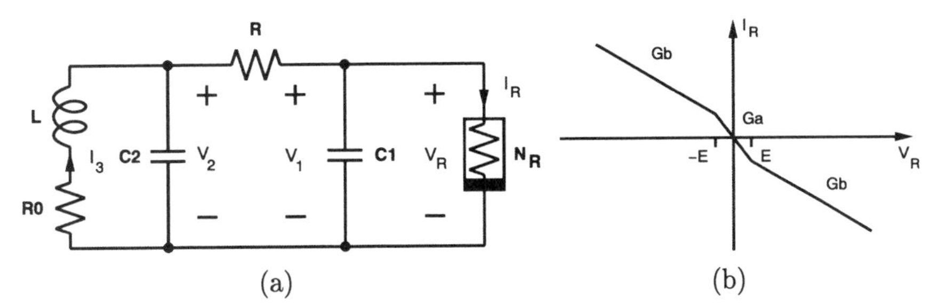

FIGURE 11.2. (a) Chua's oscillator consists of a linear inductor L with series resistance R_0, a linear resistor R, two linear capacitors C_1 and C_2, and a nonlinear resistor N_R. (b) Three-segment piecewise-linear V–I characteristic of the Chua diode N_R.

$$\begin{aligned}\frac{dV_2}{dt} &= \frac{G}{C_2}(V_1 - V_2) + \frac{1}{C_2}I_3,\\ \frac{dI_3}{dt} &= -\frac{1}{L}V_2 - \frac{R_0}{L}I_3\end{aligned}$$

where $G = 1/R$ and $f(V_R) = G_b V_R + \frac{1}{2}(G_a - G_b)(|V_R + E| - |V_R - E|)$ is the three-segment piecewise-linear V–I characteristic of the Chua diode (see Fig. 11.2(b)).

For different values of the parameters, this circuit exhibits a variety of steady-state solutions including stationary points, period and quasiperiodic trajectories, and chaos [17].

11.4 Periodicity, quasiperiodicity, and chaos

Figures 11.3(a) and (b) show periodic and quasiperiodic trajectories in Chua's oscillator. The periodic solution is characterized by a repetitive waveform in the time domain and by a discrete line spectrum in the frequency domain whose components are spaced at integer multiples of a fundamental frequency. In the state space, the periodic trajectory follows a simple closed curve.

A quasiperiodic solution is characterized by a waveform in the time domain which appears to be amplitude-modulated and by a discrete line spectrum in the frequency domain. In this case, the spectral components are not rationally related. In the state space, a quasiperiodic trajectory corresponds to motion on a torus.

A chaotic solution (such as that shown in Fig. 11.3(c)) is characterized by "apparent randomness" in the time domain and by a continuous spectrum in the frequency domain. In the state space, a chaotic trajectory moves along a complicated geometrical object called a strange attractor.

Long-term unpredictability results from persistent *stretching and folding* of bundles of trajectories in the state space. Stretching and folding also gives the strange attractor its characteristic fractal structure [23].

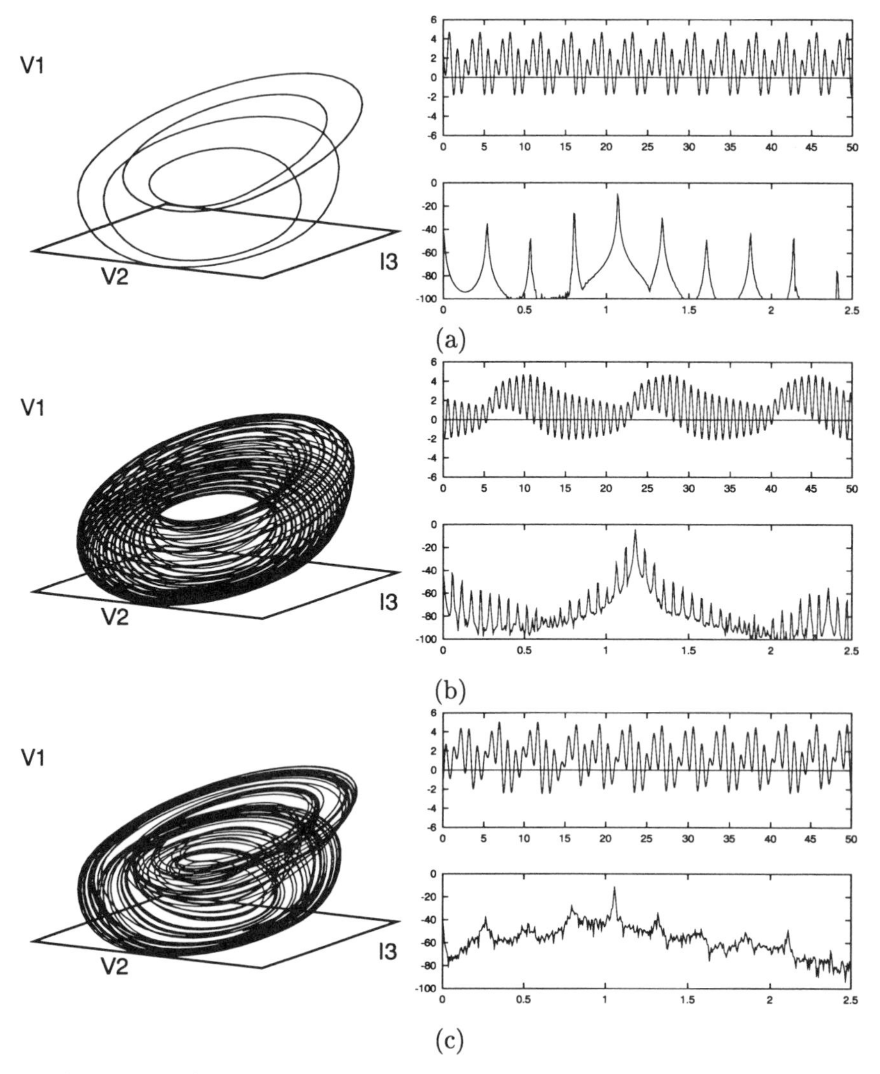

FIGURE 11.3. Simulated state space trajectory, time waveform, and power spectrum for Chua's oscillator. (a) Periodic steady-state—all spikes in the power spectrum are harmonically related to the fundamental frequency. (b) quasiperiodic steady-state—the signal is characterized by a discrete power spectrum with incommensurate frequency components; (c) chaotic steady-state—the waveform has a broadband power spectrum. Time plot: horizontal axis—t (ms); vertical axis—$V_1(t)$ (V). Power spectrum: horizontal axis—frequency (kHz); vertical axis—power of $V_2(t)$ (dB).

11.5 Applications of chaos in communications

Because chaotic systems are *deterministic*, two trajectories from two distinct but otherwise identical systems which start from *identical* initial states will follow precisely the same paths through the state space. In practice, it is impossible to construct two analog systems with identical parameters, let alone to start them from identical initial states. However, recent work has shown that it is possible to *synchronize* two analog chaotic systems so that their trajectories remain close [1, 27, 25, 10].

Figure 11.4 is a schematic representation of feed-forward synchronization. A chaotic signal $y(t)$ produced by one system (labeled T/X) is conveyed to another system (labeled R/X) and forces the state $x'(t)$ of the receiver to *synchronize* with that of the transmitter in the sense that the difference $||x'(t) - x(t)||$ between their states decays asymptotically to zero.

The possibility of synchronization has stimulated interest in applications of chaos in communications, the hope being that information can be conveyed to a remote receiver by modulating it onto a "noiselike" chaotic carrier. In addition, sensitivity to initial conditions and to parameters ensures that synchronization is possible only if the receiver knows the parameters of the transmitter. It has been suggested, therefore, that communicating with chaos might provide a degree of "security" [35].

In the remainder of this paper, we evaluate possible applications of chaos from a communications perspective. We first review the basic components of a digital communication system and identify the functional units in which chaos might play a role. We then discuss the state of the art in spreading, chaotic synchronization, and modulation, and briefly consider security aspects of proposed chaotic communication systems.

11.6 Digital communication

Communication system theory is concerned with the transmission of information from a source to a receiver through a channel [12, 30].

The goal of a digital communication system is to convey information from a digital information source (such as a computer, digitized speech or video) to a receiver as effectively as possible. This is accomplished by mapping the digital

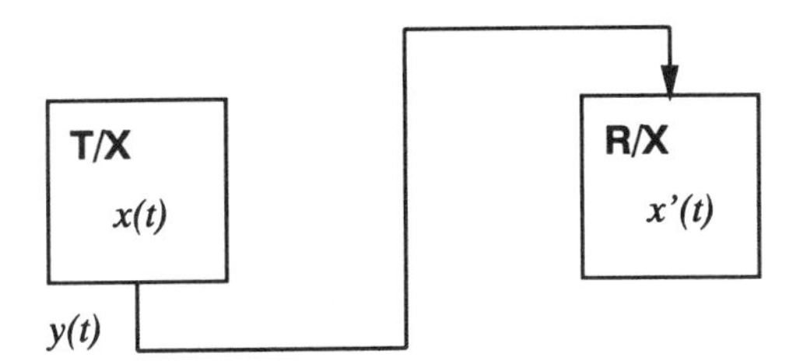

FIGURE 11.4. Feed-forward synchronization using a chaotic signal $y(t)$. Chaos communication is concerned with modulating information onto $y(t)$ and recovering it at the receiver.

information to a sequence of symbols which vary some parameter of an analog electromagnetic wave called the carrier. This process is called *modulation*. At the receiver, the received signal is demodulated, interpreted, and the information recovered.

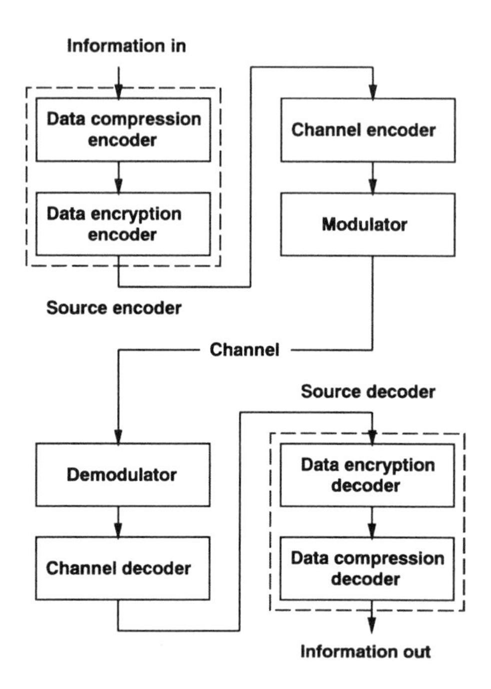

FIGURE 11.5. Digital communication system showing source and channel coding, modulation, and channel.

The mapping from *baseband* digital information to a high-frequency *passband* carrier signal may be accompanied by encryption and coding to add end-to-end security, data compression, and error-correction capability.

Built-in error-correction capability is required because real channels distort analog signals by a variety of linear and nonlinear mechanisms: attenuation, dispersion, fading, noise, interference, multipath effects, etc..

A *channel encoder* introduces algorithmic redundancy into the transmitted symbol sequence that reduces the probability of incorrect decisions at the receiver.

Modulation is the process by which a symbol is transformed into an analog waveform that is suitable for transmission. Common digital modulation schemes include Phase-Shift-Keying (PSK) and Frequency-Shift-Keying (FSK), where a one-to-one correspondence is established between phases and frequencies, respectively, of a sinusoidal carrier and the symbols.

The *channel* is the physical medium that carries the signal from the transmitter to receiver. Inevitably, the signal becomes corrupted in the channel. Hence, the receiver seldom receives exactly what was transmitted. The role of the *demodulator* in the receiver is to produce from the received corrupted analog signal an estimate of the transmitted symbol sequence. The *channel decoder* exploits redundancy in the transmitted sequence to reconstruct the original information. Because of disturbances in real communication channels, error-free transmission is not possible.

The performance of the communication system is measured in terms of the bit error rate (BER) at the receiver. In general, this depends on the coding scheme, the type of waveform used, transmitter power, channel characteristics, and demodulation scheme. The conventional graphical representation of performance in a linear channel with Additive White Gaussian Noise (AWGN) shows bit error rate versus E_b/N_0, where E_b is the energy per bit and N_0 is the power spectral density of the noise introduced in the channel.

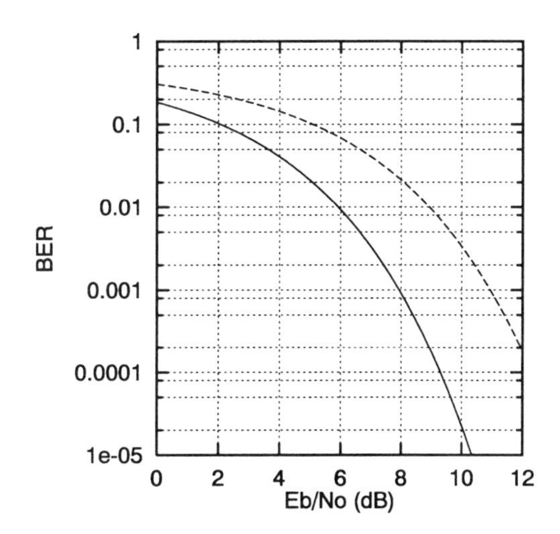

FIGURE 11.6. Comparison of the noise performances of two digital modulation schemes: differential phase shift keying (solid) and noncoherent frequency shift keying (dashed).

For a given background noise level, the BER may be reduced by increasing the energy associated with each bit, either by transmitting with higher power or for a longer period per bit. The challenge in digital communications is to achieve a specified BER with minimum energy per bit. A further consideration is *bandwidth efficiency*, defined as the ratio of data rate to channel bandwidth [12].

Nonlinear dynamics has potential applications in several of the building blocks of a digital communication system: data compression, encryption, and modulation. In this work, we focus primarily on the application of chaos as a spread spectrum modulation scheme.

11.7 Spreading

In spread spectrum communication, the transmitted signal is spread over a much larger bandwidth than is necessary to transmit the baseband information. Spread

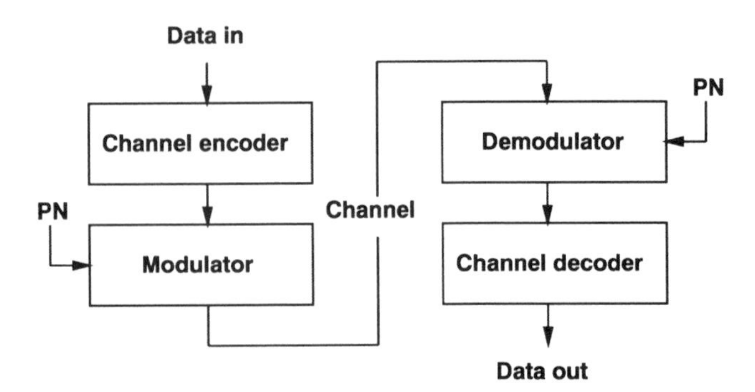

FIGURE 11.7. Spread spectrum communication systems using a conventional PN spreader. The synchronized PN despreader permits signal recovery and interference rejection.

spectrum can be used for [30]:

- combating the effects of interference due to jamming, other users, and multipath effects,
- hiding a signal "in the noise" by transmitting it at low power, and
- achieving message privacy in the presence of eavesdroppers.

Conventional spread spectrum communication systems use *pseudorandom* spreading sequences to distribute the energy of the information signal over a wide bandwidth. The transmitted signal appears similar to noise and is therefore difficult to detect by eavesdroppers. In addition, the signal is difficult to jam because its power spectral density is low. By using orthogonal spreading sequences, multiple users may communicate simultaneously on the same channel; this is called Code Division Multiple Access (CDMA).

Spread spectrum techniques are suited for applications in satellite communications (low power spectral density), mobile phones (privacy, high tolerance against multipath effects, multiple users), and military communications (anti-jamming capability, low probability of intercept).

11.7.1 Pseudorandom spreading sequences

Pseudorandom spreading sequences, also called *pseudonoise* or PN sequences, are widely used in direct sequence (DS) spread spectrum communications because their statistics and orthogonality properties are well understood, they are easy to generate, and easy to synchronize. However, the inherent periodicity of a pseudorandom sequence compromises the overall security of a spread spectrum communication system. The greater the length of the pseudorandom sequence, the higher is the level of security, but the more difficult it is to establish synchronization at the demodulator.

Figure 11.7 shows the key building blocks of a spread spectrum system using a PN sequence generator. The modulator spreads the data stream from the channel encoder, as determined by the pseudorandom spreading sequence, and transmits on a sinusoidal carrier using a variant of PSK or FSK.

11.7.2 Chaotic spreading signals

A pseudorandom sequence generator is a special case of a chaotic system, the principal difference being that the chaotic system has an infinite number of (analog) states, while the pseudorandom generator has a finite number (of digital states). A pseudorandom sequence is produced by visiting each state of the system once in a deterministic manner. With only a finite number of states to visit, the output sequence is necessarily periodic. By contrast, an analog chaos generator can visit an infinite number of states in a deterministic manner and therefore produces an output sequence which *never* repeats itself.

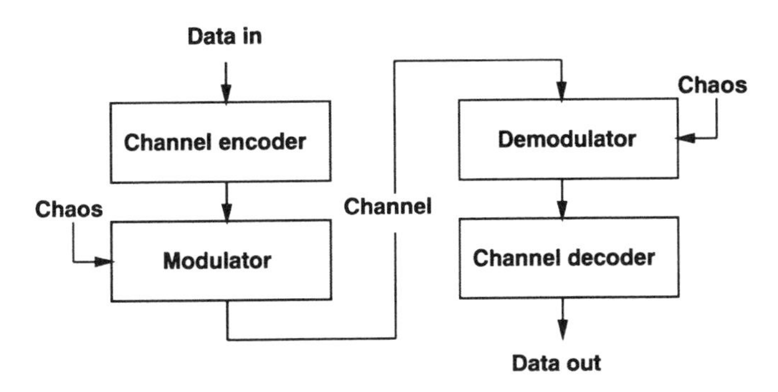

FIGURE 11.8. Spread spectrum communication systems using a chaotic spreader.

What are the potential advantages of using chaotic systems in spread spectrum communication systems? With appropriate modulation and demodulation techniques, the "random" nature and "noise-like" spectral properties of chaotic electronic circuits [19] can be used to provide *simultaneous spreading and modulation* of a transmission. The simplicity of the analog circuits involved could permit extremely high speed, low power implementations.

It should be emphasized, however, that the processing gain in a DS spread spectrum communication system results from despreading with a synchronized replica of the spreading signal [31]. Without synchronization, interference suppression due to despreading cannot be achieved.

11.8 Chaotic synchronization: state of the art

Several techniques for synchronizing chaotic systems have been proposed [10] including drive-response synchronization [27], error feedback synchronization [3], and inverse systems [14].

11.8.1 Drive-response synchronization

Drive-response synchronization is achieved by partitioning a system into two interconnected subsystems, the first of which drives local and remote copies of the second. Provided that the system has been partitioned in an appropriate manner, a chaotic output from the first subsystem causes the remote copy of the second subsystem to synchronize with the local copy.

11.8.2 Inverse systems

If a system Σ with state $x(t)$ and input $s(t)$ produces an output $y(t)$, then its inverse Σ^{-1} produces an output $y'(t) = s(t)$ when its input is $s'(t) = y(t)$ and its state $x'(t)$ has synchronized with $x(t)$ (see Fig. 11.10).

11.8.3 Error-feedback synchronization

Error-feedback synchronization is borrowed from the field of automatic control. An error signal is derived from the difference between the output of the receiver system and that which has been received from the transmitter. The error signal modifies the state of the receiver to drive the error to zero.

11.8.4 Performance evaluation

Simulations of a linear bandpass communication channel with additive white Gaussian noise suggest that drive-response synchronization is not robust (SNR $>$ 30 dB is required) and that continuous-time analog inverse systems exhibit extreme noise sensitivity (SNR $>$ 40 dB is required to maintain synchronization). By contrast, error feedback synchronization can be maintained with a SNR of less than 20 dB [15].

11.9 Chaotic modulation: state of the art

Since 1992, five chaos-based modulation and spreading techniques have been developed: chaotic masking, inverse systems, Predictive Poincaré Control (PPC) modulation, Chaos Shift Keying (CSK), and, most recently, differential CSK (DCSK).

Each of the techniques described below has been demonstrated experimentally using discrete general-purpose components [6] or dedicated circuitry [7]. More recently, prototype spreaders have been realized in integrated circuit form [8].

11.9.1 Chaotic masking

In chaotic masking [4], the information signal $s(t)$ is spread by *adding* it to the output $y(t)$ of a chaotic system. [11].

Provided that the amplitude of the information signal $s(t)$ is small compared to $y(t)$, an identical chaotic system in the receiver can be made to synchronize with $y(t)$. This permits the receiver to "filter" out the "disturbance" $s(t)$. Thus, $s(t)$ can be retrieved by simply subtracting the output of the receiver's chaotic system from the received signal. A block diagram of this scheme is shown in Fig. 11.9.

Chaotic masking suffers from the disadvantage that distortion and noise introduced by the channel are indistinguishable from the signal. Furthermore, if the amplitude of the signal is too large relative to the carrier, synchronization cannot be maintained.

11.9.2 Inverse systems

In the inverse system approach [9], the transmitter consists of a chaotic system which is excited by the information signal $s(t)$. Despite (or because of) this input, the output $y(t)$ of the transmitter is chaotic. The receiver is simply the inverse system, i.e. a system which produces $r(t) = s(t)$ as output when excited by $y(t)$ and when started from the same initial condition. If the system is properly designed, the output $r(t)$ will approach $s(t)$, regardless of the initial conditions.

Inverse systems are widely used in digital encryption and spreading schemes under the guise of self-synchronizing stream ciphers [5, 32].

11.9.3 Predictive Poincaré Control (PPC) modulation

In 1993, Hayes *et al.* proposed a method for modulating binary information directly onto a chaotic carrier by choosing initial conditions which generate a chaotic signal corresponding (in a qualitative sense) to the information signal [13]. More recently, this technique has been generalized for a broad class of chaotic transmitters [34]. This global control scheme for forcing a trajectory to follow a prescribed path is called Predictive Poincaré Control (PPC).

In PPC modulation, symbolic analysis of chaotic systems is used to encode an information signal [34]. In a suitable Poincaré section of the analog chaotic system in the transmitter, two or more disjoint regions are identified and assigned values for coding the information (for example "0" and "1"). With an appropriate control method, the trajectory of the transmitter system is successively steered through the desired regions of the Poincaré section such that the resulting succession of assigned values corresponds to the information signal that is to be transmitted.

On the receiver side, an identical chaotic system is synchronized approximately with the transmitter system. By identifying the regions of the prescribed Poincaré section as they are visited by the synchronized receiver, the information signal can be retrieved.

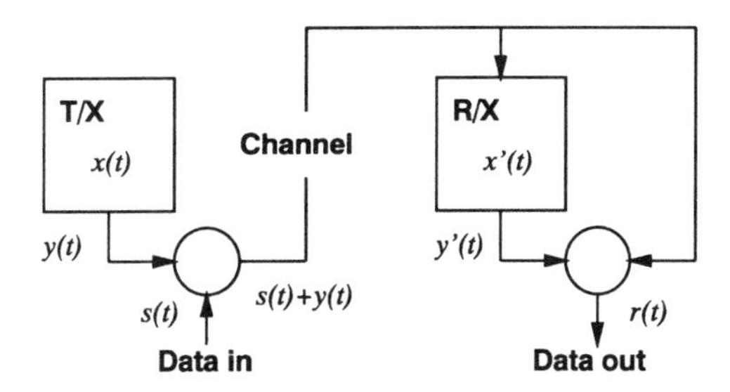

FIGURE 11.9. Chaos masking. If the systems synchronize, then $x'(t) \approx x(t)$, $y'(t) \approx y(t)$, and 'Data out' $\approx s(t)$.

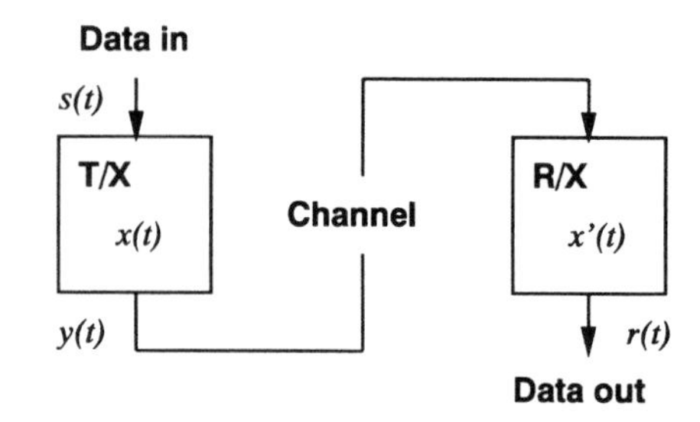

FIGURE 11.10. Inverse system.

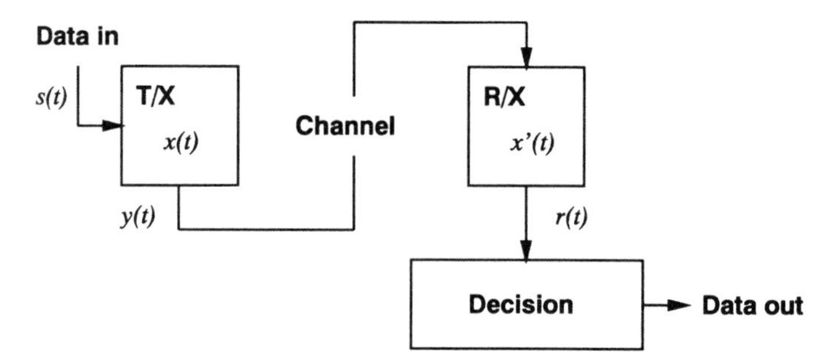

FIGURE 11.11. Predictive Poincaré Control modulation.

Although each of these techniques—chaos masking, inverse systems, and PPC—has been demonstrated in the laboratory, none has been tested experimentally using a noisy communication channel. Indeed, recent simulations [15] suggest that masking and inverse systems perform poorly if the transmitted signal is corrupted by noise. A more robust method is Chaos Shift Keying, so-called because symbols are represented as different chaotic attractors.

11.9.4 Chaos Shift Keying (CSK)

In binary Chaos Shift Keying [6, 29], an information signal is encoded by transmitting one chaotic signal for a "1" and another chaotic signal to represent the binary symbol "0". The two chaotic signals come from two different systems (or the same system with different parameters); these signals are chosen to have similar statistical properties.

Two demodulation schemes are possible: coherent and noncoherent. The coherent receiver contains copies of the systems corresponding to "0" and "1". Depending on the transmitted signal, one of these will synchronize with the incoming signal and the other will desynchronize. By detecting synchronization at the receiver, one may determine which bit is being transmitted.

In the case of non-coherent demodulation, no attempt is made to recover the carrier at the receiver; instead, one simply examines statistical attributes of the received signal.

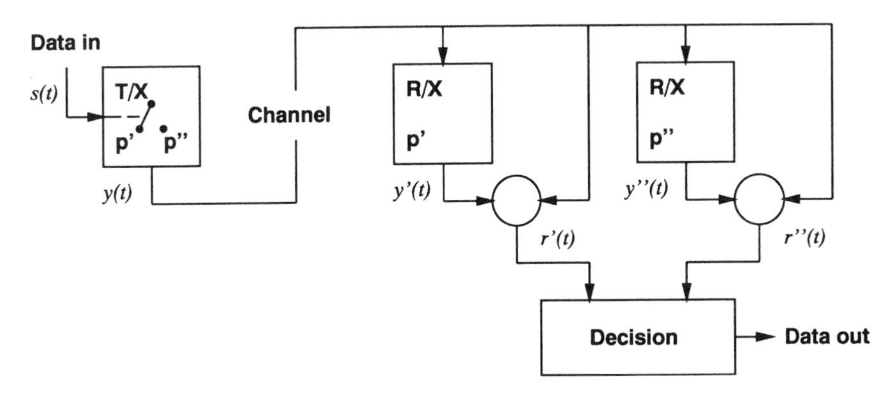

FIGURE 11.12. CSK with a coherent receiver. Either $y'(t)$ or $y''(t)$ synchronizes with $y(t)$ depending on whether the parameter in the transmitter is at p' or p''.

11.9.5 *Differential Chaos Shift Keying (DCSK)*

Differential CSK [22] is a development of CSK which exhibits lower sensitivity to channel imperfections than the techniques outlined above.

In DCSK, the modulator is a free-running chaotic generator with output $y(t)$. Each binary symbol is encoded for transmission as two bits, the first of which acts as a reference, the second carrying the information. To transmit a "1", $y(t)$ and a one-bit-delayed copy of y are applied to the channel during successive bit periods. A "0" is indicated by transmitting $y(t)$ for the first bit period, and an inverted one-bit-delayed copy of this signal during the next bit period.

The chaotic signal sent via the channel is correlated with the signal received during the preceding bit interval and a decision is made based on the output of the correlator. Since both signals presented to the correlator have passed through the same channel, DCSK exhibits robustness in the presence of channel imperfections.

11.10 Chaotic demodulation: state of the art

Digital information modulated onto a chaotic carrier may be recovered from the incoming analog waveform using either noncoherent or coherent techniques.

11.10.1 *Coherent demodulation by chaos synchronization*

CSK was originally described in terms of *synchronization* of chaotic subsystems in a receiver matched to those in the transmitter [6].

Indeed, it was the observation by Pecora and Carroll [27, 20] that two chaotic systems could be synchronized without using an external synchronizing signal which highlighted the possibility of using self-synchronizing coherent demodulators for chaotic transmissions.

While this discovery has generated a great deal of interest in exploiting the properties of chaos for spread spectrum communications, a recent study of the synchronization schemes described in section 11.8 [15] suggests that they are not sufficiently robust for use in noisy channels. In particular, chaos synchronization is not yet competitive with pseudorandom sequences for spreading and despreading.

Better synchronization methods must be developed if the goal of an efficient chaotic spread spectrum communication system with inherent interference suppression is to be realized.

11.10.2 Noncoherent demodulation

Noncoherent demodulation techniques have been evaluated for CSK [21] using a transmitter comprised of two chaotic analog phase-locked loops (APLLs) corresponding to symbols "1" and "0". Simple statistics of the received signal (mean and standard deviation) are used in the decision-making process. In this configuration, a BER of 10^{-2} requires $E_b/N_0 < 13$ dB [15].

In the case of DCSK using APLLs, a noncoherent demodulation scheme has been proposed in which the incoming signal is correlated with a delayed version of itself. The output of the correlator is positive (negative) when a "1" ("0") is transmitted and tends to zero between adjacent bits [22]. Figure 11.14 shows simulations of this system with additive white noise in the channel. Decisions are made at the positive and negative peaks of the correlator output.

Preliminary simulations of DCSK using this configuration suggest that it requires $E_b/N_0 \approx 12.5$ dB for a BER of 10^{-3} [15] compared with 10.8 dB for noncoherent FSK.

This architecture has not yet been optimized. In principle, the bandwidth efficiency could be improved if the decision-making process were accelerated by exploiting the short-term predictability of the chaotic carrier.

11.11 Additional considerations

11.11.1 Security issues

Recent studies [33, 28, 2] have shown that communication schemes using low order chaotic sources have limited security; the security of higher order hyperchaotic systems has yet not been determined.

In this work, we have adopted the viewpoint that security is an add-on feature in a digital communication system which may be implemented by adding encryption/decryption hardware at each end of the system, as shown in Fig. 11.5. Nevertheless, there exist strong similarities between the concept of discrete-time inverse systems and self-synchronizing stream ciphers [5] which may permit a hybrid approach to chaotic modulation and encryption. In addition, the cryptographic mechanisms of *diffusion* and *confusion* are closely related to sensitive dependence on initial conditions.

11.11.2 Multiple access

One of the advantages of using pseudorandom spreading sequences in a spread spectrum system is that multiple users are permitted simultaneous access to the channel provided they use uncorrelated pseudorandom sequences or *codes*. This is called code division multiple access (CDMA). Further work is required to define an equivalent concept of orthogonality for chaotic spreading signals [24].

11.12 Engineering challenges

The field of "communicating with chaos" presents many challenging research and development problems at the basic, strategic, and applied levels.

The building blocks with which to construct a practical chaos-based spread spectrum communication system already exist: chaos generators, modulation schemes, and demodulators. Nevertheless, further research and development is required in all of these subsystems.

Synchronization schemes for chaotic spreading signals are not yet sufficiently robust to be competitive with pseudorandom spreading sequences. Nevertheless, they do offer theoretical advantages in terms of security. Furthermore, an analog implementation of chaotic spreading may permit the use of simple low power, high-frequency circuitry.

The simulated performance of CSK and DCSK modulation is close to that of conventional techniques based on periodic carriers. The wideband nature of the chaotic carrier in this case may offer advantages in time-varying channels which suffer from multipath and fading effects.

Current proposals for CSK and DCSK receivers using noncoherent receivers do not exploit the fact that the carrier has been produced by a chaotic system. By exploiting the short-term predictability of this signal, improved receiver performance may result.

Acknowledgments: The author is indebted to Géza Kolumbán, TU Budapest, and Jörg Schweizer, EPFL, for extensive discussions on chaotic communications.

11.13 References

[1] V.S. Afraimovich, N.N. Verichev, and M.I. Rabinovich. Stochastic synchronization of oscillations in dissipative systems. *Izv. Vuzov. Radiofizika*, 29:795–803, 1986.

[2] T. Beth, D.E. Lazic, and A. Mathias. Cryptanalysis of cryptosystems based on remote chaos replication. In Y.G. Desmedt, editor, *Advances in Cryptology—CRYPTO '94*, pages 318–331. Springer-Verlag, 1994.

[3] G. Chen and X. Dong. Controlling Chua's circuit. *J. Circuits Syst. Computers*, 3(1):139–149, March 1993.

[4] K.M. Cuomo and A.V. Oppenheim. Circuit implementation of synchronized chaos with applications to communications. *Phys. Rev. Lett.*, 71(1):65–68, 1993.

[5] J. Daemen. *Cipher and Hash Function Design—Strategies based on linear and differential cryptanalysis*. PhD thesis, Katholieke Universiteit Leuven, Leuven, Belgium, March 1995.

[6] H. Dedieu, M. P. Kennedy, and M. Hasler. Chaos shift keying: Modulation and demodulation of a chaotic carrier using self-synchronizing Chua's circuits. *IEEE Trans. Circuits and Systems Part II: Analog and Digital Signal Processing, Special Issue on Chaos in Nonlinear Electronic Circuits—Part C: Applications*, 40(10):634–642, October 1993.

[7] A. Dmitriev, A. Panas, and S. Strakov. Transmission of complex analog signals by means of dynamical chaos. In *Proc. 3rd Int. Workshop on Nonlinear Dynamics of Electronic Systems*, pages 241–244, Dublin, 28–29 July 1995.

[8] M. Delgado-Restituto, R. López-Ahumada, and A. Rodríguez-Vázquez. Secure communication using CMOS current-mode sampled-data circuits. In *Proc. 3rd Int. Workshop on Nonlinear Dynamics of Electronic Systems*, pages 237–240, Dublin, 28–29 July 1995.

[9] U. Feldmann, M. Hasler, and W. Schwarz. Communication by chaotic signals: the inverse systems approach. *Int. J. Circuit Theory Appl.*, 24, September 1996.

[10] M. Hasler. Synchronization principles and applications. In C. Toumazou, editor, *Circuits & Systems Tutorials*, chapter 6.2, pages 314–327. IEEE ISCAS'94, London, UK, May 1994.

[11] M. Hasler. Engineering chaos for secure communication systems. *Phil. Trans. R. Soc. Lond. A*, 353(1701), 16 Oct. 1995.

[12] S.S. Haykin. *Communication Systems.* John Wiley & Sons, Inc., New York, 1994. 3rd edition.

[13] S. Hayes, C. Grebogi, and E. Ott. Communicating with chaos. *Phys. Rev. Lett.*, 70(20):3031—3034, May 1993.

[14] K.S. Halle, C.W. Wu, M. Itoh, and L.O. Chua. Spread spectrum communication through modulation of chaos. *Int. J. Bifurcations and Chaos*, 3(2):469–477, 1993.

[15] G. Kolumbán, H. Dedieu, J. Schweizer, J. Ennitis, and B. Vizvári. Performance evaluation and comparison of chaos communication systems. In *Proc. 4th Int. Workshop on Nonlinear Dynamics of Electronic Systems*, pages 105–110, Sevilla, 27–28 June 1996.

[16] M. P. Kennedy. Robust op amp realization of Chua's circuit. *Frequenz*, 46(3–4):66–80, March–April 1992.

[17] M.P. Kennedy. Three steps to chaos part II: A Chua's circuit primer. *IEEE Trans. Circuits and Systems Part I: Fundamental Theory and Applications, Special Issue on Chaos in Nonlinear Electronic Circuits—Part A: Tutorial and Reviews*, 40(10):657–674, October 1993.

[18] M.P. Kennedy. ABC (Adventures in Bifurcations and Chaos): A program for studying chaos. *J. Franklin Inst., Special Issue on Chaos in Electronic Circuits*, 331B(6):529–556, November 1994.

[19] M.P. Kennedy. Basic concepts of nonlinear dynamics and chaos. In C. Toumazou, editor, *Circuits and Systems Tutorials*, pages 289–313. IEEE Press, London, 1994.

[20] M.P. Kennedy. Bifurcation and chaos. In W.K. Chen, editor, *The Circuits and Filters Handbook*, pages 1089–1164. CRC Press, 1995.

[21] G. Kolumbán and B. Vizvári. Direct symbol generation by PLL for the chaos shift keying modulation. In *Proc. ECCTD'95*, pages 483–486, Istanbul, 27–31 August 1995.

[22] G. Kolumbán, B. Vizvári, W. Schwarz, and A. Abel. Differential chaos shift keying: A robust coding for chaotic communication. In *Proc. 4th Int. Workshop on Nonlinear Dynamics of Electronic Systems*, pages 87–92, Sevilla, 27–28 June 1996.

[23] B. Mandelbrot. *The Fractal Geometry of Nature*. Freeman, San Francisco, CA, 1982.

[24] V. Milanovic, K.M. Syed, and M.E. Zaghloul. Chaotic communications by CDMA techniques. In *Proc. 4th Int. Workshop on Nonlinear Dynamics of Electronic Systems*, pages 155–160, Sevilla, 27–28 June 1996.

[25] M.J. Ogorzałek. Taming chaos: Part I—synchronisation. *IEEE Trans. Circuits and Systems*, CAS-40(10), October 1993.

[26] A.V. Oppenheim, G.W. Wornell, S.H. Isabelle, and K.M. Cuomo. Signal processing in the context of chaotic signals. In *Proc. IEEE ICASSP*, volume IV, pages IV–117—IV–120, San Francisco, CA, 1992. IEEE.

[27] L.M. Pecora and T.L. Carroll. Synchronization in chaotic systems. *Phys. Rev. Letters*, 64(8):821—824, 1990.

[28] G. Pérez and H.A. Cerdeira. Extracting messages masked by chaos. *Phys. Rev. Lett.*, 74(11):1970–1973, 13 March 1995.

[29] U. Parlitz, L.O. Chua, L. Kocarev, K.S. Halle, and A. Shang. Transmission of digital signals by chaotic synchronization. *Int. J. Bifurcation and Chaos*, 2(4):973–977, 1992.

[30] J.G. Proakis. *Digital Communications*. McGraw-Hill, Singapore, 1983.

[31] R.L. Pickholtz, D.L. Schilling, and L.B. Millstein. Theory of spread spectrum communications. *IEEE Trans. Commun.*, COM-30:855–884, May 1982.

[32] J.B. Savage. Some simple self-synchronizing digital data scramblers. *Bell System Tech. J.*, pages 449–487, Feb. 1967.

[33] K.M. Short. Steps towards unmasking secure communications. *Int. J. Bif. Chaos*, 4:957–977, 1994.

[34] J. Schweizer and M.P. Kennedy. Predictive Poincaré control. *Phys. Rev. E*, 52(5):4865–4876, November 1995.

[35] C.W. Wu and L.O. Chua. A simple way to synchronize chaotic systems with applications to secure communication systems. *Int. J. Bifurcation and Chaos*, 3(6), Dec. 1993.

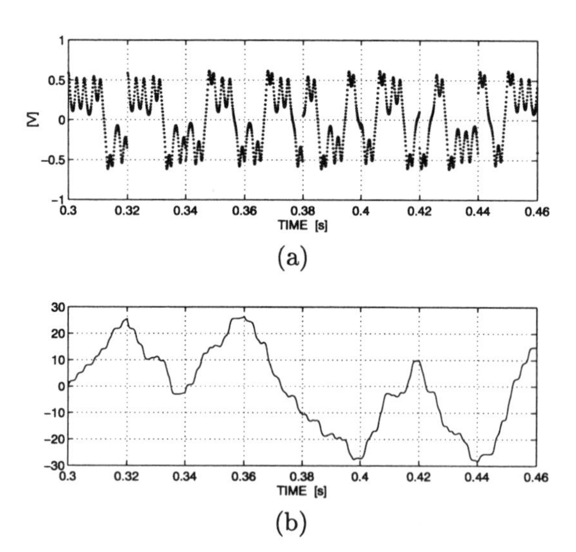

FIGURE 11.13. DCSK using a chaotic analog phase-locked loop (APLL). (a) transmitted signal corresponding to the sequence 1100; (b) output of the correlator.

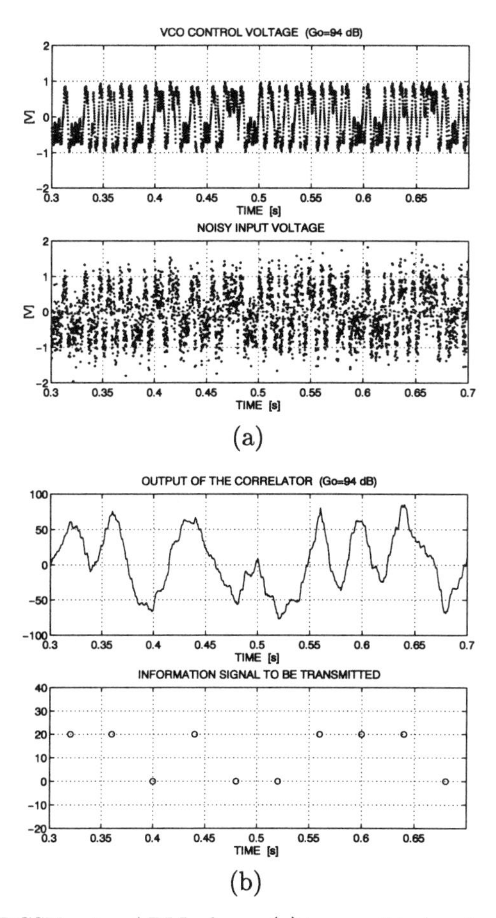

FIGURE 11.14. DCSK using APLL chaos. (a) transmitted and noisy received signals for the symbol sequence 1101001110; (b) output of correlator and received bit sequence (from [22]). Note the strong positive and negative peaks corresponding to the symbols '1' and '0' respectively.

12

Design of Near Perfect Reconstruction Non-Uniform Filter Banks

Fabrizio Argenti
Benedetto Brogelli
Enrico Del Re

ABSTRACT
Subband coding is widely used to compress speech, audio and video signals. In some applications, for example audio compression, a non-uniform splitting of the spectrum of a digital signal can be preferable to a uniform one. Some techniques that allow to implement a non-uniform width subband decomposition are here discussed, design methods based on the cosine-modulation of more than one prototype are described and some examples of filter bank design are given.

12.1 Introduction

Splitting the spectrum of a digital signal [1]-[4] can be useful in several applications, for example data compression. Most of the literature in the field of filter banks design is concerned with uniform width subbands. However, in some cases a non-uniform splitting is more suitable, for example in audio coding [5][6], where non-uniform width subbands could match better the *critical bands* of the human auditory system.

In subband coding applications an analysis filter bank is used to represent a digital signal $x(n)$ through a set of subsequences, each related to a particular frequency interval. Because of the reduced spectrum occupancy, the subband signals can be downsampled (or decimated). Quantization and coding of the subbands samples allow their transmission through the communication channel. At the receiver side, after decoding and inverse quantization, the subband signals are upsampled (or interpolated) and filtered by the synthesis bank, that eliminates the *images* of the subband spectra produced by the upsampling operation. Summing up all the filtered subbands yields the reconstructed signal $\hat{x}(n)$.

Different types of errors can affect the reconstructed signal: if a rational transfer function $T(z)$ from $x(n)$ to $\hat{x}(n)$ exists, then aliasing error is absent (in general this is not true, since the upsampling and the downsampling operators are linear but time-varying); in the absence of aliasing, the transfer function $T(z)$ should be an allpass and linear phase function, that is a delay: if these properties are not satisfied, magnitude and/or phase distortion can affect $\hat{x}(n)$.

In uniform filter banks the downsampling/upsampling factors are all equal to an integer M, so that uniform width subbands, covering $1/M$ of the global spectrum, are produced. If a non-uniform splitting is to be designed, then the scheme presented in Fig. 12.1, which shows a filter bank with rational sampling factors, can be used. At the analysis side of the *m-th* branch a fraction of the global spectrum equal to R_m/M_m is extracted. If $R_m = 1, m = 0, \ldots$,M-1, then integer decimation factors are considered.

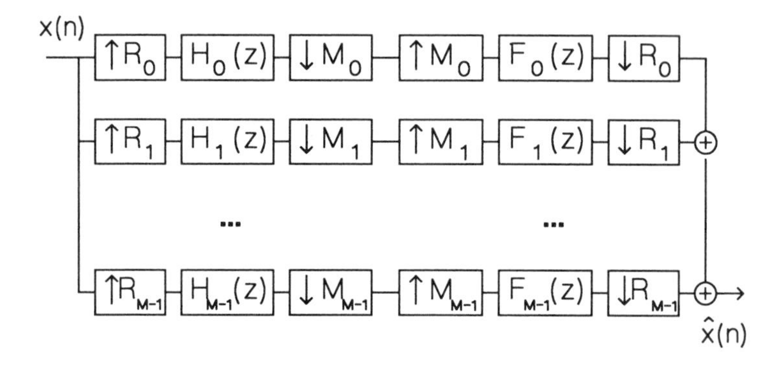

FIGURE 12.1. Non-uniform filter banks with rational decimation factors

The theory of uniform filter banks is well established and several design methods are reported in the literature. A lower number of works appears in the field of non-uniform filter bank design: this problem has been addressed, for example, in [7]-[14].

In [8] the problem of perfect aliasing cancellation in non-uniform filter banks with integer decimation factors is considered. The notion of *compatible sets* is introduced: the choice of a compatible set as integer decimation factors represents a necessary condition for the cancellation of aliasing when non-ideal filters are taken into account.

In [9] the design of non-uniform filter banks is performed in the time-domain. The reconstructed signal is expressed through the impulse responses of the analysis/synthesis filters. The filters are designed through numerical optimization that minimizes, in some sense, the reconstruction error.

In [10] the theory of Perfect Reconstruction (PR) non-uniform filter banks is described. Two transforms that allow to represent such banks through a filter bank with uniform sampling factors are described and examples of the application of these transforms are given.

In [11] cosine modulation of a low-pass prototype is used to design an uniform bank: non-uniform analysis and synthesis filters are obtained as a combination of these uniform width passband filters.

In [11] the PR property is relaxed to achieve higher stopband attenuation. Near PR or Pseudo-QMF methods are widely used in uniform filter bank design [15]-[19]. The aim of these techniques is not the complete cancellation of aliasing, but deleting only its main components. In fact, consider the subband coefficients in the *m-th* branch at the receiver side. After their interpolation, the synthesis filter extracts the band of interest - let I_m be this frequency interval - but, since a non-ideal filter is used, its transition bands overlap also the *images* of the subband spectrum produced by the upsampling operator. The overlapping with the replicas adjacent to I_m produces the *main components of aliasing*, while the overlapping with the other non-adjacent replicas produces negligible aliasing components. In Pseudo-QMF design the main components of aliasing produced in adjacent branches sum up to zero.

Cosine-modulation and Pseudo-QMF methods have been used to design non-uniform filter banks also in [12] (in [13] the technique is used to design non-uniform Short Time Fourier Transformers) and in [14]. In [12] different low-pass prototypes, with different passband widths, are cosine-modulated to yield an integer sampling factors filter bank. Rational sampling factors filter banks are also taken into account using, however, a different structure than that shown in Fig. 12.1. In [14] the analysis/synthesis filter banks are obtained from the cosine-modulation of real and complex coefficients prototypes. In fact, consider two uniform cosine-modulated filter banks with decimation factors M_1 and M_2. A non-uniform filter bank is built by selecting filters from the M_1-channels or from the M_2-channels bank. The filters of the two banks, however, have different transition bandwidths. If the *(m-1)-th* and the *(m+1)-th* branches contain filters coming from the M_1-channels and from the M_2-channels bank, respectively, then the *m-th* branch filter is obtained by the modulation of a complex coefficients low-pass prototype: the modulated filter is a real coefficients passband filter with two different transition bands that match the adjacent filters ones and, therefore, allows the cancellation of the main components of aliasing. In [14] the rational decimation factors case is also considered: the filters of the analysis/synthesis banks are obtained as a combination of passband filters.

Here a method to design filter banks with rational decimation factors is discussed. It extends the work done in [20] related only to the integer decimation factors case. It aims at the cancellation of the main component of aliasing in the output signal, so that Near PR is achieved. In the case of rational decimation factors banks, however, more than one coupling of the aliasing components of adjacent branches lead to their cancellation. If cosine-modulation of different linear phase prototypes is used, then the aliasing cancellation condition constrains the prototypes relative to adjacent branches to become dependent on each other. A procedure to design the prototypes based on these constraints is shown and examples of cosine-modulated non-uniform filter banks are given.

This work is organized as follows. In Section 12.2 the MPEG audio standard is outlined. MPEG performs compression of audio signals by using a psychoacoustic model and a uniform subband decomposition. This section aims at showing why a non-uniform subband decomposition of the input signal could be preferred to a uniform one in an audio coder. In Section 12.3 the mechanism of aliasing cancellation in non-uniform filter banks with rational sampling factors is shown and the design based on the cosine-modulation of more than one prototype is described. Finally, in Section 12.4 some examples of filter bank design are presented.

12.2 The MPEG audio coder

The MPEG audio standard [21] is a recent achievement in audio compression. Subband decomposition is used to represent the input signal by means of a set of subsequences, each of them related to a given interval of the spectrum. The sensitivity of the human auditory system is taken into account and the subbands that result to be perceptually less important can be more coarsely quantized [5][6][21]. It is well-known that the human auditory system is modeled as a bank of passband filters splitting the audio spectrum into *critical bands*: human sensitivity is approximately constant within a critical band. Critical bands have non-uniform width and, therefore, the psychoacoustic model is expected to be exploited better by non-uniform banks than by uniform ones.

A block diagram of the MPEG audio coder is shown in Fig. 12.2. The coder works on signals sampled at 32, 44.1 or 48 kHz, 16 bit/sample, and provides a compressed output at a bit rate varying from 32 to 192 kbit/s. Three layers of coding, having different efficiency and complexity are defined: in the following we will refer to the Layer II.

The filter bank used in MPEG provides a splitting of the audio spectrum into 32 uniform subbands: at a sampling frequency of 48 kHz, each subband refers to a 750 Hz wide frequency interval. Sets of 12 subband coefficients are generated from 384 input samples: for each set, a *scalefactor* is chosen from a table of predefined values.

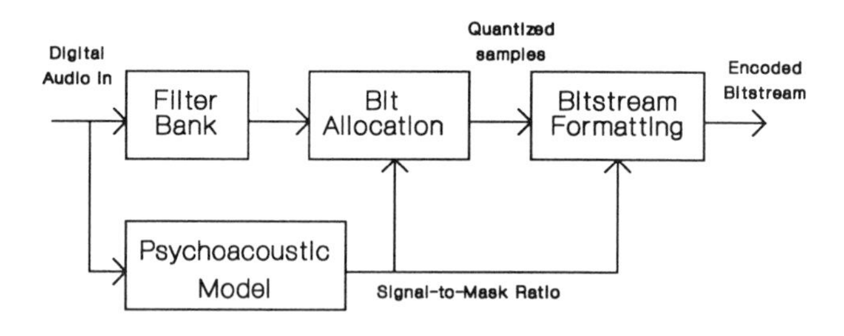

FIGURE 12.2. Block diagram of the MPEG audio coder

Once the bit rate has been fixed, the available bits are distributed among the subbands. The bit allocation is based on the psychoacoustic model and is adapted to the input signal: in fact, it is computed every 1152 samples (a *frame* of input data), i.e., it is maintained constant every 36 subband samples.

The aim of the psychoacoustic model is computing, for each frequency of the audio spectrum, the *auditory threshold*: only spectral components with an energy higher than the threshold value are actually perceived by the human auditory system. Experimental studies have shown that two main effects must be taken into account: 1) an *absolute threshold of hearing* defines the minimum energy that makes a sinusoidal component perceivable in the absence of other sounds; 2) a high energy spectral component can mask adjacent components, i.e., it raises the auditory threshold in its proximity.

The psychoacoustic model used in MPEG is defined in the frequency domain. Very roughly speaking, the main steps that lead to its computation are the following.

- The *energy* distribution of the input signal is computed. This is achieved by using a windowed FFT. An *unpredictability* measure is also computed, so that each spectral component can be classified, at the two extremes, as *tone-like* and *noise-like.*
- These two measures are transposed from the FFT domain into the *calculation partitions* domain. The calculation partitions are a refined version of the *critical bands.* The main incentive to study non-uniform banks relies on the fact that the calculation partitions have non-uniform width.
- The effect of *masking* of high energy partitions vs. low energy ones is considered: the energy and the unpredictability measure of each calculation partition are updated by applying a *spreading function* to the values of the adjacent ones.
- The unpredictability measure of each calculation partition is converted into a *tonality index.* From the tonality index a Signal-to-Noise Ratio (SNR) is derived. Since the energy of each calculation partition is known, the *masking noise*, that is the noise tolerated without being perceived, is computed from the SNR.
- The masking noise is converted again into the domain of the FFT components. These values are compared with the absolute threshold of hearing and the maximum yields the *masking threshold.*
- Energy and masking threshold are then translated into the domain of the subband intervals (the *coder partitions*): from these two variables a *Signal-to-Mask Ratio* SMR_n, referred to the *n-th* subband, is obtained.

The values of SMR_n are the input of the bit allocation procedure. Let b_n be the number of bits assigned to the *n-th* subband: a Signal-to-Noise Ratio (SNR_n), due to the quantization at b_n bits, is tabulated. The quantity $MNR_n = SNR_n - SMR_n$ defines the Mask-to-Noise Ratio, that is a measure of how much the masking noise is higher than the quantization noise. The larger these values, the better the performance of the coder. The optimal values of b_n are searched for by an iterative procedure that aims at approximately constant values of MNR_n, as large as possible within the constraint of the available global bit rate.

As can be seen a set of *coder partitions* closer to the *calculation partitions*, and therefore with non-uniform widths, would simplify, at least in principle, the coder design.

12.3 Non-uniform filter banks with rational sampling factors

12.3.1 Aliasing cancellation in non-uniform filter banks

Consider the system in Fig. 12.1. The input-output relationship in the z-domain is given by:

$$\begin{aligned}\hat{X}(z) &= \sum_{m=0}^{M-1} \frac{1}{R_m} \frac{1}{M_m} \sum_{p=0}^{R_m-1} F_m(z^{\frac{1}{R_m}} W_{R_m}^p) H_m(z^{\frac{1}{R_m}} W_{R_m}^p) X(z) + \\ &+ \sum_{m=0}^{M-1} \frac{1}{R_m} \frac{1}{M_m} \sum_{l=1}^{M_m-1} \sum_{p=0}^{R_m-1} F_m(z^{\frac{1}{R_m}} W_{R_m}^p) \cdot \\ &\cdot H_m(z^{\frac{1}{R_m}} W_{R_m}^p W_{M_m}^l) X(z W_{M_m}^{lR_m})\end{aligned} \tag{12.1}$$

where $W_M = e^{-j2\pi/M}$. The right-hand-side of eq. (12.1) highlights the reconstruction transfer function and the aliasing components.

Consider the analysis stage of each branch shown in Fig. 12.1. Real coefficients filters are taken into account and, therefore, the frequency response of each filter has passbands located at positive and negative frequencies, symmetrically with respect to the origin. The passband at positive (negative) frequencies has width π/M_m and is centered in $(k_m + 0.5)\pi/M_m$ $(-(k_m + 0.5)\pi/M_m)$, where k_m is an integer that selects which part of the spectrum of the R_m-fold upsampled input signal must be extracted. For example, R_m=3, M_m=5, $k_m = 2$ must be used to extract the spectrum in the frequency intervals $[\pi/5, 4\pi/5]$ and $[-4\pi/5, -\pi/5]$. If we consider the frequency response of each filter of the analysis/synthesis banks approximately equal to zero in their stopbands, then the filter transfer functions can be expressed as:

$$H_m(z) = U_m(z) + V_m(z) \tag{12.2}$$

$$F_m(z) = \hat{U}_m(z) + \hat{V}_m(z) \tag{12.3}$$

where $U_m(\omega)$ and $\hat{U}_m(\omega)$ have a passband for $\omega > 0$, while $V_m(\omega)$ and $\hat{V}_m(\omega)$ have a passband for $\omega < 0$.

Due to the M_m-fold upsampler at the synthesis stage, *images* of the *m-th* subband spectrum are filtered by $F_m(z)$. Let $A_m^{(low)}(z)$ and $A_m^{(high)}(z)$ be the main aliasing terms produced at the low-frequency and at the high-frequency edges of the passband of $F_m(z)$, respectively. These components have been described for a cosine-modulated uniform bank in [18]. If we consider that in rational decimation factors banks each branch operates on an R_m-fold upsampled version of $x(n)$ and if we retain only the more relevant terms, then the aliasing terms can be written as:

$$\begin{aligned}A_m^{(low)}(z) &= \frac{1}{M_m} \Big[\hat{U}_m(z) V_m(z W_{M_m}^{k_m}) X(z^{R_m} W_{M_m}^{k_m R_m}) + \\ &+ \hat{V}_m(z) U_m(z W_{M_m}^{-k_m}) X(z^{R_m} W_{M_m}^{-k_m R_m})\Big]\end{aligned} \tag{12.4}$$

$$\begin{aligned}A_m^{(high)}(z) &= \frac{1}{M_m} \Big[\hat{U}_m(z) V_m(z W_{M_m}^{(k_m+1)}) X(z^{R_m} W_{M_m}^{(k_m+1)R_m}) + \\ &+ \hat{V}_m(z) U_m(z W_{M_m}^{-(k_m+1)}) X(z^{R_m} W_{M_m}^{-(k_m+1)R_m})\Big]\end{aligned} \tag{12.5}$$

In [18] it is shown that for uniform filter banks the component $A_m^{(high)}(z)$ of the *m-th* branch is canceled by the component $A_{m+1}^{(low)}(z)$ of the *(m+1)-th* branch. In the non-uniform case, we have to consider that these components are R_m-fold downsampled and that the cancellation may occur also by coupling the *(high)-(high)*, *(low)-(low)* or *(low)-(high)* aliasing terms coming from the *m-th* and from the *(m+1)-th* branch, i.e., the following cases must be taken into account:

$$\begin{aligned} a) \quad & A_m^{(high)}(z) \downarrow R_m + A_{m+1}^{(high)}(z) \downarrow R_{m+1} = 0 \\ b) \quad & A_m^{(low)}(z) \downarrow R_m + A_{m+1}^{(low)}(z) \downarrow R_{m+1} = 0 \\ c) \quad & A_m^{(low)}(z) \downarrow R_m + A_{m+1}^{(high)}(z) \downarrow R_{m+1} = 0 \\ d) \quad & A_m^{(high)}(z) \downarrow R_m + A_{m+1}^{(low)}(z) \downarrow R_{m+1} = 0 \end{aligned} \tag{12.6}$$

where $Q(z) \downarrow M$ stands for the z-transform of the M-fold subsampled version of $q(n)$.

For example, consider the bank $\{1/5,\, 3/5,\, 1/5\}$ that can be implemented using filters having a passband equal to $\pi/5$ and centered, on the positive frequency axis, in $\pi/10$, $\pi/2$ and $9\pi/10$. The aliasing term $A_0^{(high)}(z)$ produced in the *m=0* branch at the synthesis stage must be canceled by $A_1^{(high)}(z) \downarrow 3$. This situation is depicted with more details in Fig. 12.3 and in Fig. 12.4: the normalized frequency $F = \omega/2\pi$ is used. In Fig. 12.3 the spectra relative to the first branch, $m = 0$, with $R_0/M_0 = 1/5$ are shown. The subband $W_0(z) = H_0(z)X(z)$ is subsampled by a factor 5 and, at the receiver side, it is upsampled by a factor 5, so obtaining the signal $S_0(z)$. This signal is filtered by $F_0(z)$ and the component of aliasing $A_0^{(high)}(z)$ is produced at the higher edge of the passband of $F_0(z)$. In Fig. 12.4 the spectra of the signals involved in the branch $m = 1$ are shown. Now, the analysis filter $H_1(z)$ operates on a 3-fold interpolated version of $X(z)$. The subband $W_1(z)$ is subsampled and interpolated (at the receiver side) by a factor 5. The filter $F_1(z)$ extracts the band of interest but produces also two aliasing components at the edges of its passband. After the final 3-fold subsampler the spectrum appears as mirrored and the aliasing component that cancel $A_0^{(high)}(z)$ is $A_1^{(high)}(z) \downarrow 3$, that is we have an *(high)-(high)* aliasing coupling.

Now, consider from an analytical point of view this *(high)-(high)* case: substituting (12.5) into the equation given in the case *a)* of (12.6) yields an expression that can be split into two systems: if $W_{M_m}^{(k_m+1)R_m} = W_{M_{m+1}}^{(k_{m+1}+1)R_{m+1}}$ then the following system must be solved

$$\left\{ \begin{aligned} & \tfrac{1}{M_m}[\hat{U}_m(z)V_m(zW_{M_m}^{(k_m+1)})] \downarrow R_m + \\ & + \tfrac{1}{M_{m+1}}[\hat{U}_{m+1}(z)V_{m+1}(zW_{M_{m+1}}^{(k_{m+1}+1)})] \downarrow R_{m+1} = 0 \\ & \\ & \tfrac{1}{M_m}[\hat{V}_m(z)U_m(zW_{M_m}^{-(k_m+1)})] \downarrow R_m + \\ & + \tfrac{1}{M_{m+1}}[\hat{V}_{m+1}(z)U_{m+1}(zW_{M_{m+1}}^{-(k_{m+1}+1)})] \downarrow R_{m+1} = 0 \end{aligned} \right. \tag{12.7}$$

Otherwise, if $W_{M_m}^{(k_m+1)R_m} = W_{M_{m+1}}^{-(k_{m+1}+1)R_{m+1}}$ then the following system must

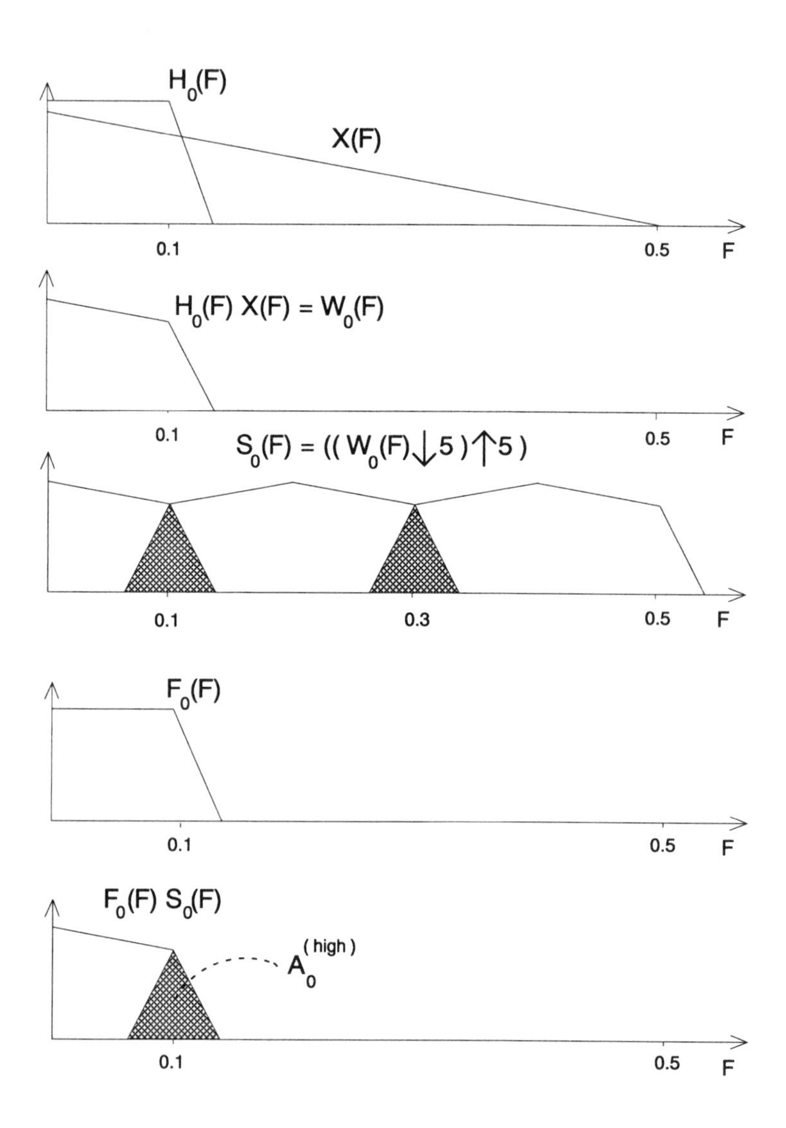

FIGURE 12.3. Bank {1/5, 3/5, 1/5 }: signals involved in the branch $m = 0$.

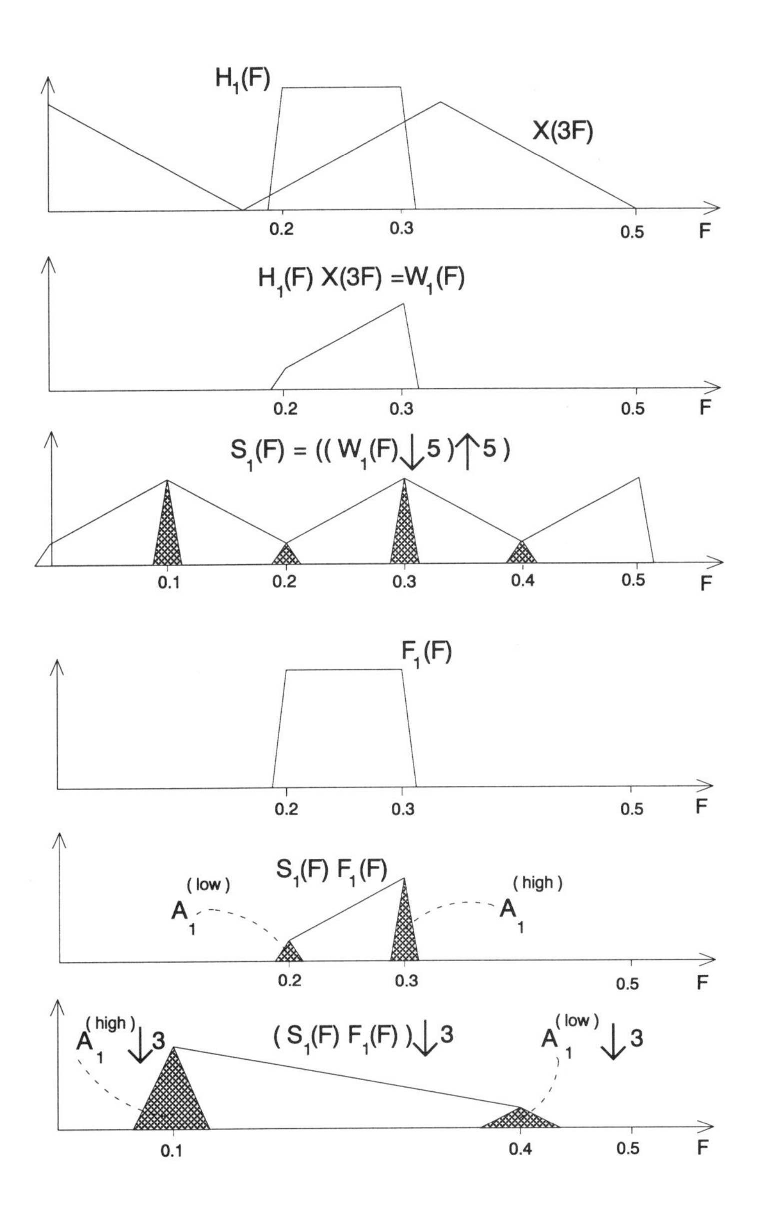

FIGURE 12.4. Bank $\{1/5,\ 3/5,\ 1/5\ \}$: signals involved in the branch $m = 1$.

be solved

$$\left\{\begin{array}{l} \frac{1}{M_m}[\hat{U}_m(z)V_m(zW_{M_m}^{(k_m+1)})]\downarrow R_m + \\ +\frac{1}{M_{m+1}}[\hat{V}_{m+1}(z)U_{m+1}(zW_{M_{m+1}}^{-(k_{m+1}+1)})]\downarrow R_{m+1} \;=0 \\ \\ \frac{1}{M_m}[\hat{V}_m(z)U_m(zW_{M_m}^{-(k_m+1)})]\downarrow R_m + \\ +\frac{1}{M_{m+1}}[\hat{U}_{m+1}(z)V_{m+1}(zW_{M_{m+1}}^{(k_{m+1}+1)})]\downarrow R_{m+1} \;=0 \end{array}\right. \tag{12.8}$$

Similar systems can be written also for the other possible couplings of the aliasing components.

12.3.2 Use of cosine-modulated filter banks

The use of cosine-modulation simplifies the fulfillment of the aliasing cancellation condition. Suppose that each filter of the analysis/synthesis bank is obtained as follows:

$$\begin{aligned} h_m(n) &= 2g_m(n)\cos((2k_m+1)\tfrac{\pi}{2M_m}(n-\tfrac{N_m-1}{2})+\theta_m) \\ f_m(n) &= 2g_m(n)\cos((2k_m+1)\tfrac{\pi}{2M_m}(n-\tfrac{N_m-1}{2})-\theta_m) \\ &= h_m(N_m-1-n) \end{aligned} \tag{12.9}$$

for $m = 0, 1, \ldots, M-1$. N_m is the length of $g_m(n)$. The prototypes $g_m(n)$ have a linear phase and satisfy $g_m(n) = g_m(N_m - 1 - n)$. The phase terms θ_m are chosen to satisfy the aliasing cancellation constraint, as described later. In the case of cosine-modulated banks, the following relationships hold:

$$\begin{aligned} U_m(z) &= G_m(zW_{2M_m}^{(k_m+(1/2))})a_m b_m \\ V_m(z) &= G_m(zW_{2M_m}^{-(k_m+(1/2))})a_m^* b_m^* \\ \hat{U}_m(z) &= G_m(zW_{2M_m}^{(k_m+(1/2))})a_m b_m^* \\ \hat{V}_m(z) &= G_m(zW_{2M_m}^{-(k_m+(1/2))})a_m^* b_m \end{aligned} \tag{12.10}$$

where $a_m = W_{2M_m}^{(k_m+(1/2))(N_m-1)/2}$ and $b_m = e^{j\theta_m}$. Consider, for example, the *(high)-(high)* case. Substituting the expressions shown above into the aliasing cancellation constraints (12.7) and (12.8) yields a relationship between the prototypes of adjacent branches. In [22] it is shown that for $W_{M_m}^{(k_m+1)R_m} = W_{M_{m+1}}^{\pm(k_{m+1}+1)R_{m+1}}$ the choice $e^{-j2\theta_m} + e^{\mp j2\theta_{m+1}} = 0$ allows aliasing cancellation if the following relationship holds:

$$\begin{aligned} &\frac{1}{M_m}\frac{1}{R_m}\sum_{p=0}^{R_m-1} G_m(z^{1/R_m}W_{R_m}^p)G_m(z^{1/R_m}W_{R_m}^p W_{2M_m}) = \\ &= \frac{1}{M_{m+1}}\frac{1}{R_{m+1}}\sum_{p=0}^{R_{m+1}-1} G_{m+1}(z^{1/R_{m+1}}W_{R_{m+1}}^p W_{4M_m}^{R_m/R_{m+1}} W_{4M_{m+1}}^{-1})\cdot \\ &\cdot G_{m+1}(z^{1/R_{m+1}}W_{R_{m+1}}^p W_{4M_m}^{R_m/R_{m+1}} W_{4M_{m+1}}) \end{aligned} \tag{12.11}$$

The same result is obtained also by considering the cases *b)-d)* given in (12.6). The constraints on the phase terms θ_m that lead to the cancellation of aliasing are [22]:

$$\begin{array}{llll} b) & W_{M_m}^{k_m R_m} = W_{M_{m+1}}^{\pm k_{m+1} R_{m+1}} & \Rightarrow & e^{-j2\theta_m} + e^{\mp j2\theta_{m+1}} = 0 \\ c) & W_{M_m}^{k_m R_m} = W_{M_{m+1}}^{\pm (k_{m+1}+1) R_{m+1}} & \Rightarrow & e^{-j2\theta_m} + e^{\mp j2\theta_{m+1}} = 0 \\ d) & W_{M_m}^{(k_m+1) R_m} = W_{M_{m+1}}^{\pm k_{m+1} R_{m+1}} & \Rightarrow & e^{-j2\theta_m} + e^{\mp j2\theta_{m+1}} = 0 \end{array}$$

Moreover, it is possible to demonstrate [22] the following facts, which outline also the steps of the procedure to design rational sampling factors filter banks.

By using the zero-phase representation of the prototypes, i.e.,

$$G_m^{(zp)}(\omega) = G_m(\omega) e^{j\frac{N_m-1}{2}\omega} \tag{12.12}$$

and by imposing the condition

$$\frac{N_m - 1}{R_m} = \frac{N_{m+1} - 1}{R_{m+1}} \tag{12.13}$$

on the lengths of the prototypes, the constraint of aliasing cancellation reduces to the following relationship between the zero-phase frequency responses of the prototype filters

$$\begin{aligned} &\frac{1}{M_m}\frac{1}{R_m} G_m^{(zp)}\left(\frac{\omega}{R_m}\right) G_m^{(zp)}\left(\frac{\omega}{R_m} - \frac{\pi}{M_m}\right) = \\ &\frac{1}{M_{m+1}}\frac{1}{R_{m+1}} G_{m+1}^{(zp)}\left(\frac{\omega}{R_{m+1}} - \frac{\pi R_m}{2M_m R_{m+1}} + \frac{\pi}{2M_{m+1}}\right) \cdot \\ &\cdot G_{m+1}^{(zp)}\left(\frac{\omega}{R_{m+1}} - \frac{\pi R_m}{2M_m R_{m+1}} - \frac{\pi}{2M_{m+1}}\right) \end{aligned} \tag{12.14}$$

Let $\omega_{c,m} = \pi/(2M_m)$ be the cut-off frequency of $G_m(\omega)$. Suppose the transition band, having width $\Delta\omega_m$, is centered in $\omega_{c,m}$ and let $\omega_{p,m} = \omega_{c,m} - (\Delta\omega_m/2)$ and $\omega_{s,m} = \omega_{c,m} + (\Delta\omega_m/2)$ be the upper bound of the passband and the lower bound of the stopband, respectively, of $G_m(\omega)$. Therefore, the constraint (12.14) is satisfied if $R_m\Delta\omega_m = R_{m+1}\Delta\omega_{m+1}$ and if $G_{m+1}^{(zp)}(\omega)$ is chosen as follows

$$G_{m+1}^{(zp)}(\omega) = \begin{cases} 0 & -\pi < \omega \le -\omega_{s,m+1} \\ \sqrt{\frac{R_{m+1}M_{m+1}}{R_m M_m}}\, G_m^{(zp)}\left(-\omega_{p,m} + (\omega + \omega_{p,m+1})\frac{\Delta\omega_m}{\Delta\omega_{m+1}}\right) & -\omega_{s,m+1} < \omega \le -\omega_{p,m+1} \\ \sqrt{R_{m+1}M_{m+1}} & -\omega_{p,m+1} < \omega \le \omega_{p,m+1} \\ \sqrt{\frac{R_{m+1}M_{m+1}}{R_m M_m}}\, G_m^{(zp)}\left(\omega_{p,m} + (\omega - \omega_{p,m+1})\frac{\Delta\omega_m}{\Delta\omega_{m+1}}\right) & \omega_{p,m+1} < \omega \le \omega_{s,m+1} \\ 0 & \omega_{s,m+1} < \omega \le \pi \end{cases} \tag{12.15}$$

Assuming the aliasing components have been completely eliminated, the input-output relationship shown in (12.1) can be expressed as

$$\begin{aligned}\hat{X}(z) &= \left[\sum_{m=0}^{M-1} \frac{1}{R_m} \frac{1}{M_m} \sum_{p=0}^{R_m-1} F_m(z^{\frac{1}{R_m}} W_{R_m}^p) H_m(z^{\frac{1}{R_m}} W_{R_m}^p)\right] X(z) \\ &= T(z)X(z)\end{aligned} \quad (12.16)$$

Phase error is absent since the synthesis filters are a time reversed version of the analysis filters, while the magnitude error is maintained at low levels if $T(z)$ is approximately allpass. The reconstruction error is reduced choosing prototype filters with high stopband attenuation and with a proper behavior in the transition band.

The design procedure can be outlined as follows. A first prototype is designed: for example, the methods proposed in [23][19][24][25][14] for uniform filter banks can be used. This prototype is relative to the *m-th* branch, where m must be chosen so that R_m/M_m=min$\{R_k/M_k$, k=0,...,M-1$\}$. Its cut-off frequency is $\omega_{c,m} = \frac{\pi}{2M_m}$ and $\sqrt{R_m M_m}$ is the gain in the passband. $G_m(\omega)$ must have a *power complementary* transition band, i.e., satisfies

$$| G_m(\omega) |^2 + | G_m(\frac{\pi}{M_m} - \omega) |^2 = R_m M_m \quad (12.17)$$

for $\omega_{p,m} < \omega < \omega_{s,m}$. The prototypes in the other branches are obtained by using (12.15) and by adding the correct linear phase term to determine $G_{m+1}(\omega)$; $g_{m+1}(n)$ is obtained by the inversion of $G_{m+1}(\omega)$.

12.4 Examples of non-uniform filter banks design

To show the effectiveness of the design procedure described in the previous section we consider three examples of non-uniform banks. Example 1 and 2 are relative to banks with rational sampling factors, while Example 3 refers to an integer sampling factors bank, suitable for audio coding applications, that has been proposed in [26]. We indicate with K and Θ the sets $\{ k_m$, m=0,...,M-1 $\}$ and $\{ \theta_m$, m=0,...,M-1 $\}$, respectively.

Example 1: Bank { 2/9, 2/9, 2/9, 2/9, 1/9 }. Two prototypes need to be designed ($g_0(n) = g_1(n) = g_2(n) = g_3(n)$). In this example more than one choice is possible for K. If we choose K={ 0, 7, 6, 3, 8 } then all possible types of aliasing couplings (*(high)-(high)*, *(low)-(high)*, *(low)-(low)*, *(high)-(low)* in the order) occur; in this case $\Theta = \{\pi/4, \pi/4, -\pi/4, -\pi/4, \pi/4\}$.

Example 2: Bank { 1/2, 5/16, 3/16 }. Three prototypes have to be designed. All the aliasing couplings are of the type *(high)-(low)*. In this case K={ 0, 8, 15 } and $\Theta = \{ \pi/4, -\pi/4, \pi/4 \}$.

Example 3: Non-uniform bank having 16, 32 and 64 as possible decimation factors and allowing the splitting of an audio signal sampled at 48 kHz as shown in Fig. 12.5.

The performance of the presented design method is evaluated in terms of both the overall distortion function $T(\omega)$ and the residual aliasing error. As to the latter error, a global measure relative to the whole structure is used. According to the input-output relationship in (12.1), the aliasing contribution relative to

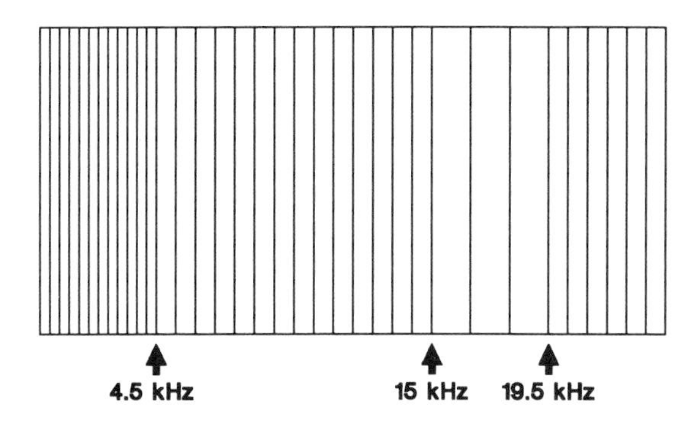

FIGURE 12.5. Subband splitting relative to Example 3

$X(zW_{M_m}^{lR_m})$ can be written as

$$A_{l,m}(z) = \frac{1}{R_m}\frac{1}{M_m}\sum_{p=0}^{R_m-1} H_m(z^{\frac{1}{R_m}} W_{R_m}^{p} W_{M_m}^{l}) F_m(z^{\frac{1}{R_m}} W_{R_m}^{p}) \tag{12.18}$$

with m=0,...,M-1, l=0,...,$M_m - 1$. The functions $A_{l,m}(\omega)$ are 2π-periodic functions. All the aliasing terms $A_{l,m}(z)$ that refer to the same shifted version of $X(z)$, i.e., having the same value of the factor $W_{M_m}^{lR_m}$ in the argument, must be summed up, so that the following aliasing error can be defined:

$$E_a(\omega) = \sqrt{\sum_{r=1}^{M_{max}-1} \left| \sum_{m=0}^{M-1} \sum_{l=1,(lR_m) \bmod M_m = r}^{M_m-1} A_{l,m}(\omega) \right|^2} \tag{12.19}$$

where $M_{max} = \max\{M_m,\ m\text{=0,...,M-1}\}$ and where the inner summation in (12.19) is evaluated only for the values of l and m satisfying the condition $(lR_m) \bmod M_m = r$.

Therefore,

$$E_{p-p} = \max_{0\leq\omega\leq\pi} |\, T(\omega) \,| - \min_{0\leq\omega\leq\pi} |\, T(\omega) \,| \tag{12.20}$$

$$E_{a,max} = \max_{0\leq\omega\leq\pi} E_a(\omega) \tag{12.21}$$

can be used as measures of the quality of the designed banks.

Tables 12.1 and 12.2 report the results obtained for Example 1 and 2, respectively. As can be seen, both the magnitude distortion and the aliasing error are kept small.

In Fig. 12.6 the frequency responses of the final cosine modulated analysis filters, the overall distortion function and the aliasing error relative to Example 1 and obtained with prototypes having 90 and 179 coefficients are shown. The results relative to Example 2 are shown in Fig. 12.7. From the inspection of these figures it can be seen that the design based on (12.15) does not degrade the passband and the stopband characteristics of the new prototypes.

TABLE 12.1. Results relative to Example 1

N_k, $k = 0, 1, 2, 3$	N_4	E_{p-p}	$E_{a,max}$
87	44	6.28 E−02	2.90 E−02
107	54	2.72 E−02	1.37 E−02
179	90	3.60 E−03	6.26 E−03

TABLE 12.2. Results relative to Example 2

N_0	N_1	N_2	E_{p-p}	$E_{a,max}$
100	496	298	1.21 E−02	9.51 E−03

In Fig. 12.8 the final bank relative to Example 3 and obtained with filter lengths equal to 512 (the same number of coefficients as in MPEG) is shown: the reconstruction and the aliasing error are $E_{p-p} = 3.88E - 03$ and $E_{a,max} = 8.99E - 03$, respectively.

12.5 Conclusions

In this work the problems related to the design of non-uniform filter banks have been discussed. This issue can be of interest in some applications, such as audio coding. The MPEG audio standard has also been outlined in order to show why and how non-uniform banks could be exploited.

A method based on the cosine-modulation of more than one prototype has been described. To achieve the cancellation of the main aliasing component, the prototypes cannot be independent of each other. A simple procedure, that requires numerical optimization of only one prototype, being the others derived in a straightforward way from this one, has been described. Even if only Near PR is achieved, the results seem to give a quality of the reconstructed signal sufficiently good for most of the applications.

As a drawback, more than one prototype must be used and only few channels of the uniform bank related to each of them is actually selected. These two facts imply that prototype convolution followed by DCT, as done in the uniform case, can not be used in the implementation.

12.6 References

[1] M. J. Smith and T. P. Barwnell, "Exact reconstruction techniques for tree-structured subband coders", *IEEE Trans. Acoust., Speech, Signal Processing*, vol. ASSP-34, n. 4, pp. 434-441, June 1986.

[2] M.Vetterli, "A theory of multirate filter banks", *IEEE Trans. Acoust., Speech, Signal Processing*, vol. ASSP-35, n. 3, pp. 356-372, March 1987.

[3] P.P.Vaidyanathan, "Theory and design of M-Channel maximally decimated quadrature mirror filters with arbitrary M, having the perfect reconstruction property", *IEEE Trans. Acoust., Speech, Signal Processing*, vol. ASSP-35, n. 4, pp. 476-492, Apr. 1987.

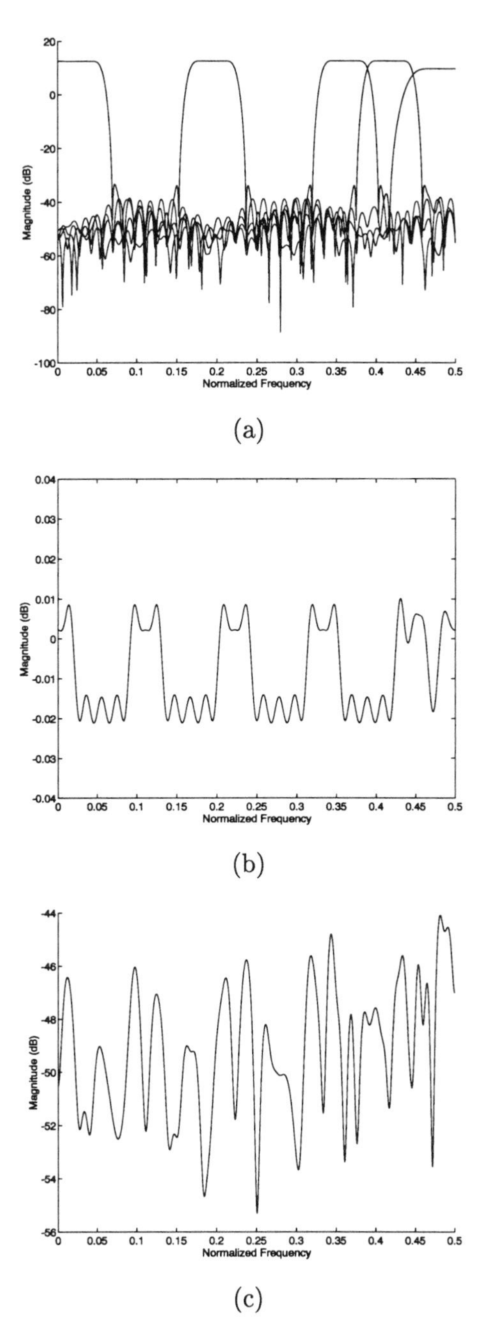

FIGURE 12.6. Results relative to Example 1 (bank {2/9, 2/9, 2/9, 2/9, 1/9 }): (a) final cosine-modulated bank; (b) direct transfer function $| T(\omega) |$; (c) aliasing error

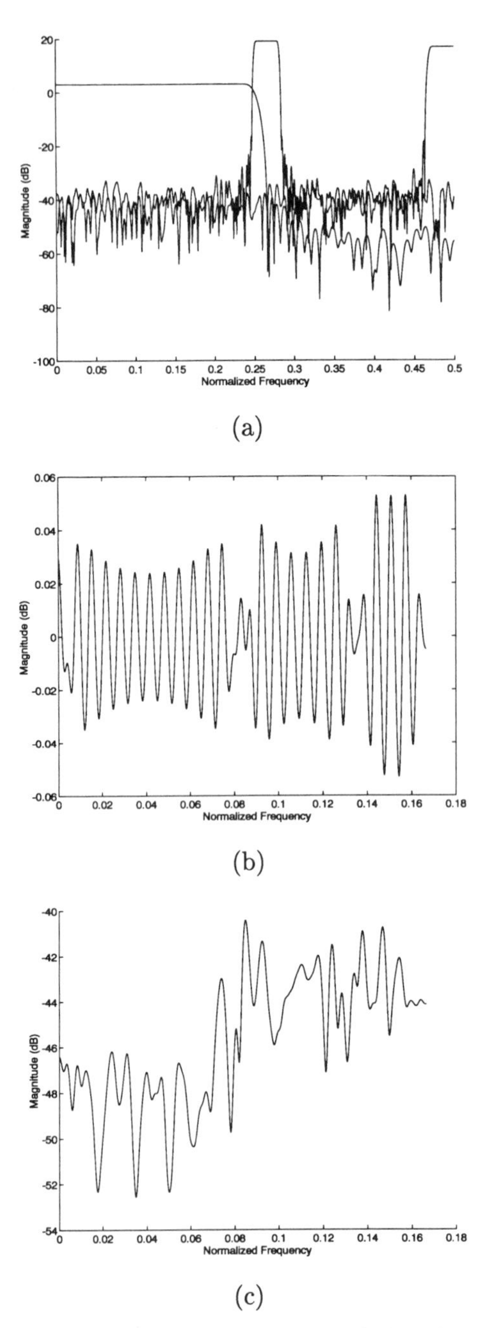

FIGURE 12.7. Results relative to Example 2 (bank {1/2, 5/16, 3/16 }): (a) final cosine-modulated bank; (b) direct transfer function $| T(\omega) |$; (c) aliasing error

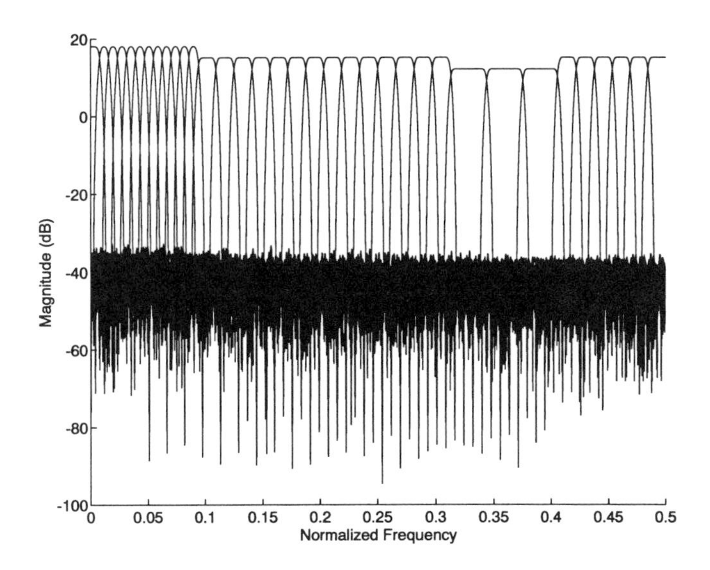

FIGURE 12.8. Filter bank relative to Example 3

[4] P.P.Vaidyanathan, *Multirate Systems and Filter Banks*, Englewood Cliffs, NJ, Prentice-Hall, 1992.

[5] N. Jayant, J. Johnston and R. Safranek, "Signal Compression Based on Models of Human Perception", *Proceedings of the IEEE*, Vol. 81, no. 10, pp. 1385-1422, Oct. 1993.

[6] P. Noll, "Wideband Speech and Audio Coding", *IEEE Commun. Mag.*, pp. 34-44, Nov. 1993.

[7] R.V. Cox, "The Design of Uniformly and Nonuniformly Spaced Pseudo-quadrature Mirror Filters", *IEEE Trans. Acoust., Speech, Signal Processing*, Vol. ASSP-34, no. 5, pp. 1090-1096, Oct. 1986.

[8] P.Q. Hoang and P.P. Vaidyanathan, "Non-uniform Multirate Filter Banks: Theory and Design", *Proc. Int. Symp. Circuits Syst.*, pp. 371-374, May 1986.

[9] K. Nayebi, T.P. Barnwell III and M.J.T. Smith, "Nonuniform Filter Banks: A Reconstruction and Design Theory", *IEEE Trans. Signal Processing*, Vol. 41, no. 3, pp. 1114-1127, Mar. 1993.

[10] J. Kovacevic and M. Vetterli, "Perfect Reconstruction Filter Banks with Rational Sampling Factors", *IEEE Trans. Signal Processing*, Vol. 41, no. 6, pp. 2047-2066, Jun. 1993.

[11] J. Li, T.Q. Nguyen and S. Tantaratana, " A Simple Design Method for Nonuniform Multirate Filter Banks", *28th Asilomar Conf. on signal, systems and computers*, Pacific Grove, CA, Oct. 1994, pp. 1015-1019.

[12] S. Wada, "Design of Nonuniform Division Multirate FIR Filter Banks", *IEEE Trans. Circuits Syst. II*, Vol. 42, no. 2, pp. 115-121, Feb. 1995.

[13] S.Wada, "Generalized Short-Time Fourier Transforms Based on Nonuniform Filter Bank Structure", *Trans. IEICE*, Vol. E78-A, no. 3, pp. 431-436, March 1995.

[14] J. Princen, "The Design of Nonuniform Modulated Filterbanks", *IEEE Trans. Signal Processing*, Vol. 43, no. 11, pp. 2550-2560, Nov. 1995.

[15] H. Nussbaumer, "Pseudo QMF Filter Banks", *IBM Technical Disclosure Bulletin*, Vol. 24, No. 6, pp. 3081-3087, Nov. 1981.

[16] J.H. Rothweiler, "Polyphase Quadrature Filters - A New Subband Coding Technique", *Proc. IEEE ICASSP 83*, pp. 1280-1283, Boston, Apr. 1983.

[17] P.L. Chu, "Quadrature Mirror Filter Design for an Arbitrary Number of Equal Bandwidth Channels", *IEEE Trans. Acoust., Speech, Signal Processing*, Vol. ASSP-33, no. 1, pp. 203-218, Feb. 1985.

[18] R.D Koilpillai and P.P. Vaidyanathan, "A Spectral Factorization Approach to Pseudo-QMF Design", *IEEE Trans. Signal Processing*, Vol. 41, no. 1, pp. 82-92, Jan. 1993.

[19] T.Q. Nguyen, "Near-Perfect-Reconstruction Pseudo-QMF Banks", *IEEE Trans. Signal Processing*, Vol. 42, no. 1, pp. 65-76, Jan. 1994.

[20] F. Argenti and E. Del Re, "Non-uniform filter banks based on a multi-prototype cosine modulation", *IEEE ICASSP'96*, Atlanta, May 1996, pp. 1511-1514.

[21] ISO/IEC, "Information technology - Coding of moving pictures and associated audio for digital storage media up to about 1.5 Mbit/s", 1994.

[22] F. Argenti, B. Brogelli and E. Del Re, "Design of filter banks with rational sampling factors based on a multi-prototype cosine modulation", Tech. Rep. n. 960701, Dept. of Elec. Eng., University of Florence (Italy).

[23] R.D Koilpillai and P.P. Vaidyanathan, "Cosine-Modulated FIR Filter Banks Satisfying Perfect Reconstruction", *IEEE Trans. Signal Processing*, Vol. 40, no. 4, pp. 770-783, Apr. 1992.

[24] T.Q. Nguyen, "Digital Filter Bank Design Quadratic-Constrained Formulation", *IEEE Trans. Signal Processing*, Vol. 43, no. 9, pp. 2103-2108, Sept. 1995.

[25] C.D. Creusere, S.K. Mitra, "A Simple Method for Designing High-Quality Prototype Filters for *M*-Band Pseudo QMF Banks", *IEEE Trans. Signal Processing*, Vol. 43, no. 4, pp. 1005-1007, Apr. 1995.

[26] F.Argenti, V.Cappellini, E.Del Re and A.Fiorilli, "Non-uniform subband analysis banks for the compression of audio signals", *Proc. 1st Workshop on Sampling Theory and Applications*, Jurmala, Latvia, 20-22 Sept. 1995, pp. 285-289.

13

Source Coding of Stereo Pairs

Halûk Aydınoğlu
Monson H. Hayes

ABSTRACT [1] Due to recent advances in display technology, three dimensional (3-D) imaging systems are becoming increasingly more common in applications such as computer vision, virtual reality, terrain mapping, navigation, and image understanding. To achieve 3-D perception, these systems use a stereo pair, which is a pair of images of the same scene acquired from different perspectives. Since there is an inherent redundancy between the images of a stereo pair, data compression algorithms can be employed to transmit and store these images efficiently.
In this chapter, we consider the problem of stereo image coding. We begin with a description of the stereo coding problem, and survey the current approaches to stereo coding. Then, we describe a new coding algorithm that is based on disparity compensation and subspace projection. This algorithm, called the Subspace Projection Technique (SPT), is an incomplete local transform with a data-dependent space-varying transform matrix. The advantage of the SPT approach over other techniques is that it is able to adapt to changes in the cross-correlation characteristics of stereo pairs locally.

13.1 Introduction

A pair of images of a scene that is acquired from two different perspectives is called a *stereo pair*. When each image of a stereo pair is presented simultaneously to the left and right eyes, the brain is able to *fuse* the images and perceive depth[2]. This ability is called stereo vision. Stereoscopic systems that enable three-dimensional viewing are becoming increasingly more common in applications such as computer vision, virtual reality, terrain mapping, navigation, and image understanding. With the use of stereo pairs in these vision systems, it is becoming important to be able to transmit and store them efficiently. Although it is possible to code the left and right images of a stereo pair independently, this approach does not exploit the redundancies that exist between the two images. Therefore, to achieve higher compression, a number of stereo image coding systems have been developed

[1]This material is based upon work supported by the U.S. Army Research Office under grant number DAAH04-96-1-0161.

[2]The stereo images presented in the appendix can be perceived in 3-D with free viewing techniques.

to exploit the correlation between the images [1, 2, 3, 4]. In the design of a stereo compression system, it is often necessary to impose certain restrictions on the encoder in order to make it suitable for its intended application. For example, 3-D television broadcasting will most likely be considered within the context of an add-on feature to currently existing systems. Therefore, the receiver should maintain compatibility with normal video decoding and display systems and be able to decode one of the images separately [5, 6]. On the other hand, in applications such as terrain mapping and autonomous piloting of unmanned vehicles, the compression system should be designed to minimize the errors in the 3-D information [7], i.e., preserve depth information.

In this chapter, we consider the problem of very low-bit compression of stereo pairs, and present a new approach for low-bit rate stereo image coding. In this approach, it is assumed that stereo is an extension to a basic monocular transmission channel and, therefore, a *conditional coder* is used as the framework for the proposed stereo compression system. This structure ensures the compatibility of the coder with a monocular decoding system. We begin, in the following section, with an overview of stereo image compression, and a summary of current approaches to stereo image coding.

13.2 Stereo Image Coding

A stereo image pair is produced by a pair of cameras that image a given scene from two different perspectives. Typically, it is assumed that the optical axes of these cameras are parallel, so that the image planes are coplanar and co-linear. In addition, it is assumed that the cameras are displaced along the horizontal axis, and that the focal lengths and the magnification factors of the lenses are the same. This camera geometry is illustrated in Figure 13.1. The parallel optical axes geometry and the horizontal camera displacement are convenient for several reasons. First, this geometry is appropriate to model the human visual system [8, 9]. Second, it ensures that the positions of *corresponding* pixels in the two images will differ only in the horizontal direction. This horizontal displacement, denoted by d, varies from one pixel to the next, and is referred to as the *disparity* (See Figure 1). The distance to an object, $f-z$, is inversely proportional to its disparity and may be computed as follows

$$f - z = \frac{f \cdot s}{d}, \tag{13.1}$$

where f is the focal length of the cameras, and s is the distance between the two camera centers (note that z is always negative for objects in the scene). Finally, the parallel axis geometry ensures that a rectangular surface lying in a plane that is parallel to the image plane will have the same area and aspect ratio in both images (rectangular surfaces that are not parallel to the image plane will suffer from perspective distortion).

Due to the different camera positions, there are two effects that need to be understood in stereo compression systems. The first effect is that of *occlusion*. As illustrated in Fig. 13.2, when a 3-D object is imaged by two different cameras, a part of the object will only be visible by one of the cameras. This region is

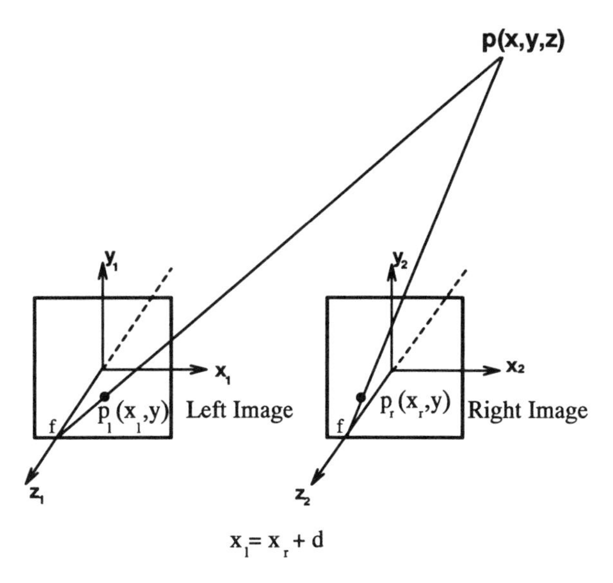

FIGURE 13.1. The assumed geometry for a stereo acquisition system.

said to be occluded. Occurrence of a depth discontinuity may yield occlusion. The border of the image can also act as an occluding edge. The difficulty with occluded regions is that the depth cannot be determined (it could be predicted, however, from neighboring regions), and occluded regions cannot be estimated from the second image. The second effect is the observation that the recorded intensity of an object by the left and right cameras will generally not be the same. These *photometric variations* are a result of the fact that the light that is reflected from an object and recorded by a camera depends on the position of that camera relative to the object [10]. Other sources of photometric variations include noise, and different photometric characteristics of the camera systems.

13.2.1 Theory of Stereo Image Coding

A stereo pair (X, Y) may be modeled as a vector-valued outcome of two correlated discrete random processes. According to Shannon's source coding theorem [11], if $H(X)$ is the entropy of X and $H(Y|X)$ is the conditional entropy of Y given X, then an average rate that is greater than $H(X,Y) = H(X) + H(Y|X)$ is sufficient to recover a stereo pair with arbitrarily low probability of error, provided that they are encoded together. This implies that a stereo pair may be optimally noiselessly encoded using a *conditional coding* structure (referred to as the CONCOD structure), that involves independent coding of the first image, and then coding the second image, given the coded first image. What makes the CONCOD structure attractive is that it allows the decoder to independently decode a single image, thus making it compatible with non-stereo decoders.

In many applications, the goal is not to design a lossless coder, but rather to design one that allows some distortion in exchange for a decrease in the number

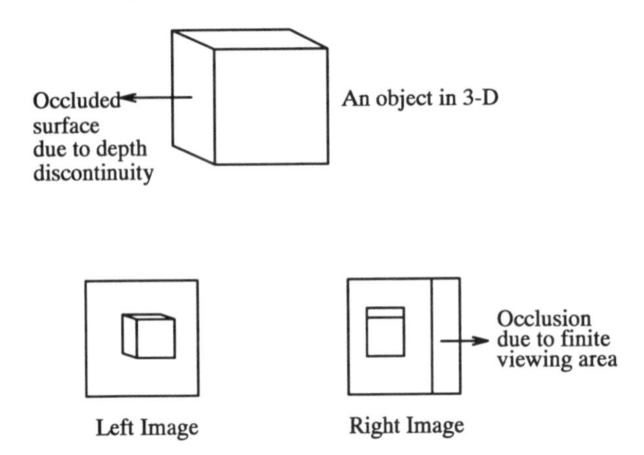

FIGURE 13.2. Occlusion in a stereo imaging system.

of bits. An important question to ask, therefore, is the following: "What is the minimum number of bits necessary to code the stereo source (X, Y) such that the distortion for X is D_X and the distortion for Y is D_Y?" Although the answer to this question is unknown, a candidate system would be to use the CONCOD structure to independently encode image X with a distortion D_X, and conditionally encode image Y with a distortion D_Y. The disadvantage of this approach is that the CONCOD structure has been shown to be suboptimal over some regions of the rate-distortion surface [2] in the sense that.

$$R_X(D_X) + R_{Y|X}(D_Y) \leq R_{X,Y}(D_X, D_Y) \leq R_X(D_X) + R_{Y|\hat{X}}(D_Y), \quad (13.2)$$

where $R_X(D_X)$ is the rate-distortion function for X, $R_{X,Y}(D_X, D_Y)$ is the joint rate-distortion function for X and Y (the optimal coder), $R_{Y|X}(D_Y)$ is the conditional rate-distortion function for Y given the original X, and $R_{Y|\hat{X}}(D_Y)$ is the conditional rate-distortion function for Y given the encoded X (the conditional coder) [2]. An interesting property for very low bit rate coding of the second image follows from this inequality if we rewrite Eq.(13.2) as follows

$$R_X(D_X) \leq R_{X,Y}(D_X, D_Y) \leq R_X(D_X) + R_{Y|\hat{X}}(D_Y). \quad (13.3)$$

In the extreme case, in which we allocate zero bits for the second image (the distortion D_Y for the second image is equal to its maximum value $D_{Y_{max}}$, i.e., $R_{Y|\hat{X}}(D_{Y_{max}}) = 0$), the CONCOD structure is optimal since the lower and upper bounds of Eq. (13.3) are identical. Due to the continuity of the rate-distortion functions, for any $\epsilon > 0$ there exists a δ such that if $|D_{Y_\delta} - D_{Y_{max}}| < \delta$ then

$$|R_X(D_X) + R_{Y|\hat{X}}(D_{Y_\delta}) - (R_X(D_X) + \underbrace{R_{Y|\hat{X}}(D_{Y_{max}})}_{0})| = R_{Y|\hat{X}}(D_{Y_\delta}) < \epsilon. \quad (13.4)$$

However, since

$$R_{X,Y}(D_X, D_{Y_\delta}) > R_{X,Y}(D_X, D_{Y_{max}}) = R_X(D_X) + R_{Y|\hat{X}}(D_{Y_{max}}), \quad (13.5)$$

(a) The difference between the left and right images.

(b) The difference due to global translation method.

we can rewrite Eq. (13.4) in the following form[3]:

$$|R_X(D_X) + R_{Y|\hat{X}}(D_{Y_\delta}) - R_{X,Y}(D_X, D_{Y_\delta})| < \epsilon, \qquad (13.6)$$

which implies that the CONCOD structure is performing arbitrarily close to the optimal solution given that $R_{Y|\hat{X}}(D_Y)$ is small. Based on this observation, several stereo coding systems have been proposed that use a conditional coder that minimizes the bit rate for the second image by exploiting the stereo redundancy, while preserving a required quality [12, 13]. Note that the operational rate-distortion curve of any DC scheme is bounded from below by the theoretical rate-distortion curve of the CONCOD structure.

13.2.2 Historical Perspective

In this section, we provide an overview of the basic approaches that have been proposed for stereo image coding. Each of these methods is designed to exploit the redundancy that exists between the images in a stereo pair.

One of the first techniques proposed for stereo image coding is to code the sum and difference of the two images [8]. If the left and right images have the same first and second order statistics, i.e., $E\{X\} = E\{Y\} = m$ and $E\{X^2\} = E\{Y^2\}$, then the sum difference images will be uncorrelated and may therefore be independently coded. Shown in Figure 13.2.2(a) is the difference image for the "Lab" stereo pair (see appendix for the original images). The bit rate reduction that is obtained by sum-difference coding over independent coding is approximately 15% for the Lab stereo pair.

An improvement in sum-difference coding is to horizontally shift one image to the point where the cross-correlation between the images of the stereo pair reaches its maximum value (the shift approximately equals to the mean disparity of the scene). The sum and difference images are then encoded [1]. This method

[3]Note that $R_X(D_X) + R_{Y|\hat{X}}(D_{Y_\delta}) - R_{X,Y}(D_X, D_{Y_\delta})$ is always positive

is referred to as the *global translation method*, and works well when all of the objects in a scene have similar disparity values. Unfortunately, however, this is not generally the case. In Figure 13.2.2(b), the prediction error due to the global translation method is shown for the Lab stereo pair (the translation amount is 29 pixels which is equal to the mean disparity of the scene). The peak signal to noise ratio (PSNR) is equal to 22.18dB for the non-occluded region. Note that the entropy of the prediction error is 4.02 bpp, which is equivalent to 47 % bit rate reduction over independent coding and 30 % bit rate reduction over sum-difference coding of the second image.

A refinement to the global translation method is to individually shift each row. The amount of translation is determined either by cross-correlation statistics (correlation enhanced technique) [1], or by using prominent features in each row [14].

Both the global translation and correlation enhanced techniques are examples of disparity compensated prediction [15]. The basic idea of disparity compensation (DC) is as follows. First, one of the images (say the right) is partitioned or segmented into subimages, $\mathbf{b}_r$. Then, a subimage in the left image that most closely resembles $\mathbf{b}_r$ is found (the relative displacement between the two subimages is the average disparity). This subimage is then used as an estimate, $\hat{\mathbf{b}}_r$, of $\mathbf{b}_r$. Thus, the first step in disparity compensated stereo compression is to estimate the *disparity field*, i.e., disparity for each subimage. There are two basic approaches to disparity estimation. The first are the *feature-based* methods that attempt to match significant features, such as edges, to estimate the disparity field. Selected features are generally those that are not sensitive to noise and calibration errors. Since feature matching generates a sparse disparity map (disparity values are only extracted for the features), the disparity field is then interpolated to form a dense disparity field [12]. Since this is a nontrivial procedure, feature-based disparity estimation is typically only used in applications where very accurate disparity information is required.

The second approach to disparity estimation are the *intensity-based* methods that establish a correspondence between points in the image pair based on the intensities of corresponding areas in the right and left images. A disparity value is typically assigned to each point even if a real match does not exist (occlusions).

Once the disparity field has been estimated, there is a variety of stereo coding strategies that involve prediction based on the disparity information. The first technique is to perform blockwise prediction using intensity-based disparity estimation, and to transmit the prediction error along with the disparities for each block [1]. This technique, referred to as disparity compensated prediction with residual coding (DC-RC), encodes the prediction error using a lossy scheme and encodes the disparity without distortion.

Another approach is to use a mixed inter/intra frame coder. Instead of coding the residual signal, either the disparity information is sent and the subblock is replaced by the approximating subblock at the decoder, or the subblock is independently coded using an adaptive discrete cosine transform [4]. Which approach is used for a block depends on the accuracy of the subblock approximation. The approach is a special case of DC-RC. In some applications, this algorithm is used only for the lowest subband to decrease the computational complexity and to exploit the fact that the correlation between the high frequency components of the images of a stereo pair is low [3].

Disparity compensated prediction in the transform domain was proposed by Perkins [2]. In this approach, the transform of a right image block, $R = T(\mathbf{b}_r(i,j))$, is estimated from the transform of the matching left image block, $L = T(\mathbf{b}_l(i + d, j))$, using $\hat{R} = A \bigotimes L + B$, where the array multiplication operation $\bigotimes$ is performed by multiplying the $(i,j)^{th}$ element of the first matrix by $(i,j)^{th}$ element of the second matrix. The matrices A and B are chosen based on the average statistics of a training set, so that each coefficient is predicted by a minimum variance predictor. After estimating the transform domain coefficients, the transform domain error is quantized and sent through the channel. The minimum variance predictor performs better than the simple predictor proposed by [1] for high bit rates. However, the coding gain is very small for low bit rates [2].

Block-based disparity compensation can be implemented using either fixed block sizes (FBS) or variable block sizes (VBS). The advantage of FBS schemes is that there is no overhead required to specify the block locations. However, if there are multiple objects or an occluded object within a block, then FBS-DC cannot fully exploit the stereo redundancy. Although the block size could be reduced to minimize the effect of this problem, this would result in an increase in the overall bit rate for the disparity field. A quadtree decomposition, on the other hand, provides a relatively economical and effective solution to this problem. The basic idea is to adaptively decrease the block size until the prediction error falls below some threshold. A regular quadtree approach, where a block can be divided only at the midpoint of its sides has been implemented in [16, 17]. Also, a generalized quadtree can be considered where a block can be divided at $2^K - 1$ locations (K is the number of bits that can be allocated per side per node) [12]. The division takes place at the permitted location that lies closest to a sharp intensity discontinuity. In general, VBS schemes outperform the FBS schemes [12, 16, 17].

Conventional disparity compensation algorithms using rectangular blocks often result in discontinuities between neighboring disparity compensated blocks in the predicted images. In order to alleviate this blocking effect, a windowed disparity compensation (WDC) scheme was recently proposed [17]. This approach involves the use of a data window that overlaps a portion of the conventional disparity compensation block.

Other approaches to disparity estimation include a multiresolution hierarchical disparity estimation technique [18] to reduce the complexity of the search, and a genetic block matching algorithm to obtain a smooth disparity field at a small block size without sacrificing its accuracy [19].

13.3 The Subspace Projection Technique

Conventional stereo image coders employ disparity compensated prediction followed by the transform coding of the residual to compress the stereo information. Disparity compensation removes the stereo redundancy between the images of a stereo pair, and the transform coder exploits the correlation between the samples of the residual signal. Although transform coders, such as the DCT, are generally efficient in terms of packing a large fraction of the energy of an image into relatively few transform coefficients, this is not always the case for residual coding. Therefore, in order to improve the coding gain, a framework has been proposed for combining disparity compensation and residual coding [20]. The idea is to apply a transformation, $\mathbf{T}$, to each $m \times m$ block, $\mathbf{b}_r$, of the right image in such a way that it exploits the stereo and spatial redundancy and allows the block to be represented with fewer bits.

The transformation that is applied to each image block is motivated by the following two observations. First, the correlation between the low frequency components in a pair of matching blocks is greater than the correlation between the high frequency components [3]. Second, if one of the images of the stereo pair is encoded with little distortion, thereby resulting in a high-resolution decoded image, then it is the low frequency components of the second image that are perceptually more important than the high frequency components [2]. Based on these observations, the transformation $\mathbf{T}$ is chosen to be a projection operator that projects an $m \times m$ block $\mathbf{b}_r$ onto a subspace $\mathcal{S}$ of R^{m^2}. This subspace is spanned by a set of N vectors. The first vector in the spanning set is a block-dependent vector and is equal to $\hat{\mathbf{b}}_r$, the estimate of the block $\mathbf{b}_r$ that is obtained from the left image through disparity compensation. The remaining $N-1$ vectors are fixed, $\mathbf{F} = \{\mathbf{f}_i\}_1^{N-1}$, and are selected so that they approximate the low frequency components of the block by forming a polynomial approximation to the vector $\mathbf{b}_r$. The number of these polynomial vectors define the order of the approximation. Generally, either one (zero-order approximation), three (first-order approximation), or six (second-order approximation) fixed vectors are used. For example, with $m = 4$, the fixed vectors for a first-order approximation would be

$$\mathbf{f}_1 = \begin{bmatrix} 1 & 1 & 1 & 1 \\ 1 & 1 & 1 & 1 \\ 1 & 1 & 1 & 1 \\ 1 & 1 & 1 & 1 \end{bmatrix}, \quad \mathbf{f}_2 = \begin{bmatrix} 1 & 1 & 1 & 1 \\ 2 & 2 & 2 & 2 \\ 3 & 3 & 3 & 3 \\ 4 & 4 & 4 & 4 \end{bmatrix}, \quad \mathbf{f}_3 = \begin{bmatrix} 1 & 2 & 3 & 4 \\ 1 & 2 & 3 & 4 \\ 1 & 2 & 3 & 4 \\ 1 & 2 & 3 & 4 \end{bmatrix}.$$

Thus, the projection of $\mathbf{b}_r$ onto the subspace $\mathcal{S}$ is given by

$$\mathbf{T}(\mathbf{b}_r) = \beta_0 \hat{\mathbf{b}}_r + \sum_{i=1}^{N-1} \beta_i \cdot \mathbf{f}_i, \tag{13.7}$$

where β_i are the least-square coefficients that minimize the mean-square error

$$\mathcal{E} = \|\mathbf{b}_r - \mathbf{T}(\mathbf{b}_r)\|^2. \tag{13.8}$$

Generally, the spanning vectors are first orthogonalized so that the projection coefficients are uncorrelated and easier to evaluate. Using the Gram-Schmidt orthogonalization procedure, the orthogonalized basis vectors are

$$\mathbf{v}_1 = \mathbf{f}_1, \tag{13.9}$$

$$\mathbf{v}_k = \mathbf{f}_k - \sum_{i=1}^{k-1} \frac{<\mathbf{f}_k, \mathbf{v}_i>}{<\mathbf{v}_i, \mathbf{v}_i>} \cdot \mathbf{v}_i \quad ; \quad k = 2, 3, \ldots, N-1, \tag{13.10}$$

$$\mathbf{b}_0 = \hat{\mathbf{b}}_r - \sum_{i=1}^{N-1} \frac{<\hat{\mathbf{b}}_r, \mathbf{v}_i>}{<\mathbf{v}_i, \mathbf{v}_i>} \cdot \mathbf{v}_i, \tag{13.11}$$

where $< \cdot, \cdot >$ is the inner product. For example, the first-order orthogonalized polynomial basis vectors are

$$\mathbf{v}_1 = \mathbf{f}_1 \quad \mathbf{v}_2 = \begin{bmatrix} -1.5 & -1.5 & -1.5 & -1.5 \\ -0.5 & -0.5 & -0.5 & -0.5 \\ 0.5 & 0.5 & 0.5 & 0.5 \\ 1.5 & 1.5 & 1.5 & 1.5 \end{bmatrix}, \quad \mathbf{v}_3 = \mathbf{v}_2'.$$

Note that this orthogonalization procedure is computationally efficient and, except for $\mathbf{b}_0$, these vectors may be calculated off-line. In addition, since the vectors $\mathbf{v}_i$ will be known by both the encoder and decoder, they do not need to be encoded or transmitted. Finally, note that although $\mathbf{b}_0$ is an orthogonalized block-dependent vector, it is not necessary to transmit this vector either. Specifically, since $\hat{\mathbf{b}}_r$ may be derived by the decoder from the disparity field and the decoded left image, then $\mathbf{b}_0$ may be recovered using Eq. (13.11).

In terms of the orthogonalized spanning vectors, the transform domain representation for the block $\mathbf{b}_r$ is

$$\mathbf{T}(\mathbf{b}_r) = \gamma \cdot \mathbf{b}_0 + \sum_{i=1}^{N-1} \alpha_i \cdot \mathbf{v}_i. \tag{13.12}$$

Due to the orthogonality of the basis vectors, the projection coefficients, γ and α_i, are easily evaluated as follows,

$$\alpha_i = \frac{< \mathbf{b}_r, \mathbf{v}_i >}{< \mathbf{v}_i, \mathbf{v}_i >} \quad i = 1, 2, \ldots, N-1, \tag{13.13}$$

$$\gamma = \frac{< \mathbf{b}_r, \mathbf{b}_0 >}{< \mathbf{b}_0, \mathbf{b}_0 >}. \tag{13.14}$$

Note that the fixed subspace component of $T(\mathbf{b}_r)$,

$$\mathbf{b}_f = \sum_{i=1}^{N-1} \alpha_i \cdot \mathbf{v}_i \tag{13.15}$$

captures the low frequency components of the block $\mathbf{b}_r$, whereas the block-dependent component

$$\mathbf{b}_v = \gamma \cdot \mathbf{b}_0 \tag{13.16}$$

exploits the stereo redundancy of the block and contains the high frequency stereo information.

The use of polynomial vectors in the fixed subspace approximation is motivated by the smooth variations that are characteristics of the image pixel intensities found in most natural images. Specifically, the image intensity at a given pixel is generally close to the average intensity in a small neighborhood of that pixel. The coefficient of the fixed vector $\mathbf{f}_1$ represents the average intensity or *dc* component of each block. Similarly, the coefficients of $\mathbf{f}_2$ and $\mathbf{f}_3$ provide a measure of the slope in image pixel intensities in the vertical and horizontal directions, respectively. Two properties of the polynomial vectors are worth mentioning. First, the orthogonal basis vector $\mathbf{b}_0$ and the transform coefficients α_i can be calculated with multiplication-free operations, i.e., by using adds and shifts. Second, the first three basis vectors of the discrete cosine transform and polynomial vectors result in similar approximations [21].

An important observation to make at this point is that, due to the stereo redundancy, there is a similarity between the image block $\mathbf{b}_r$ in the right image, and the disparity compensated prediction $\hat{\mathbf{b}}_r$ that is formed from the left image. Therefore, if we expand $\hat{\mathbf{b}}_r$ in terms of the orthogonalized basis vectors $\mathbf{b}_0$ and $\mathbf{v}_i$ then we have, from Eq. (13.11),

$$\hat{\mathbf{b}}_r = \mathbf{b}_0 + \sum_{i=1}^{N-1} \hat{\alpha}_i \cdot \mathbf{v}_i, \tag{13.17}$$

where

$$\hat{\alpha}_i = \frac{< \hat{\mathbf{b}}_r, \mathbf{v}_i >}{< \mathbf{v}_i, \mathbf{v}_i >} \quad i = 1, 2, \ldots, N-1. \tag{13.18}$$

Thus, it follows that the coefficients α_i and $\hat{\alpha}_i$ will be correlated. Consequently, it is the coefficient *prediction errors*, $\epsilon_i = \alpha_i - \hat{\alpha}_i$ that are encoded and transmitted instead of the transform coefficients α_i themselves.

Since the operator $\mathbf{T}$ projects the block $\mathbf{b}_r$ onto a lower-dimensional subspace, $\mathbf{T}(\mathbf{b}_r)$ is not generally equal to $\mathbf{b}_r$. However, among all vectors $\mathbf{b} \in \mathcal{S}$, $\mathbf{T}(\mathbf{b}_r)$ is the best estimate of the vector $\mathbf{b}_r$, i.e.,

$$\min_{\mathbf{b} \in \mathcal{S}} \|\mathbf{b}_r - \mathbf{b}\| = \|\mathbf{b}_r - \mathbf{T}(\mathbf{b}_r)\|. \tag{13.19}$$

In fact, since we adaptively change the subspace $\mathcal{S}$, the projection error, $\mathbf{e} = \mathbf{b}_r - \mathbf{T}(\mathbf{b}_r)$, which is orthogonal to $\mathcal{S}$ is small, in general. The subspace approximation that is generated by the fixed vectors and the block varying vector $\mathbf{b}_0$ is illustrated in Figure 13.3.

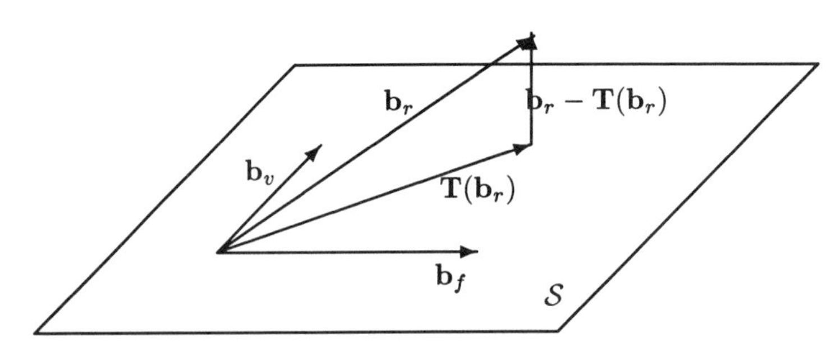

FIGURE 13.3. Approximation of the vector $\mathbf{b}_r$ by an element of the subspace generated by the fixed vector set $\mathcal{F}$ and the block varying vector $\mathbf{b}_0$.

In summary, for the SPT encoder, we need to transmit the following components for each block:

- The disparity vector d of the block.

- The residual coefficients ϵ_i of the fixed vectors $\mathbf{v}_i$ for $i = 1, \ldots, N-1$.
- The projection coefficient γ of the block-varying basis vector $\mathbf{b}_0$.

The disparity vector field is coded using a lossless differential pulse code modulation followed by a modular warping scheme as proposed in the MPEG coding standard [22]. A near-lossless scheme has also been considered for coding the disparity field [23], but this approach requires further studies, and is not discussed in this chapter.

Before discussing how to quantize the coefficients, we first describe the operation of the decoder. With the CONCOD structure, the first step is to decode the left image. Then, for each block in the right image, the disparity compensated prediction $\hat{\mathbf{b}}_r$ is recovered from the decoded left image. Finally, the subspace approximation for $\mathbf{b}_r$ is given by Eq. 13.12 However, note that since $\alpha_i = \hat{\alpha}_i + \epsilon$ then $T(\mathbf{b}_r)$ is, in practice, evaluated as follows,

$$T(\mathbf{b}_r) = \gamma \cdot \mathbf{b}_0 + \sum_{i=1}^{N-1} (\epsilon_i + \hat{\alpha}_i) \cdot \mathbf{v}_i. \tag{13.20}$$

Note that since γ and the residual coefficients ϵ_i are transmitted by the decoder, this computation is straightforward.

Let us now consider the quantization of the coefficients ϵ_i and γ. Empirical work has shown that efficient quantization of the coefficients is a key to higher compression rates [24]. At the decoder, we have

$$\mathbf{Q}[\mathbf{T}(\mathbf{b}_r)] = \gamma_q \cdot \mathbf{b}_0 + \sum_{i=1}^{N-1} (\epsilon_{i_q} + \hat{\alpha}_i) \cdot \mathbf{v}_i, \tag{13.21}$$

where $\mathbf{Q}$ is the quantization operator and where γ_q and ϵ_{i_q} are the quantized coefficients. The overall coding distortion, $\mathcal{E}$, is

$$\mathcal{E} = \mathbf{b}_r - \mathbf{Q}[\mathbf{T}(\mathbf{b}_r)] = \mathbf{T}(\mathbf{b}_r) + \mathbf{e} - \mathbf{Q}[\mathbf{T}(\mathbf{b}_r)], \tag{13.22}$$

where $\mathbf{e}$ is the subspace projection error. Therefore, it follows that

$$\mathcal{E} = (\gamma - \gamma_q) \cdot \mathbf{b}_0 + \sum_{i=1}^{N-1} (\epsilon_i - \epsilon_{i_q}) \cdot \mathbf{v}_i + \mathbf{e}. \tag{13.23}$$

Since we have an orthogonal decomposition, then

$$||\mathcal{E}||^2 = (\gamma - \gamma_q)^2 \cdot ||\mathbf{b}_0||^2 + \sum_{i=1}^{N-1} (\epsilon_i - \epsilon_{i_q})^2 \cdot ||\mathbf{v}_i||^2 + ||\mathbf{e}||^2. \tag{13.24}$$

In order to improve the perceptual quality, the subspace projection operation is performed on 4-by-4 blocks, which results in a small projection error vector $\mathbf{e}$. We can then construct the operational rate-distortion function $\mathcal{R}(\mathcal{D})$, where $\mathcal{D}$ is given by

$$\mathcal{D} = \sum_{\text{all blocks}} \left((\gamma - \gamma_q)^2 \cdot ||\mathbf{b}_0||^2 + \sum_{i=1}^{N-1} (\epsilon_i - \epsilon_{i_q})^2 \cdot ||\mathbf{v}_i||^2 \right). \tag{13.25}$$

Since the distortion is additive in each term, we can find a locally optimal bit allocation scheme using the operational rate-distortion curves as proposed in [25]. The proposed allocation scheme is as follows.

We have N different projection coefficients. For each of these coefficients, we use n_i different quantizers $Q_{i,1}$, $Q_{i,2}$, $Q_{i,3}$, $\cdots$, Q_{i,n_i}, where $1 \leq i \leq N$. The quantizers are ordered so that the distortion, d_i is decreasing and the rate b_i is increasing along the quantizer sequence. The distortion for each component of the error term is given by

$$d_i = \sum_{\text{all blocks}} (\epsilon_i - \epsilon_{i_q})^2 \cdot \|\mathbf{v}_i\|^2 \quad 1 \leq i < N, \tag{13.26}$$

$$d_N = \sum_{\text{all blocks}} (\gamma - \gamma_q)^2 \cdot \|\mathbf{b}_0\|^2. \tag{13.27}$$

We add the N distortion components for each term to find the actual distortion, $\mathcal{D}$ for the image. The rate, $r_i(k_i)$, is the actual bit rate for the ith coefficient (summed over all blocks) using the kth quantizer. Our aim is to find the optimal set of r_i's that minimize $\mathcal{D}$, subject to the constraint that $\sum_{i=1}^{N} r_i \leq \mathcal{R}_t$. We follow an iterative four step algorithm to achieve this objective.

1. Determine an initial bit allocation using the n_i^{th} quantizer for each coefficient.

2. Calculate all possible values of the slope function

$$s(j_i, k_i) = \frac{r_i(j_i) - r_i(k_i)}{d_i(j_i) - d_i(k_i)} \tag{13.28}$$

   for each class i. Here, $r_i(j_i)$ is the current bit assignment. For each class, find the quantizer for which the slope function $s(j_i, k_i)$ is maximum.

3. Determine the class for which the maximum values of $s(j_i, k_i)$ is largest and assign to this class the quantizer for which this maximum is obtained.

4. Calculate the new bit rate $\mathcal{R}$ and distortion $\mathcal{D}$ and check if $\mathcal{R} \leq \mathcal{R}_t$; if so, then stop the algorithm. If not go back to step 2.

The bit allocation algorithm is locally optimal. That is, given the points of the operational rate distortion curve, one can find the best bit allocation among the coefficients. Note that it is possible to prepare lookup tables for operational rate-distortion curves.

The actual quantizer choice is crucial for the best coding performance. Instead of the standard zig-zag scanning, uniform quantization, and entropy coding schemes, we construct coefficient images for each coefficient and utilize a subband coder for each of these coefficient images. This approach exploits the spatial correlation between the coefficients of the neighboring blocks. The gain achieved with the proposed approach becomes significant at very low bit rates. In addition, it has been shown that exploiting and alleviating interblock correlation of the transform domain coefficients diminish the visibility of the blocking artifacts [26].

method	f.garden	lab	room
DC	22.97	29.83	24.00
DC-RC	22.00	30.02	22.68
SPT	23.70	32.00	26.83

TABLE 13.1. Rate-distortion performance of the stereo image coding algorithms. The PSNRs of the encoded images (in dB) are presented at a fixed bit rate (0.06bpp for the lab and room pairs and 0.05bpp for the flower garden pair).

13.4 Experimental Results

In this section we present a comparison of stereo image coding algorithms. In these experiments, the left image is independently coded, and the right image is coded based on the coded left image.

Three different stereo pairs are used in these experiments. The first pair was obtained from the "flower garden" video sequence (frame numbers 0 and 2). These images are 352-by-240. The second is the "Lab" pair (512-by-480), which was obtained by shifting a video camera and taking two pictures of a stationary scene from two horizontally separated locations. Finally, the third one is the "Room" pair (256-by-256), which was obtained using the same procedure as the "Lab" pair. These images are given in the appendix.

First, the performance of stereo image coding algorithms at very low bit rates (around 0.05 bpp for the right image) are compared. To operate at this bit rate is very challenging. The bits spent on the disparity field constitute an important portion of the total bit allocation. On the other hand, the prediction suffers from blocking artifacts and mismatches (especially for occlusion regions). The advantage of this lower rate is that the transmission cost for the stereo pair is only slightly more than that of a single mono image. Three algorithms are compared: disparity compensated prediction with 8-by-8 block size (Figure 13.5), disparity compensated prediction (16-by-16 block size) followed by transform domain residual coding (DC-RC) (Figure 13.6), and SPT is implemented with 4-by-4 projection block size using the fixed block size (16-by-16) DC estimates (Figure 13.7). The subspace is generated with three fixed vectors and one block-varying vector. An optimal bit allocation scheme is employed to determine the quantization levels for each coefficient.

The distortion for the non-occluded regions at a fixed bit rate are presented in Table 13.1. SPT is significantly better than the other two implementations of the disparity compensation. Note that the standard disparity compensation techniques may be improved if modified predictors (such as generalized block matching [16], windowed disparity compensation [17], or variable block size disparity compensation [12]) are employed. On the other hand, if the results of these modified estimates are used in the design of the local transforms, similar gains can be achieved by the SPT[17].

Second, the operational rate distortion curves of the DC and the SPT algorithms are compared for the "Lab" stereo pair. The SPT is implemented using a 4-by-4 projection block size and 16-by-16 disparity block size. Disparity compensation is implemented using 16-by-16 blocks for estimation. The residual due to the disparity compensation is coded with an optimized JPEG coder. The rate-

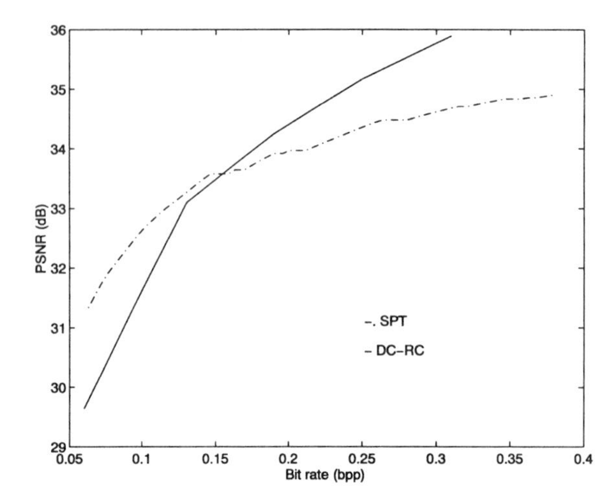

FIGURE 13.4. A comparison of the rate distortion performance of the SPT and DC-RC algorithms.

distortion curves are presented in Figure 13.4. For low bit rates, the SPT is 1.7 dB better than disparity compensation. Another interpretation of this result is that we can achieve a bit rate reduction of 40% over DC. The minimum distortion achieved by the SPT is bounded from below since this implementation cannot perform exact reconstruction. The maximum achievable PSNR may be increased by decreasing the block size or increasing the number of the fixed vectors. The rate-distortion performances of these different implementations (different block size or fixed vector size) are, however, almost identical for low bit rates.

13.5 Conclusion

This chapter summarizes our recent work on stereo image coding. We propose a new coding technique based on subspace projection. The novelty of the approach is that the transformation matrix of the projection operation adaptively changes to exploit the inherent stereo redundancy and non-stationary cross-correlation characteristics between the images of a stereo pair. In addition, we used a combined transform-subband coding scheme that is very efficient for coding transform domain coefficients. The subspace projection technique is appealing since its performance at very low bit rates was found to be superior to the standard stereo coding algorithms. The proposed coder is flexible and can operate over a broad range of the rate-distortion curve and does not require training.

FIGURE 13.5. Image coded at 0.06 bpp using DC ($PSNR = 29.83$dB)

FIGURE 13.6. Image coded at 0.06 bpp using DC-RC ($PSNR = 30.02$dB)

FIGURE 13.7. Image coded at 0.06 bpp using the SPT. ($PSNR = 32.00$dB)

APPENDIX A

(a) Left flower garden image.

(b) Right flower garden image.

(c) Left lab image.

(d) Right lab image.

(e) Left room image.

(f) Right room image.

13.6 References

[1] H. Yamaguchi, Y. Tatehira, K. Akiyama, and Y. Kobayashi, "Stereoscopic Images Disparity for Predictive Coding," in *ICASSP*, pp. 1976–1979, IEEE, 1989.

[2] M. G. Perkins, "Data Compression of Stereopairs," *IEEE Trans. on Communications*, vol. 40, pp. 684–696, April 1992.

[3] T. Ozkan and E. Salari, "Coding of Stereoscopic Images," in *Proc. SPIE, Image and Video Processing*, vol. 1903, pp. 228–235, 1993.

[4] I. Dinstein, G. Guy, J. Rabany, J. Tzelgov, and A. Henik, "On Stereo Image Coding," in *Proc. Int. Conf. on Pat. Recog.*, pp. 357–359, IEEE, 1988.

[5] H. Aydınoğlu and M. H. Hayes, "Source Coding of Stereo Image Pairs," in *Proc. 4th Bayona Workshop on Intelligent Methods in Signal Processing and Communication*, June 1996.

[6] A. Puri, R. V. Kollarits, and B. G. Haskell, "Stereoscopic video compression using temporal scalability," in *Proc. Visual Communications and Image Processing*, pp. 745–756, SPIE, May 1995.

[7] D. S. Kauffman and S. A. Wood, "Digital Elevation Model Extraction from Stereo Satelite Images," in *Proc. Int. Geoscience and Remote Sensing Symp*, vol. 1, pp. 349–352, 1987.

[8] M. Perkins, *Data Compression of Stereopairs.* PhD thesis, Stanford University, Stanford, CA, 1988.

[9] V. S. Grinberg, G. Podnar, and M. Siegel, "Geometry of Binocular Imaging," in *Proc of the IS&T/SPIE Symp on Electronic Imaging, Streoscopic Displays and Applications*, vol. 2177, 1994.

[10] S. D. Cochran and G. Medioni, "3-D Surface Description from Binocular Stereo," *IEEE Trans. on PAMI*, vol. 14, pp. 981–994, Oct 1992.

[11] T. Cover and J. Thomas, *Elements of Information Theory.* Telecommunications, Wiley, 1991.

[12] S. Sethuraman, A. Jordan, and M. Siegel, "A Multiresolutional Region Based Hierarchical Segmentation Scheme for Stereoscopic Image Compression," in *Proc. Digital Video Compression: Algorithms and Technologies*, vol. 2419, SPIE, 1995.

[13] H. Aydınoğlu and M. H. Hayes, "Stereo Image Coding: A Subspace Approach," *submitted to IEEE Trans. on Image Processing*, 1996.

[14] E. Salari and W. Whyte, "Compression of Stereoscopic Image Data," in *Proc. Data Compression Conference*, p. 425, 1991.

[15] M. E. Lukacs, "Predictive Coding of Multi-Viewpoint Image Sets," in *Proc. Int. Conf. on Acoustics, Speech, and Signal Processing*, pp. 521–524, IEEE, 1986.

[16] V. E. Seferidis and D. V. Papadimitriou, "Improved Disparity Estimation in Stereoscopic Television," *Electronics Letters*, vol. 29, pp. 782–783, April 1993.

[17] H. Aydınoğlu and M. H. Hayes, "Performance Analysis of Stereo Image Coding Algorithms," in *Proc. Int. Conf. on Acoustics, Speech, and Signal Processing*, vol. IV, pp. 2191–2195, 1996.

[18] S. Sethuraman, A. Jordan, and M. Siegel, "Multiresolution Based Hierarchical Disparity Estimation for Stereo Image Compression," in *Proc. Symposium on Application of Subbands and Wavelets* (A. Akansu, ed.), IEEE, March 1994.

[19] R. Franich, R. Lagendijk, and J. Biemond, "Stereo-enhanced Displacement Estimation by Genetic Block Matching," in *Proc. SPIE, Visual Communication and Image Processing*, vol. 2094, pp. 362–371, 1993.

[20] H. Aydınoğlu, F. Kossentini, and M. H. Hayes, "A New Framework for Multi-View Image Coding," in *Proc. Int. Conf. on Acoustics, Speech, and Signal Processing*, pp. 2173–2176, May 1995.

[21] H. Aydınoğlu and M. H. Hayes, "Image Coding with Polynomial Transforms," in *30th Asilomar Conference on Signals, Systems, and Computers*, November 1996.

[22] R. Jonsson, *Adaptive Subband Coding of Video Using Probability Distribution Models.* PhD thesis, Georgia Institute of Technology, Atlanta, GA, 1994.

[23] H. Aydınoğlu, F. Kossentini, Q. Jiang, and M. H. Hayes, "Region-Based Stereo Image Coding," in *Proc. Int. Conf. on Image Processing*, vol. II, pp. 57–61, October 1995.

[24] H. Aydınoğlu and M. H. Hayes, "Stereo Image Coding," in *Proc. Int. Symp. on Circuits ans Systems*, vol. I, pp. 247–250, April 1995.

[25] A. Gersho and R. Gray, *Vector Quantization and Signal Compression.* Kluwer Academic Publishers, 1992.

[26] K. W. Lim, K. W. Chun, and J. B. Ra, "Improvements on Image Transform Coding by Reducing Interblock Correlation," *IEEE Trans. on Image Proc.*, vol. 4, pp. 1146–1150, August 1995.

14

Design Methodology for VLSI Implementation of Image and Video Coding Algorithms – A Case Study

Javier Bracamonte
Michael Ansorge
Fausto Pellandini

ABSTRACT In this chapter a methodology for the design of VLSI circuits for image and video coding applications is presented. In each section a different phase of the design procedure is discussed, along with a description of the involved software environments. An example of an area efficient single-chip implementation of a JPEG coder is presented to illustrate the methodology.

14.1 Introduction

To produce an image in digital form, an enormous amount of data is required. Due to the explosive proliferation of digital images in today's world, the prodigality of digital image data has three direct economic and technical implications: a) the storage of digital images demands a significant amount of memory media, b) the transmission of digital images requires long transmission times, c) image processing applications need a large computational power.

In their original form (just after they have been digitized), two images with the same spatial resolution require exactly the same number of data (bits) regardless of their information content. For example, an image of a cloudless sky would require the same number of bits as an image of a crowd in a stadium, even though the latter contains much more information than the former. Technically speaking, it is said that the former image contains a high degree of redundancy, in the sense that, in exchange for some processing, the same image could be equally described with less data.

The kind of redundancy pointed out in the previous paragraph is only one among others [1]. Image coding—or compression—is the art or science of eliminating all possible kinds of redundancies in a digital image, so as to represent it with the least possible data, while retaining most of the original information. The same definition is valid for video coding, where another and exploitable kind of redundancy is present, given the high degree of similarity between two successive images in a video sequence.

Several image compression techniques are currently available [2][3]. In general, the original image is processed with the selected coding algorithm, to produce a compressed version. Depending on the application, the compressed version is either stored, or transmitted via a communication channel. Then, when the original image is required, the decoding algorithm is applied to the compressed version to recover an exact copy of the original image (lossless compression), or a very good replica (lossy compression). In effect, to increase the compression efficiency, most of the coding algorithms are allowed to introduce some errors.

In order to produce a compressed (or decompressed) image or video sequence, a significant amount of computational resources are needed. For some applications, a general purpose personal computer would suffice to execute the coding/decoding algorithms. However, when the process must be performed in real time, the required computational power might be so high that it can only be achieved by means of particular VLSI circuits. These circuits may correspond to special programmable video signal processors (VSP) of the last generation [4][5], or to application specific integrated circuits (ASICs). The ASIC solution, towards which the methodology presented in this chapter is oriented, is essential for particularly demanding applications such as miniaturized and/or portable equipment.

The process of materializing the mathematical description of an image compression algorithm into a chip implies the solution of a relatively complex problem. The complete design process extends over many levels of abstraction; each level requiring the use of different methods and technologies, whose intermediate results should be useful to facilitate the tasks of the next abstraction level. An effective methodology is thus indispensable to orchestrate the multiple involved procedures in order to reduce time, effort and design errors, while constraining the final result for a particular feature such as high speed, small area or low power consumption.

The different steps of the methodology reported in this chapter are: a) high-level modeling of the algorithm, b) VLSI architecture design, c) bit-true level modeling, d) layout design, and e) verification and testing. Figure 14.1 shows the process stream of the methodology, along with the supporting software environments. To illustrate the methodology, a case study consisting of the VLSI implementation of a JPEG coder will be described.

The remainder of this chapter is organized as follows. A brief description of the JPEG algorithm is presented in section 2. The high level modeling is described in section 3. The VLSI architectures are presented in section 4. The bit-true level modeling of the algorithm is described in section 5, and the layout design is discussed in section 6. Finally, the results are given in section 7, just before the conclusions in section 8.

14.2 JPEG Baseline Algorithm

The Joint Photographic Expert Group (JPEG) algorithm defines an international standard for compression of continuous-tone still images. It specifies four modes of operation: sequential, progressive, lossless and hierarchical [6]. Thus, rather

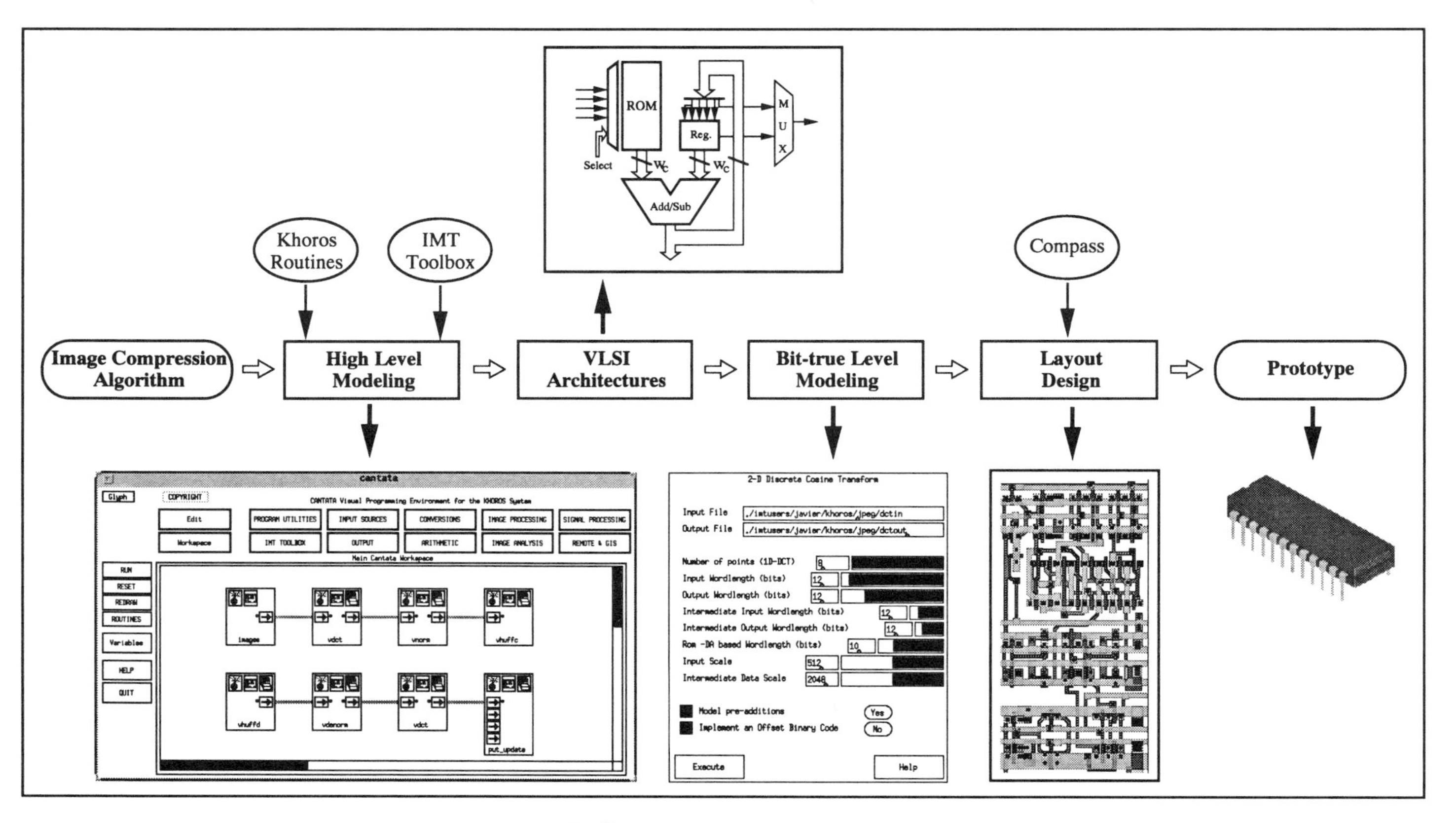

FIGURE 14.1. Methodology.

than a single image compression method, JPEG defines a family of application-dependent coding schemes. Among them, the baseline system of the sequential mode is by far the most used, since it covers a wide range of applications. It is hence the implementation system we address in this chapter.

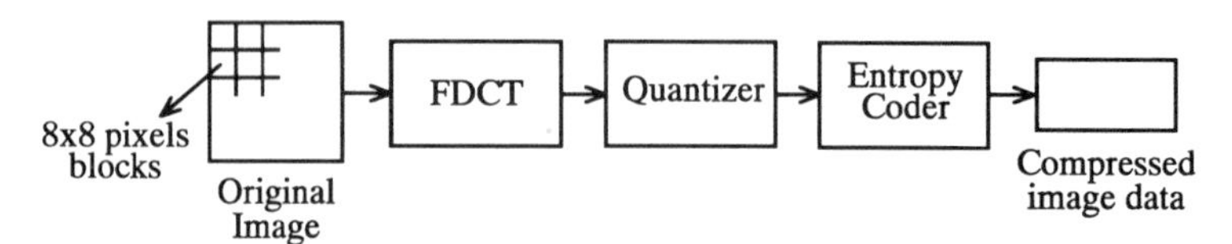

FIGURE 14.2. Baseline JPEG algorithm.

The baseline sequential JPEG algorithm is depicted in Figure 14.2. It is a Discrete Cosine Transform (DCT) -based process which transforms blocks of 8x8 pixels sequentially from the original image, into 8x8 blocks of coefficients in the frequency domain [7]. The goal of this transformation is to decorrelate the original data and redistribute the signal energy among only a small set of transform coefficients in the low frequency zone. For illustration, an array of 8x8 pixels, corresponding to a region of the eye in an image of a person's face, is given in Equation (14.1). The result after a 2-D DCT has been applied to the array $\mathbf{x}$ (whose elements are previously offset by -128, as required by JPEG) is given in Equation (14.2). It can be seen that most of the energy is concentrated in only a few low frequency (i.e., the top left corner region) coefficients.

$$\mathbf{x} = \begin{bmatrix} 77 & 69 & 69 & 70 & 71 & 67 & 67 & 67 \\ 74 & 70 & 72 & 67 & 64 & 66 & 66 & 65 \\ 67 & 72 & 68 & 68 & 64 & 66 & 68 & 66 \\ 65 & 72 & 71 & 71 & 62 & 65 & 69 & 90 \\ 68 & 76 & 79 & 65 & 61 & 56 & 58 & 99 \\ 78 & 93 & 95 & 69 & 59 & 56 & 56 & 71 \\ 73 & 108 & 106 & 92 & 68 & 60 & 60 & 83 \\ 70 & 105 & 110 & 105 & 90 & 84 & 83 & 102 \end{bmatrix} \tag{14.1}$$

$$\mathbf{X} = \begin{bmatrix} -431 & 25 & 11 & -48 & -3 & -16 & 4 & -6 \\ -52 & -16 & 6 & 40 & 8 & 15 & 0 & 5 \\ 35 & 9 & -20 & 4 & -5 & 4 & -6 & 0 \\ -18 & 19 & 11 & -2 & 2 & -3 & 6 & 6 \\ 15 & -26 & 2 & -3 & 11 & -3 & 2 & -1 \\ -1 & 6 & -3 & -2 & -1 & 1 & -5 & -4 \\ -1 & 1 & -4 & 9 & -4 & 2 & -1 & 3 \\ -2 & -1 & 0 & -4 & 4 & -4 & -1 & -1 \end{bmatrix} \tag{14.2}$$

Based on a psychovisual analysis, a normalization array can be defined. Its purpose is to quantize those DCT coefficients that are visually significant with relative short quantization steps, while using large quantization steps for those coefficients which are less important. This large-step quantization, associated with the energy packing effect of the transformation, results in general in the zeroing out of many DCT coefficients. Equation (14.3) gives the resulting array $\mathbf{Q}$, when the matrix $\mathbf{X}$ in Equation (14.2) has been quantized by using the normalization matrix given in Equation (14.4) on page 308.

$$\mathbf{Q} = \begin{bmatrix} -27 & 2 & 1 & -3 & 0 & 0 & 0 & 0 \\ -4 & -1 & 0 & 2 & 0 & 0 & 0 & 0 \\ 2 & 1 & -1 & 0 & 0 & 0 & 0 & 0 \\ -1 & 1 & 0 & 0 & 0 & 0 & 0 & 0 \\ 1 & -1 & 0 & 0 & 0 & 0 & 0 & 0 \\ 0 & 0 & 0 & 0 & 0 & 0 & 0 & 0 \\ 0 & 0 & 0 & 0 & 0 & 0 & 0 & 0 \\ 0 & 0 & 0 & 0 & 0 & 0 & 0 & 0 \end{bmatrix} \tag{14.3}$$

A long sequence of zero-valued coefficients can be effectively abridged by runlength coding. In order to increase the efficiency of the runlength coding, the array of normalized DCT coefficients is reordered by following a zigzag pattern. The reordering has the effect of producing longer runs of zeros. The resulting zigzag-reordered, 64-element 1-D array is: $\{-27, 2, -4, 2, -1, 1, -3, 0, 1, -1, 1, 1, -1, 2, 0, 0, 0, 0, 0, -1, 0, \ldots, 0\}$, where the ellipsis replaces a sequence of 42 zeros.

The top-left coefficient (element $(0,0)$) of the array $\mathbf{X}$ in Equation (14.2), is called the DC coefficient (the remaining 63 coefficients are referred to as AC coefficients). Its value represents the (upscaled) average value of the 64 pixels of the original image (this value in Equation (14.2) is negative due to the offset of -128, as pointed out above). Since the average value of two adjacent 8x8 blocks of pixels is, with a high probability, very similar, a straight DPCM (differential pulse code modulation) scheme can be applied to the DC coefficients, using the DC coefficient of the previous coded block as the prediction value.

Though the quantization of the DCT coefficients is the main mechanism of data compression (and also of information loss), additional data compression can be obtained by entropy coding the output of the DPCM and runlength coders. While both Huffman and arithmetic coding are defined by the general JPEG standard for the entropy coding stage, the baseline JPEG algorithm uses Huffman coding only. This fact presents two important advantages: first, Huffman coding requires a less complex hardware implementation; and second, many algorithms on arithmetic coding are patented, and their commercial implementation, in consequence, requires the payment of license fees [8].

14.3 High Level Modeling

A modular high level implementation of the JPEG algorithm is shown in Figure 14.3. The top and bottom flowgraphs represent the encoder and decoder respectively. The runlength coding and DPCM operations, described in the previous section, were merged into a single module with the Huffman coder/decoder (i.e., *vhuffc/vhuffd*). From an image processing point of view, this high-level modeling is an application program by itself [9]. Typical compression ratios of images are around 10, depending on the image activity and spatial resolution.

In the high level model, each process of the algorithm was implemented with the C programming language, using floating point precision for the arithmetic operations. No concerns were raised at this point on issues as to how the operations will be carried out by the hardware (e.g., hardware multiplexing, parallel processing, bit-serial architectures, etc.). With respect to our methodology, de-

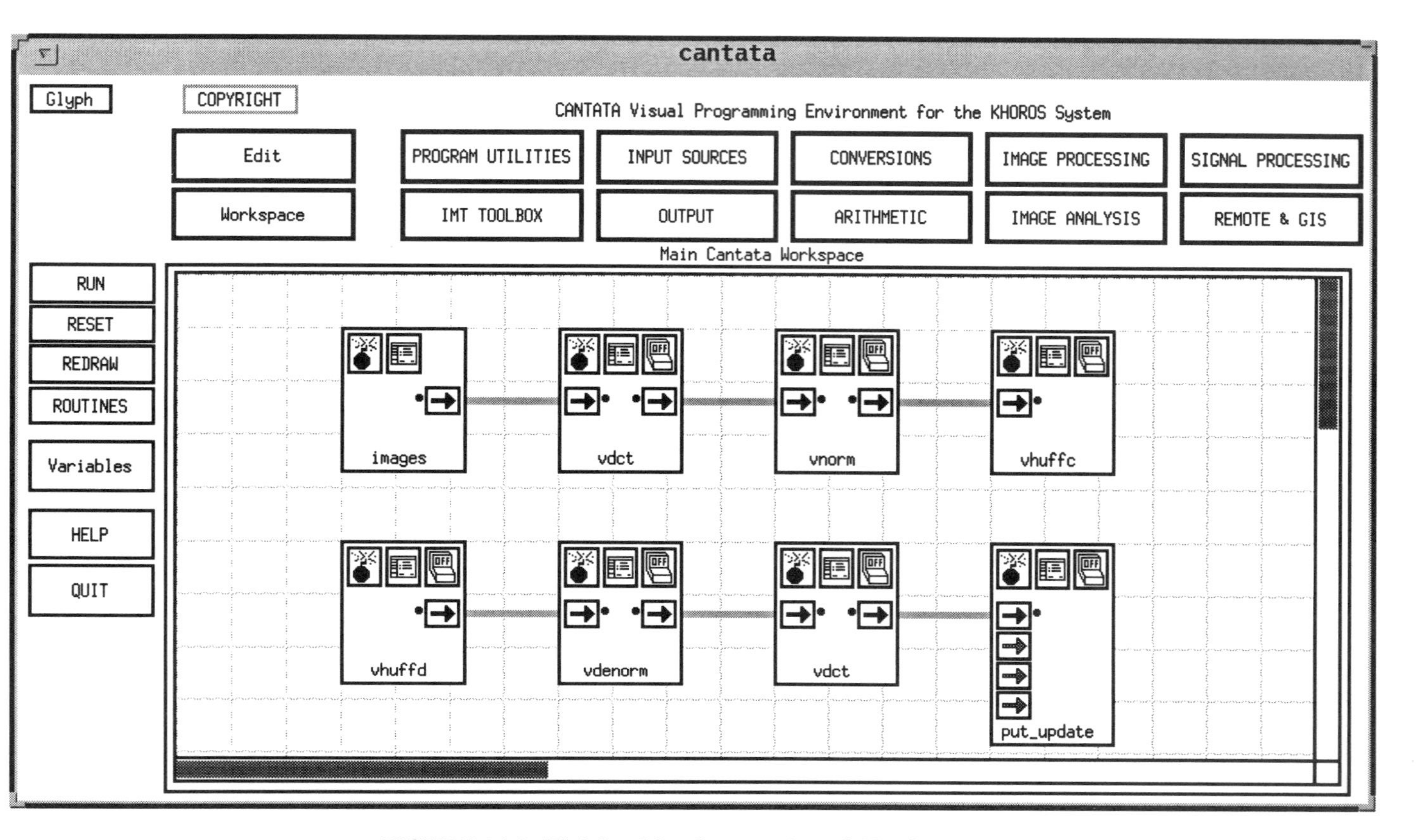

FIGURE 14.3. High level implementation of JPEG in Khoros.

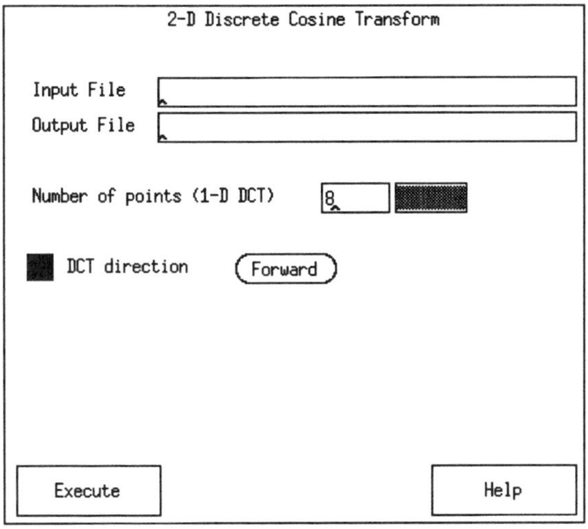

2-D Discrete Cosine Transform

Input File

Output File

Number of points (1-D DCT) 8

DCT direction Forward

Execute

Help

FIGURE 14.4. GUI of the 2-D DCT.

veloping a high level model of an algorithm allows us to have both a reference and a framework to support the bit-true level modeling (section 5).

The Khoros software [10] has been chosen as the environment for developing the high and bit-true level modeling. Due to its user-friendliness and visual programmable capability, it permits a rapid evaluation of the results of a given configuration. Each glyph (or icon) in Figure 14.3 corresponds to a C program that has been converted into a Khoros routine. The glyph named *images* addresses a database and provides the image to be coded. The glyph *put_update* has the task of displaying the decompressed image. Both routines are part of the original Khoros software.

The graphical user interface (GUI) of the routine *vdct* is shown in Figure 14.4. Though JPEG specifies a fixed block size of 8x8 pixels, the number of points of the inherent 1-D DCT has been left as a parameter for experimental purposes. The same GUI is used for both the forward and inverse 2-D DCT.

The GUI of the routine *vnorm* is shown in Figure 14.5. It performs the quantization of the 2-D DCT coefficients with a normalization array that is user selectable. By default the normalization array given in Equation (14.4) has been programmed; it corresponds to the array proposed in Table K.1 in Annex K of reference [11]. The *Scale Factor* parameter allows the user to control the compression ratio. The higher the setting of this scale factor the lower the amount of the resulting compressed image data. This kind of improvement of the compression ratio (CR) is made to the detriment of the quality of the reconstructed image. By default, *Scale Factor* is set to 1, which offers a good compromise: image quality/CR.

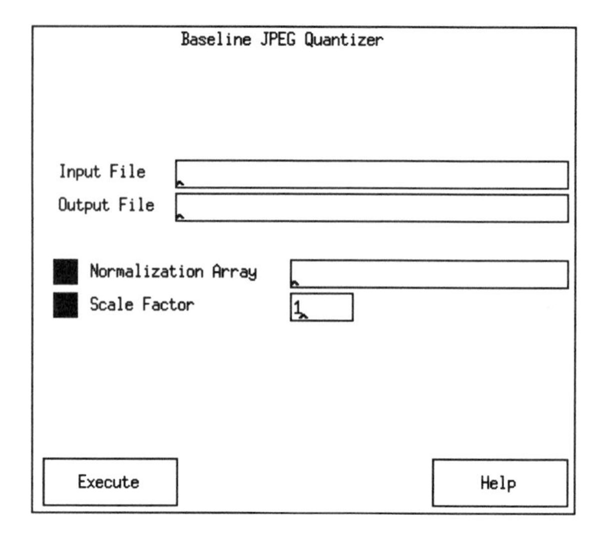

FIGURE 14.5. GUI of the quantizer.

$$\mathbf{N} = \begin{bmatrix} 16 & 11 & 10 & 16 & 24 & 40 & 51 & 61 \\ 12 & 12 & 14 & 19 & 26 & 58 & 60 & 55 \\ 14 & 13 & 16 & 24 & 40 & 57 & 69 & 56 \\ 14 & 17 & 22 & 29 & 51 & 87 & 80 & 62 \\ 18 & 22 & 37 & 56 & 68 & 109 & 103 & 77 \\ 24 & 35 & 55 & 64 & 81 & 104 & 113 & 92 \\ 49 & 64 & 78 & 87 & 103 & 121 & 120 & 101 \\ 72 & 92 & 95 & 98 & 112 & 100 & 103 & 99 \end{bmatrix} \tag{14.4}$$

The *vhuff* routine implements the Huffman coding. The user is given the possibility of selecting a Huffman table customized to the application. By default, the Huffman codes proposed in Table K.5 in Annex K of reference [11] have been programmed.

As the number of new developed Khoros routines grows, it is convenient to integrate them into a toolbox. This provides a software structure that is easier to reuse, maintain, document, and distribute [12]. The created IMT Toolbox, contains the JPEG related routines shown in Figure 14.3, and some routines for other methods for image compression.

14.4 VLSI Architectures

To introduce this section, it is worth making a point regarding the design flowgraph in Figure 14.1. Although it is not explicitly shown in Figure 14.1, there is a bidirectional interaction among the different processes. When a proposed scheme turns out not to be a good solution, the designer should go back and forth in the design process while exploring other approaches. The previous remark is particularly pertinent to the design of the VLSI architectures. Thus, the reader should be aware that the solutions presented in this section are the result of several evaluations and trade-offs.

Taking into consideration the constraints of the target application, a VLSI architecture for each block in Figure 14.2 is to be developed. As pointed out in the introduction, given their associated high data throughput, image and video coding applications require particular hardware implementations that operate at high clocking frequencies. If the compression circuits are intended to be used in portable equipment, then the power consumption is also of paramount importance.

Even though the relentless advance in the development of new semiconductor technologies makes feasible the production of devices that run faster or consume less power, a good architecture is nevertheless the key to achieving the specifications of a given application. Low power VLSI for image processing applications is currently a hot research topic. Though it was an issue that was considered during the design of the architectures below, the main objective remained the realization of high speed and small silicon area circuits.

Bit-serial architectures present several advantages [13]: a) the size of bit-serial structures is considerably smaller than their bit-parallel counterparts; b) wiring, which in a VLSI circuit requires a large part of silicon area, is dramatically reduced, since a single wire is required for each input word; c) communication between bit-serial modules is easier to implement, and d) bit-serial architectures lead to tightly pipelined structures at the bit-level, which implies that a maximum clock-rate can be achieved [14]. Thus, whenever it was possible, we have chosen a fully pipelined bit-serial approach.

In the following paragraphs we describe the VLSI architecture for each of the main modules of the encoder in Figure 14.2.

14.4.1 FDCT

The forward 2-D DCT appropriate for JPEG is defined in [11] as:

$$X_{uv} = \frac{1}{4} C_u C_v \sum_{i=0}^{7} \sum_{j=0}^{7} x_{ij} \cos[(2i+1)u\pi/16] \cos[(2j+1)v\pi/16]; \qquad (14.5)$$

for $u, v = 0, 1, \ldots, 7$, where $C_u, C_v = 1/2$ for $u, v = 0$; $C_u, C_v = 1$ otherwise. X_{uv} represents the value of the 2-D DCT coefficient at the point (u, v) in the frequency domain, and x_{ij} represents the value of the image pixel at the point (i, j). Given that the 2-D DCT is separable, it can be reformulated as two successive 1-D DCTs: the first one being applied to the rows of the block of 8x8 pixels, whereas the second is applied to the columns of the resulting matrix (or to the rows of its transpose) [7]. This decomposition leads to a simpler hardware implementation.

Using matrix notation, the N-point 1-D DCT $\mathbf{X}$, of an input vector $\mathbf{x}$, can be expressed as:

$$\mathbf{X} = \mathbf{A}\,\mathbf{x}, \qquad (14.6)$$

where $\mathbf{A}$ represents the cosine transformation matrix. For $N = 8$ (as required by the JPEG standard) each element of $\mathbf{X}$ is obtained by means of an 8-point inner product of the following form:

$$X_k = \sum_{l=0}^{7} A_{kl}\, x_l, \quad \text{for } k = 0, 1, \ldots, 7. \tag{14.7}$$

Distributed arithmetic [15] is appropriate for the calculation of inner products. It has the interesting property of circumventing all the multiplications, which are elegantly replaced by the less complex addition and shift operations.

Let us consider the evaluation of the following inner product:

$$y = \mathbf{a}\mathbf{x} = \sum_{i=0}^{N-1} a_i\, x_i, \tag{14.8}$$

where $\mathbf{x}$ and $\mathbf{a}$ denote an input signal vector and a fixed coefficient vector, respectively.

If x_i is represented as a W_d bits binary number in 2's complement form, then it can be expressed as:

$$x_i = \sum_{j=1}^{W_d-1} x_{ij} 2^{-j} - x_{i0}, \tag{14.9}$$

where x_{ij} represents the jth bit of x_i. Substituting Equation (14.9) into (14.8) and interchanging the order of the summations, we obtain:

$$y = \sum_{j=1}^{W_d-1} \left[\sum_{i=0}^{N-1} a_i\, x_{ij} \right] 2^{-j} - \sum_{i=0}^{N-1} a_i\, x_{i0}. \tag{14.10}$$

Let us define F_j as:

$$F_j = \sum_{i=0}^{N-1} a_i\, x_{ij}, \tag{14.11}$$

since the coefficients a_i are fixed, and x_{ij} is a binary number, then any evaluation of F_j must be a number among only 2^N possible values. In a hardware implementation, these 2^N values can be simply stored in memory (e.g., a ROM).

By substituting Equation (14.11) into (14.10), the original expression of the scalar product of Equation (14.8) can be rewritten as:

$$y = \sum_{j=1}^{W_d-1} F_j 2^{-j} - F_0, \tag{14.12}$$

which can be implemented by a ROM for storing the values of F_j, an accumulator to implement the sum, and a shifter for executing the multiplications by 2^{-j}.

The architecture of the 2-D DCT is shown in Figure 14.6. Each 1-D DCT is executed by a single, multiplexed, distributed arithmetic processor (DAP) [16], whose architecture is shown in Figure 14.7. For the computation of each of the eight coefficients X_k, in Equation (14.7), the input vector $\mathbf{x}$, is applied eight consecutive times to the DAP. Each coefficient is computed by using a different lookup table. Thus, the ROM depicted in Figure 14.7, contains eight different ROMs which are chosen with the signal *Select.* The shift register bank (SRB) in

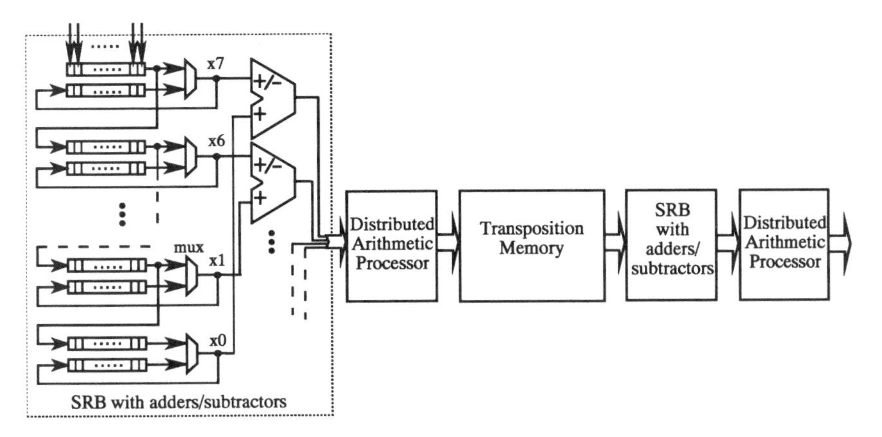

FIGURE 14.6. Architecture of the 2-D DCT.

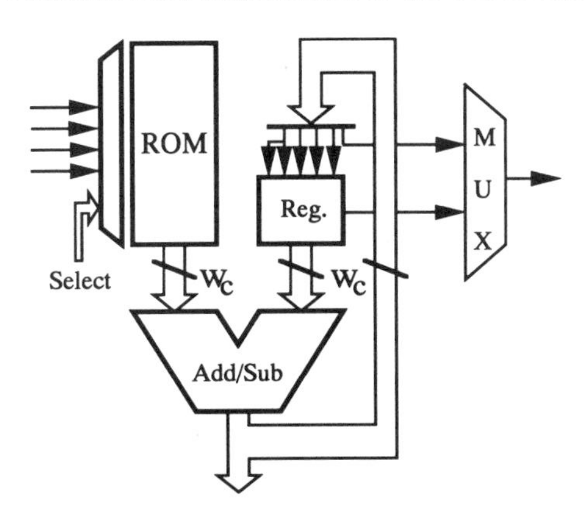

FIGURE 14.7. Distributed arithmetic processor.

Figure 14.6 provides the storage and sequential dataflow required for the multiple applications of the input vector.

It turns out that the cosine transformation matrix in Equation (14.6) presents a symmetry on its even rows and an antisymmetry on its odd rows. This implies that Equation (14.7) can be rewritten as:

$$X_k = \sum_{l=0}^{3} A_{kl}\,(x_l \pm x_{8-l}), \quad \text{for } k = 0, 1, \ldots, 7 \tag{14.13}$$

where the plus sign is used for the even values of k, and consequently, the minus sign for the odd values. The importance of Equation (14.13) is evident: the number of multiplications has been halved. In terms of the DAP's hardware, and in accordance with Equation (14.11), this means that the size of the DAP's ROMs in Figure 14.7, can been reduced from 2^N to $2^{N/2}$ words. For $N = 8$, eight small ROMs of 16 (instead of 256) words each, are required. Due to this significant memory saving, Equation (14.13) was retained for implementation. The adders/subtractors in Figure 14.6 perform the extra additions/subtractions in accordance with Equation (14.13).

14.4.2 Quantizer

The second operation of the JPEG coder is the quantization of the 2-D DCT coefficients and it is defined as: $Cq_{ij} = round(C_{ij}/N_{ij})$ for $i, j = 0, 1, ..., 7$, where C_{ij} denotes the ij-th element of an 8x8 DCT coefficients matrix, whereas N_{ij} denotes the ij-th element of an 8x8 normalization matrix. Reformulating the previous equation as $Cq_{ij} = round(C_{ij} * (1/N_{ij}))$ allows the replacement of the circuitry of a divisor with that of a simpler multiplier. The VLSI architecture of the quantizer is shown in Figure 14.8. Since the output of the 2-D DCT proces-

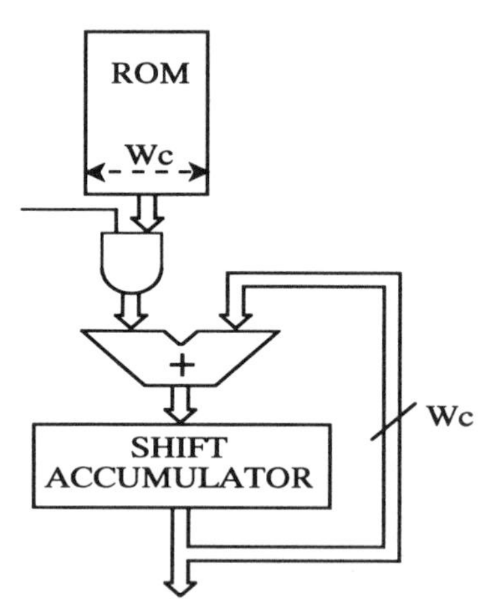

FIGURE 14.8. Architecture of the quantizer.

sor is bit-serial, a serial-parallel multiplier was implemented. The parallel input to this multiplier being the output of a ROM that contains the inverse of the normalization coefficients.

14.4.3 Entropy coder

The last operation of the baseline JPEG algorithm is entropy coding. Its goal is to increase the compression performance of the encoder by taking advantage of the statistics from the data at the output of the quantizer.

The entropy coder of JPEG can be better explained by means of an example. Let us consider that the zigzag-reordered, 64-element array containing the quantized DCT coefficients, is: $A = \{a_0, a_1, 0, a_3, 0, 0, 0, a_7, 0, 0, a_{10}, 0, \ldots, 0\}$, for $a_i \neq 0$. The first operation consists in the DPCM applied to the DC coefficient. Supposing that the value of the DC coefficient of the previous 8x8 pixels block was b_0, then the DC value to be coded will be: $(a_0 - b_0)$. Runlength coding is applied to the AC coefficients; its result is a sequence of 2-element arrays $(r; a_i)$, where r represents the number of zeros preceding a non-zero a_i

coefficient. Thus, after the DPCM and runlength coding, the sequence A, becomes: $B = \{(a_0 - b_0), (0; a_1), (1; a_3), (3; a_7), (2; a_{10}), EOB\}$, where EOB is a special code which indicates that the rest of the values, until the end of the sequence, are zeros.

For all the possible values that an a_i coefficient can take, different categories C_i are defined. Each category indicates a range to which the magnitude of a_i belongs. An extra number D_i is thus needed, in addition to C_i, to completely specify the value of a_i. Each pair of values $(r; C_i)$ is used to address a table that contains the Huffman codes. The splitting operation of each a_i into two numbers: C_i and D_i, is made in order to reduce the number of entries in the Huffman table.

The architecture of the entropy coder is shown in Figure 14.9. The zigzag reordering, required prior to the entropy coding, was implemented by a RAM with different read and write sequences. The DPCM and runlength operations are implemented by means of a subtractor and a zero detection counter, respectively, plus additional control logic and registers. The ROM contains the Huffman codes; since they are inherently of variable length (minimum 2 bits and maximum 16 bits, for the table that was chosen), a bit packer [17] has been added at the output of the circuit. The function of the bit packer is to concatenate consecutive codewords in order to output, at irregular intervals, words of fixed length.

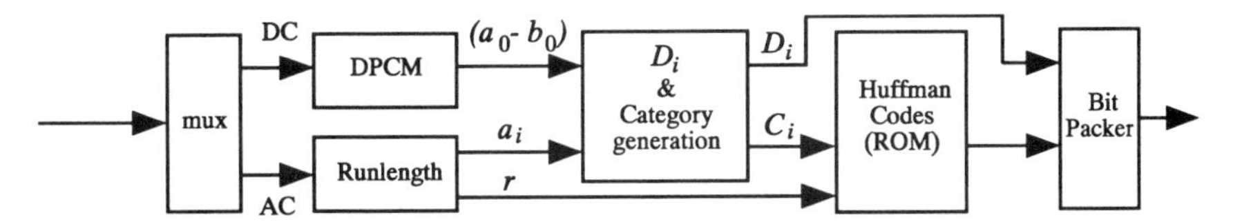

FIGURE 14.9. Architecture of the entropy coder.

14.5 Bit-true Level Modeling

The motivation for implementing a bit-true level model, is to assure the functionality of the architectures, the scheduling, the resource allocation and to find the optimum wordlength of the signals and coefficients of the algorithm.

Based on the programming framework of the high level implementation, a bit-true level model of the baseline JPEG algorithm was built. Each process of the algorithm was again implemented with C code and converted into a Khoros routine. However this time, the arithmetic operations are all carried out in binary arithmetic, modeling accurately the same processing and dataflow as would be executed by the corresponding processors and architectures described in section 4.

The GUI corresponding to the bit-true level model of the 2-D DCT is shown in Figure 14.10. Both rounding and truncation can be modeled. *Input Wordlength* refers to the number of bits per pixel required at the input of the 2-D DCT circuit. The original number of bits of an original image (e.g., 8 bits for an image of 256 gray levels) might be increased by padding trailing or leading zeros, to comply with the control signals in bit-serial structures.

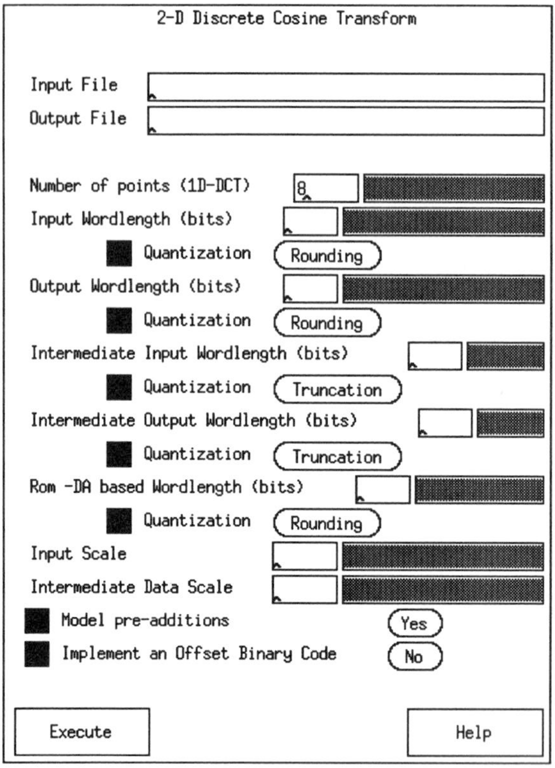

2-D Discrete Cosine Transform

Input File
Output File
Number of points (1D-DCT) 8
Input Wordlength (bits)
Quantization Rounding
Output Wordlength (bits)
Quantization Rounding
Intermediate Input Wordlength (bits)
Quantization Truncation
Intermediate Output Wordlength (bits)
Quantization Truncation
Rom -DA based Wordlength (bits)
Quantization Rounding
Input Scale
Intermediate Data Scale
Model pre-additions Yes
Implement an Offset Binary Code No
Execute Help

FIGURE 14.10. GUI of the 2-D DCT bit-true level model.

Output Wordlength refers to the number of bits required to represent the 2-D DCT coefficients. As pointed out in section 2, the 2-D DCT is computed by means of two successive 1-D DCTs. The *Intermediate* parameters in Figure 14.10 refer to the data between the two 1-D DCTs. The wordlength of the DAP's ROM refers to the number of bits required to represent the lookup table coefficients given by Equation (14.11). In Figure 14.7 this wordlength is denoted by W_c.

The *Scale* parameters are required to fix the binary point to the same position for all the data registers. In general the scale factors are chosen to be a positive integer power of 2, in order to facilitate the implementation by means of simple shifts.

The *Model pre-additions* parameter allows the user to decide whether the implementation of Equation (14.7) or (14.13) should be modeled. As the adder/subtractors in Figure 14.6 confirm, this option has been set to *Yes*, for our JPEG circuit.

The Offset Binary Coding (OBC) technique allows a binary table to be reduced from 2^N to 2^{N-1} elements, as explained in [18]. Though this option was not selected for our circuit (the reduction of a DAP's ROM size from 16 to 8 words was not very significant, due to the required additional logic and control), its model was implemented for experimental purposes.

For the 2-D DCT circuit, the optimum DAP's data- and ROM-wordlength found was 12 and 10 bits respectively. The intermediate input and output wordlength was evaluated to 12 bits. Simulations showed no difference in the results between using rounding or truncation. Thus, truncation mode was used, due to its simpler implementation.

The simulations of the bit-true level model of the quantizer circuit gave as a result a ROM wordlength value of 9 bits and an input/output data wordlength of 12 bits. No bit-true level model was developed for the entropy coder, due to the fact that the Huffman coder does not involve arithmetic operations.

14.6 Layout Design

By layout design we refer to the translation of the architectures into a CMOS circuit, which can be characterized by its speed, power consumption, etc.. Several software environments exist on the market today, to design circuits at any level (i.e., behavioral, logic, or transistor level). In our case, we have used the Compass environment [19], which contains a set of tools that aids the designer during the different stages of the circuit development.

When a module of an architecture is highly regular [21], the layout editor tool was used for developing full custom modules. The layout editor allows the design of circuits at the transistor level. For highly regular structures, this does not penalize the time for development since only a few cells have to be designed. On the other hand, minimum silicon area, minimum power consumption and high clock frequency circuits are achieved.

For less regular structures, the standard cell approach was applied. A cell library containing a collection of logic and arithmetic circuits (e.g., inverters, ANDs, NORs, latches, flip-flops, adders) are available to the user. By selection, arrangement and interconnection, the circuit designer is able to build circuits of higher complexity. The Logic Assistant tool of Compass is well suited to this task.

Cell compilers were also used during the datapath development of the JPEG circuit, mainly to generate the RAM for matrix transposition in Figure 14.6, and the ROM containing the Huffman tables [20] of the entropy coder in Figure 14.2.

At each hierarchical level of the design, extensive simulations were carried out with both the Mixed-Mode and SPICE simulators. This allows verification of the functional correctness of the circuit, and the extraction of characteristics such as the maximum attainable working frequency. The layout of the JPEG coder circuit is shown in Figure 14.11. The area of the chip is 4.6 x 3.1 $mm^2 \approx 14.5$ mm^2. It was implemented in the $1.2\mu m$ 5V CMOS CMN12 process from VLSI Technology Inc. On the left part of the layout are located the SRBs, the quantizer and the two DAPs. All these circuits were built with full-custom layouts, a fact that is reflected by the very small area of the modules. The reutilization of the basic layout cells was efficiently maximized. The RAM for executing the matrix transposition and the zigzag reordering is located at the bottom-left corner of the circuit. The Huffman coder was realized by using the standard cell approach. It corresponds to the regular structure composed of rows of cells in the right part of the layout. The two blocks in the middle of the layout correspond to the ROMs containing the Huffman codes tables.

14.7 Results

At a clock frequency of 36 MHz, this circuit is able to process 25 CIF (Common Intermediate Format: 352 x 288 pixels) images per second. Thus, it is also suitable for motion JPEG (MJPEG) or for the nonrecursive path of the H.261 [22] low-bit rate video coder. With a negligible increase in the silicon area, this circuit has been modified to feature four power management modes of operation [23]. Each mode of operation allows the user to trade image quality for power consumption; an option that can be very useful in portable multimedia equipment.

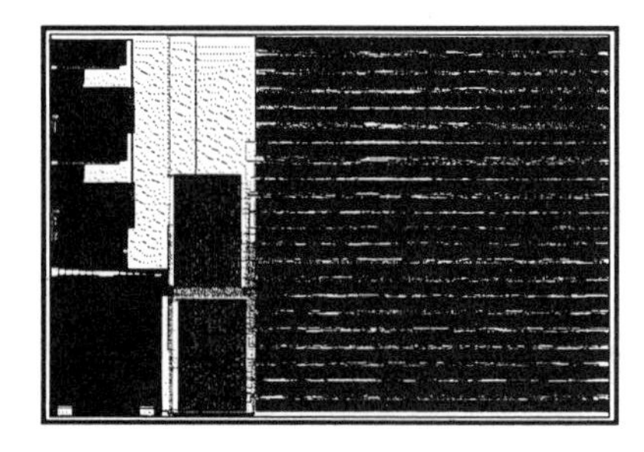

FIGURE 14.11. JPEG coder circuit.

14.7.1 Extensions for Video Coding

Motion JPEG, though formally not supported by any coding standard, represents one practical solution for DCT-based video coding. That is, if an available JPEG encoder is fast enough, then one could use it to compress a video sequence by coding each frame as an independent still image. On the other hand, a more effective video coding system can be seen as a still coding scheme, to which a module to reduce the frame-to-frame redundancies has been added. Motion compensation is a very well-suited technique for reducing the interframe redundancies, related to the movements of objects across consecutive frames in a video sequence. For example, in the standards H.261, MPEG-1 [24] and MPEG-2 [25], the principle is to use a frame, plus the motion information, as a prediction to the immediately following frame(s). The prediction error frame—the current frame minus its prediction frame—is then coded as in a classic DPCM scheme.

The estimation of the motion vectors is the most computationally intensive operation in a video coder. In accordance with architectures that were discussed in section 4, a bit-serial VLSI architecture for a motion estimation algorithm was reported in [26]. In the frame of the methodology, the incorporation of the study of the interframe redundancy reduction modules has to be added, in order to evaluate the implementation of video coding systems.

14.8 Conclusions

A semi-custom methodology for the VLSI implementation of image compression systems was reported. The software environments were described along with an example of the implementation of a JPEG coder circuit. Though image compression was addressed in this chapter, the methodology could equally be applied to other image processing applications. The intermediate results of this methodology can also be used to develop solutions for other kinds of technologies (DSP, FPGA, etc.).

The main objective of the chapter was to give a general overview of a methodology, and corresponding CAD framework, that transforms an image processing algorithm into an integrated circuit. Readers unfamiliar with VLSI circuits are reassured that some of the descriptions and options, in the previous sections, were not described in detail, due to unavoidable space limitations. For further descriptions the reader is invited to consult the given references.

14.9 Acknowledgements

This work was supported by the Swiss National Science Foundation under Grant FN 2000-40'627.94, and by the Laboratory of Microtechnology (LMT EPFL). The latter is an entity common to the Swiss Federal Institute of Technology, Lausanne, and the Institute of Microtechnology, University of Neuchâtel.

The support provided by the Swiss Federal Office for Education and Science under Grant 874-01-49 is also acknowledged.

14.10 References

[1] R. C. Gonzalez and R. E. Woods, *"Digital Image Processing"*, Addison-Wesley, Reading, MA, USA, 1992.

[2] M. Rabbani and P.W. Jones, *"Digital Image Compression Techniques"*, Vol. TT 7, SPIE Optical Engineering Press, Bellingham, WA, USA, 1991.

[3] M. Kunt, G. Granlund and M. Kocher, *"Traitement Numérique des Images"*, Presses Polytechniques et Universitaires Romandes, Collection Electricité, Traitement de l'Information, Vol. 2, Lausanne, Switzerland, 1993. (In French).

[4] Philips, *TriMedia VLIW-Based PCI Multimedia Processor System (TriMedia White Paper)*, 1995.

[5] Texas Instruments, *TMS320C8x*, Technical Documentation, 1995.

[6] W.B. Pennebaker and J.L. Mitchell, *"JPEG Still Image Data Compression Standard"*, Van Nostrand Reinhold, New York, NY, USA, 1993.

[7] K. R. Rao, and P. Yip, *"Discrete Cosine Transform: Algorithms, Advantages, Applications"*, Academic Press, Boston, MA, USA, 1990.

[8] V. Bhaskaran and K. Konstantinides, *"Image and Video Compression Standards. Algorithms and Architectures"*, Kluwer Academic Publishers, Boston, MA, USA, 1995.

[9] J. Bracamonte, *"A high and bit-true level implementation of the baseline JPEG image compression algorithm"*, Internal Report IMT, Institute of Microtechnology, University of Neuchâtel, Switzerland, 1996.

[10] D. Rasure, D. Arguiro, T. Sauer and C. William, "A visual language and software development environment for image processing", *Int'l J. of Imaging Systems and Technology*, Vol. 2, 1990, pp. 183–199.

[11] ITU-T Recommendation T.81 *"Digital compression and coding of continuous-tone still images"*, September, 1992.

[12] Khoral Research, Inc., *Khoros Programmer's Manual*, 1995.

[13] R. I. Hartley and K. K. Parhi, *"Digit-Serial Computation"*, Kluwer Academic Publishers, Boston, MA, USA, 1995.

[14] P. Denyer and D. Renshaw, *"VLSI Signal Processing: A Bit-serial Approach"*, Addison-Wesley, VLSI System Series, 1985.

[15] A. Peled and B. Liu, "A new hardware realization of digital filters", *IEEE Trans. on Acoustics, Speech, and Signal Processing*, Vol. ASSP-22, No. 6, Dec. 1974, pp. 456–462.

[16] U. Sjöström, *"On the design and implementation of DSP algorithms: An approach using wave digital state-space filters and distributed arithmetic"*, Ph.D. Thesis, University of Neuchâtel, Switzerland, 1993.

[17] S-M. Lei and M-T. Sun, "An entropy coding system for digital HDTV applications", *IEEE Trans. on Circuits and Systems for Video Technology*, Vol. 1, No. 1, March, 1991, pp. 147–155.

[18] S. G. Smith and P. B. Denyer, *"Serial-Data Computation"*, Kluwer Academic Publisher, Boston, MA, USA, 1988.

[19] COMPASS Design Automation, Inc.: Manuals, COMPASS, San Jose, CA, USA, 1993.

[20] C. Henny, *"A VLSI implementation of a Huffman Coder"*, Diploma Project, University of Neuchâtel, Switzerland, August, 1995.

[21] N. Weste and K. Eshraghian. *"Principles of CMOS VLSI design: A Systems Perspective"*, Addison-Wesley, VLSI System Series, 2nd Edition, 1993.

[22] ITU-T Recommendation H.261 *"Video codec for audiovisual services at $p \times 64$ kbits"*, March, 1993.

[23] J. Bracamonte, M. Ansorge and F. Pellandini. "VLSI systems for image compression. A power-consumption/image-resolution trade-off approach", *Proc. Conf. on Digital Compression Technologies & Systems for Video Communications*, Berlin, Germany, Oct. 7-11, 1996.

[24] ISO/IEC JTC1 CD 11172, "Coding of moving pictures and associated audio for digital storage media up to 1.5 Mbits/s", International Organization for Standardization (ISO), 1992.

[25] ISO/IEC JTC1 CD 13818, "Generic coding of moving pictures and associated audio", International Organization for Standardization (ISO), 1994.

[26] J. Bracamonte, I. Defilippis, M. Ansorge and F. Pellandini. "Bit-serial parallel processing VLSI architecture for a block matching motion estimation algorithm", *Proc. of the International Picture Coding Symposium PCS'94*, Sacramento, CA, USA, Sept. 21-23, 1994, pp. 22–25.